Plant genetic resources: A review of current research and future needs

It is widely recognised that agriculture is a significant contributor to global warming and climate change. Agriculture needs to reduce its environmental impact and adapt to current climate change whilst still feeding a growing population, i.e. become more 'climate-smart'. Burleigh Dodds Science Publishing is playing its part in achieving this by bringing together key research on making the production of the world's most important crops and livestock products more sustainable.

Based on extensive research, our publications specifically target the challenge of climate-smart agriculture. In this way we are using 'smart publishing' to help achieve climate-smart agriculture.

Burleigh Dodds Science Publishing is an independent and innovative publisher delivering high quality customer-focused agricultural science content in both print and online formats for the academic and research communities. Our aim is to build a foundation of knowledge on which researchers can build to meet the challenge of climate-smart agriculture.

For more information about Burleigh Dodds Science Publishing simply call us on +44 (0) 1223 839365, email info@bdspublishing.com or alternatively please visit our website at www.bdspublishing.com.

Related titles:

Achieving sustainable cultivation of wheat Volume 1: Breeding, quality traits, pests and diseases
Print (ISBN 978-1-78676-016-6); Online (ISBN 978-1-78676-018-0, 978-1-78676-019-7)

Advances in breeding techniques for cereal crops
Print (ISBN 978-1-78676-244-3); Online (ISBN 978-1-78676-246-7, 978-1-78676-247-4)

Achieving sustainable cultivation of barley
Print (ISBN 978-1-78676-308-2); Online (ISBN 978-1-78676-310-5, 978-1-78676-311-2)

Genome editing for precision crop breeding
Print (ISBN 978-1-78676-447-8); Online (ISBN 978-1-78676-449-2, 978-1-78676-450-8)

Chapters are available individually from our online bookshop:
https://shop.bdspublishing.com

climate•SMART•publishing
IN AGRICULTURAL SCIENCE

BURLEIGH DODDS SERIES IN AGRICULTURAL SCIENCE
NUMBER 100

Plant genetic resources: A review of current research and future needs

Edited by Dr M. Ehsan Dulloo, Bioversity International, Italy

burleigh dodds SCIENCE PUBLISHING

Published by Burleigh Dodds Science Publishing Limited
82 High Street, Sawston, Cambridge CB22 3HJ, UK
www.bdspublishing.com

Burleigh Dodds Science Publishing, 1518 Walnut Street, Suite 900, Philadelphia, PA 19102-3406, USA

First published 2021 by Burleigh Dodds Science Publishing Limited

Library of Congress Control Number: 2021931056

British Library Cataloguing in Publication Data
A catalogue record for this book is available from the British Library

ISBN 978-1-78676-451-5 (Print)
ISBN 978-1-78676-454-6 (PDF)
ISBN 978-1-78676-453-9 (ePub)
ISSN 2059-6936 (print)
ISSN 2059-6944 (online)

DOI 10.19103/AS.2020.0085

Typeset by Deanta Global Publishing Services, Dublin, Ireland

Contents

Series list

Title	Series number
Achieving sustainable cultivation of maize - Vol 1 From improved varieties to local applications Edited by: Dr Dave Watson, CGIAR Maize Research Program Manager, CIMMYT, Mexico	001
Achieving sustainable cultivation of maize - Vol 2 Cultivation techniques, pest and disease control Edited by: Dr Dave Watson, CGIAR Maize Research Program Manager, CIMMYT, Mexico	002
Achieving sustainable cultivation of rice - Vol 1 Breeding for higher yield and quality Edited by: Prof. Takuji Sasaki, Tokyo University of Agriculture, Japan	003
Achieving sustainable cultivation of rice - Vol 2 Cultivation, pest and disease management Edited by: Prof. Takuji Sasaki, Tokyo University of Agriculture, Japan	004
Achieving sustainable cultivation of wheat - Vol 1 Breeding, quality traits, pests and diseases Edited by: Prof. Peter Langridge, The University of Adelaide, Australia	005
Achieving sustainable cultivation of wheat - Vol 2 Cultivation techniques Edited by: Prof. Peter Langridge, The University of Adelaide, Australia	006
Achieving sustainable cultivation of tomatoes Edited by: Dr Autar Mattoo, USDA-ARS, USA & Prof. Avtar Handa, Purdue University, USA	007
Achieving sustainable production of milk - Vol 1 Milk composition, genetics and breeding Edited by: Dr Nico van Belzen, International Dairy Federation (IDF), Belgium	008
Achieving sustainable production of milk - Vol 2 Safety, quality and sustainability Edited by: Dr Nico van Belzen, International Dairy Federation (IDF), Belgium	009
Achieving sustainable production of milk - Vol 3 Dairy herd management and welfare Edited by: Prof. John Webster, University of Bristol, UK	010
Ensuring safety and quality in the production of beef - Vol 1 Safety Edited by: Prof. Gary Acuff, Texas A&M University, USA & Prof. James Dickson, Iowa State University, USA	011
Ensuring safety and quality in the production of beef - Vol 2 Quality Edited by: Prof. Michael Dikeman, Kansas State University, USA	012
Achieving sustainable production of poultry meat - Vol 1 Safety, quality and sustainability Edited by: Prof. Steven C. Ricke, University of Arkansas, USA	013
Achieving sustainable production of poultry meat - Vol 2 Breeding and nutrition Edited by: Prof. Todd Applegate, University of Georgia, USA	014

Foreword

After several promising years, hunger and malnutrition are again on the rise. The latest edition of '*The State of Food Security and Nutrition in the World*', published jointly by five UN agencies[1], estimated that in 2019 almost 690 million people went hungry, an increase of nearly 60 million over the previous five years. This occurred against a complex and alarming backdrop of growing economic, racial, and social inequality, persistent and widespread environmental degradation, continually expanding human populations, rapidly changing climates and a weakening of the architecture of international cooperation. Most recently, the global disruption caused by COVID-19 has further exacerbated the situation.

Within this web of inter-connected challenges, biodiversity continues to be lost at an alarming rate in spite of the best efforts of many individuals, organizations and nations. A report published by the Convention on Biological Diversity in September 2020, '*Global Biodiversity Outlook 5*', concluded that none of the 20 Aichi biodiversity targets agreed to by the parties to the Convention in 2010 had been fully achieved, and only six had been partially achieved. The failure to adequately stem the loss of biodiversity has significant implications not only for the environment but also for human wellbeing, even survival.

A critically important component of biodiversity is that which feeds and clothes us, in particular the genetic diversity within and among our domesticated animals, cultivated crops and their wild relatives. This vast diversity underpins agriculture, and its deployment helps us to meet the challenges and opportunities of today's world and as well as the unknown circumstances that will prevail tomorrow.

In 2019, the Food and Agriculture Organization of the United Nations (FAO), published '*The State of the World's Biodiversity for Food and Agriculture*'. The report pointed out that as in the case of biodiversity in general, many components of agricultural biodiversity remain under threat. For crops, neglected and underused species in particular remain under-represented in *ex situ* collections and in many parts of the world the use of landraces and traditional farmer varieties continues to decline in local farming systems. However, the situation is highly complex, and the news is not all bad, especially for those crops for which past conservation efforts are paying off handsomely. Although a great deal remains to be done, many countries are expanding the range of genetic resources that they conserve *ex situ* and many farmers who have adopted modern varieties also continue to maintain traditional landraces.

1 FAO, IFAD, UNICEF, WFP and WHO

But there is no room for complacency and the situation for crop wild relatives is even more concerning. This is particularly so in regions experiencing severe climate change and widespread degradation of habitats, and where species migration is limited by ecogeographical and other barriers.

Why should we be concerned? For the simple reason that agricultural genetic diversity represents a very powerful tool in our efforts to meet a wide variety of current and future challenges. The genetic resources of our crops underpin our ability to breed new cultivars that can help make agriculture more sustainable and productive within a rapidly changing world. Such cultivars could, for example, provide more nutritious food, novel products, be more resilient to pests, disease and environmental stresses, or have the ability to sequester more carbon.

Thus, crop genetic diversity provides a basis for making significant inroads into improving the lives and wellbeing of humanity as well as the environment in which we all live. However, in order to play this role effectively, genetic diversity must be safely conserved and wisely used, and in both of these areas we can do better. Although we know much about how best to safeguard and use genetic resources there are still major areas where further research is needed. This book makes a very important contribution to this effort through providing an up-to-date overview of the current state of scientific research and provides some invaluable pointers to future research needs and opportunities.

This book does not go into detail about the current state of scientific plant breeding as this is a topic that has been well covered in other recent publications. What if provides is an extensive, much-needed and detailed coverage of a broad range of topics, from the importance and value of plant genetic resources and key methods for conserving them, through to their international exchange and local dissemination through more effective seed systems.

The senior editor of the book, Ehsan Dulloo, is to be congratulated on such a timely and useful publication. Furthermore, the range of authors of the individual chapters and the institutions they represent is truly impressive and reads like a Who's Who of plant genetic resources!

At the time of writing, the parties to the Convention on Biological Diversity are working to finalize a '*post 2020 global biodiversity framework*' to supersede the Aichi targets. The framework aims to act as a steppingstone on the way to achieving the Convention's 2050 Vision: '*By 2050, biodiversity is valued, conserved, restored and wisely used, maintaining ecosystem services, sustaining a healthy planet and delivering benefits essential for all people.*' The current version of the framework envisages several long-term goals for biodiversity in general, including maintaining genetic diversity, and the fair and equitable sharing of benefits arising from its use. It is greatly to be hoped that in the final, agreed text this will be further built upon to ensure that the importance of genetic resources for food and agriculture, and their potential contribution to achieving the Sustainable Development Goals are fully recognized.

I commend this book to all who share these aims.

Geoffrey Hawtin OBE PhD
Former Director General, Bioversity International and
Centro Internacional de Agricultura Tropical (CIAT)
Founding CEO, Global Crop Diversity Trust

January 2021

Acknowledgements

We wish to acknowledge the following for their help in reviewing particular chapters:

- Chapter 1: Dr Geoffrey Hawtin, Consultant, UK; and Dr Jean-Louis Pham, IRD, France
- Chapter 2: Dr Theo van Hintum, Wageningen University & Research, The Netherlands
- Chapter 3: Dr Michele Walters, Council for Scientific and Industrial Research (CSIR), South Africa
- Chapter 4: Professor Stephen Smith, Iowa State University, USA; and Dr Mario Marino, FAO, Italy
- Chapter 13: Professor Rod Snowdon, Justus Liebig University, Germany

Introduction

Plant genetic diversity is fundamental for ensuring global food and nutrition security. Its importance cannot be underestimated as it provides the genetic materials for humankind to continue produce food sustainably in the face of global challenges such as pest and diseases, land degradation, climate change and other socio-economic changes that are occurring today and that will take place in future. While at the global level, their importance is widely recognised as evidenced by the Convention on Biological Diversity and the International Treaty on Plant Genetic Resources for Food and Agriculture, and political processes established for monitoring the conservation and use of genetic resources, we find at the national and local levels that the conservation and use of plant genetic diversity are not getting the attention they deserve with the result that much of the diversity are eroding away, thereby undermining agriculture as a whole. Some key challenges include a narrowing genetic base for many key crops, the loss of landraces and wild relatives, due to agricultural intensification and urbanisation, land and soil degradation, the lack of support to minor and under-utilized crops that are locally and regional important. There is a need to develop the role of existing gene banks from repository of genetic diversity to enablers of the flow of germplasm and genetic information for breeding more robust varieties. There is equally an urgent need to stop the erosion of genetic diversity from their sites where they have developed their distinctive characteristics (in situ) such that the diversity among them can continue to evolve and develop traits for future use.

This volume examines the wealth of research addressing these challenges and the opportunity for a more integrated, global approach to protecting and leveraging plant genetic diversity for a more sustainable agriculture. Part 1 assesses the importance of valuing conservation and the use plant genetic diversity. Part 2 discusses protecting plant genetic diversity with *in-situ* and on-farm strategies. The final section reviews ways of enhancing conservation and the use of plant genetic diversity, including community-based conservation of crop genetic resources, participatory plant breeding programmes and more effective seed systems.

Part 1 Importance and value of conservation and use of plant genetic diversity

Part 1 of the book begins with a discussion of plant genetic resources for food and agriculture for sustainable development. Chapter 1 reviews relevant international agreements, instruments and mechanisms, which address the conservation, sustainable use and access and benefit-sharing for these resources along with their remarkable contributions to food security and nutrition. The chapter also highlights the state-of-the-art for the scientific and technological methods used to conserve and add value through genetic gains to these resources. Underscoring the importance of collaborations at various scales, the authors call for continued global coordination and partnerships on the internationally agreed activities for conserving effectively and deriving the most benefits sustainably from these irreplaceable resources.

Chapter 2 examines valuing past, present and future plant genetic resources in genebanks. The chapter summarises the main messages from applied economics literature on genebank values that began in the late 1990s, a recent set of studies undertaken by the CGIAR Genebank Platform and Crop Diversity Trust, and assessments of national genebanks. Three points are salient: the value of information; the demand for a diversity of approaches to document the diversity of ways that genebanks benefit society; and the need to establish priorities for collecting and conservation based on cost-effectiveness. Cost-effectiveness will require a) carefully targeted investments in characterization, evaluation, and genotyping, along with management of the resulting data, and its use to develop packages of breeder- and researcher-oriented materials, and b) investment in supportive research and various organizational optimization procedures, including automation.

The subject of Chapter 3 is monitoring plant genetic resources for food and agriculture. The chapter reviews the methodologies developed to assess the extent of diversity of PGRFA at species, variety, and genetic level and examines the efforts made at global level in monitoring them at different scales. Efforts have been made to halt the loss of biodiversity (including genetic diversity) by United Nations Organizations (FAO, UNEP, UN) at setting of global targets (second Global Plan of Action on conservation and sustainable use of PGRFA, Aichi Targets and Sustainable Development Goals) and indicators have been established to monitor progress towards them. Yet none of the targets on genetic diversity have been achieved, due to a lack of implementable monitoring system that would allow progress to be accurately monitored. Further research is needed to improve the methodologies for monitoring plant genetic resources, particularly at the varietal and genetic level. The chapter discusses the opportunities and challenges as well as provides recommendations for future conservation and monitoring strategies that may safeguard PGRFA for posterity.

The final chapter of Part 1 reviews improving the global exchange of germplasm for crop breeding. Different norms, rules and practices organize the exchange of germplasm to address broader global challenges such as advancement of science and innovation, food security, sustainable agriculture or global equity. Some of these institutions are now embedded in various treaties and national regulations. Chapter 4 claims that these regulations are not as successful as they could be because they fail to effectively integrate the complexity of the exchange environment. In order to better understand how exchange could be improved, it is important to go beyond a legalistic approach to exchange and look at the broader socio-technical context in which these exchanges take place. By developing an analytical framework that includes several dimensions beyond the regulatory one, this contribution creates the foundation for a more comprehensive approach that acts upon the relationship between germplasm and the diversity of institutional logics, germplasm and its resource system, and germplasm and its social environment.

Part 2 Protecting plant genetic diversity: in-situ and on-farm strategies

The first chapter of Part 2 assesses key steps in conservation and use of plant genetic resources: an overview. Chapter 5 begins by providing an example of a model which includes a series of steps starting with the full range of genetic diversity for all the target plant taxa, through the prioritisation of target taxa, the planning and the implementation of conservation action, leading through characterisation and evaluation, and utilisation in the development of novel crop varieties by farmers and/or breeders. The chapter then goes on to discuss conservation planning which is then followed by a review of the different strategies and techniques that are used in conservation. A section on the link between conservation and utilization is also included, before providing a conclusion that emphasises the importance of maintaining the current wealth of natural plant diversity.

Chapter 6 addresses key issues facing genebanks in preserving crop genetic diversity ex situ: an overview of the range of challenges. Local crop genetic diversity is challenged with changes in land use, urbanization, land degradation, changes in agricultural practises, availability of improved varieties, changes in market preference, and the impact of climate change. Efforts have been made to secure plant genetic resources ex situ for future use but there are significant issues related to cost effective, efficient, secure, rational, and sustainable long-term ex situ conservation. The chapter begins by addressing issues for the composition of ex situ collections and moves on to discuss issues for routine operations for conservation. It also highlights issues for the use of conserved genetic resources, before concluding with a summary of why the development of sustainable genebank systems is so important.

The next chapter examines the techniques and key issues in collecting Crop Wild Relatives (CWR). The genetic diversity found in populations of crop wild relatives is an essential resource for future crop breeding, but populations are at risk of loss before germplasm has been fully conserved in genebanks. Chapter 7 describes best practice for targeting and identifying species, and review knowledge about the variation in wild plant populations to guide the timing of collecting and approaches for genetic sampling. Indicators are presented for seed quality, ripeness and dispersal. Techniques for collection of seed, herbarium vouchers and associated data are reviewed with examples drawn from the Adapting Agriculture to Climate Change (Crop Wild Relative) project. Further research is needed to find optimal approaches for handling of seed to ensure high longevity of seed collections, and improved tools are needed to guide sampling of genetic diversity of crop wild relatives.

Moving on to Chapter 8, this chapter discusses new technologies to improve the *ex-situ* conservation of plant genetic resources. Access to plant genetic resources is fundamental to the development of more resilient and nutritious crops. The efficient and effective conservation of plant genetic resources is therefore key to ensuring global food security. There are more than 1750 genebanks around the world, storing various types of plant germplasm including tissue cultures, seeds, embryos and pollen. The chapter describes some of the recent applications of science and technology to improve the management of genebank collections of orthodox seeds, not least through the introduction of automation. In addition, the chapter looks at how routine cryopreservation procedures are now possible for many species, and how storage procedures for recalcitrant-seeded and vegetatively propagated species continue to evolve given the challenges they present in terms of long-term ex situ germplasm conservation.

Part 2 concludes with an analysis of the role of the Svalbard Global Seed Vault in preserving crop genetic diversity. Chapter 9 provides an overview of this facility in Svalbard and describes its operations as well as its functions and status. Genetic material in individual genebanks is potentially vulnerable to being lost (e.g. through conflict or a natural disaster). One important and simple security measure is to ensure that samples of these valuable genetic materials are conserved in more than one place. In this context, the Svalbard Global Seed Vault offers a free of charge service to store duplicate samples of seeds that are conserved in genebanks world-wide.

Part 3 Enhancing conservation and use of plant genetic diversity

The final part of the book begins with a discussion on community-based conservation of crop genetic resources. Chapter 10 explores the current state of research, knowledge and practice of community-based conservation in the context of continued farmer-driven processes and the emergence of diverse external interventions. Autonomous community-based conservation is largely powered by farmer demand for crop diversity. Cultural, provisioning and regulating ecosystem services play an essential role. It is argued that an enhanced understanding of the drivers and rationales influencing farmers decision-making continues to be essential under the current scenario of accelerated global change. A large and diverse portfolio of exogenous interventions have emerged since on-farm conservation has become a development issue. A variety of different interventions designed to support on-farm conservation for are reviewed. The chapter also highlights the need for an effective monitoring framework for community-based management of crop genetic resources. Such a systems could enhance the measurement, metrics and intelligence underlying the on-farm conservation status of crops and landraces.

Chapter 11 reviews participatory plant breeding programmes to optimize use of crop genetic resources. The chapter summarises a sample of variety evaluation, experimental design, and breeding method innovations that have served as solid approaches for participatory plant breeding (PPB) efforts. With success in PPB comes success in conservation at a local level of useful alleles and allele assemblages in the form of on-farm crop genetic resources. PPB programs of this sort have the potential to add value to local or traditional varieties that might otherwise be abandoned, thus promoting their *in-situ* conservation. The chapter briefly touches on methodologies to assess farmers' variety preferences. This is followed by sections that highlight some experimental designs for on-farm variety evaluation and farmer-participatory breeding methods for combining in-situ conservation with genetic improvement. Finally, some of the challenges that may limit genetic gain from PPB programs are noted – problems that increase the risk of wholesale replacement of on-farm genetic diversity rather than conservation through improvement.

Moving on to Chapter 12, this chapter examines seed systems and diversity. Farmers use various seed systems to obtain seeds for their next crop. The diversity of seeds determines, together with other aspects of the farming system, the biodiversity in farmed areas. Conservation strategies for crop genetic resources distinguish ex situ, on-farm and in situ components. The on-farm management of such resources is highly influenced by the seed systems that farmers use. Next to primary and secondary centres of diversity, this chapter introduces a tertiary source of diversity, based on the creation of new diversity through modern breeding. The impact of different seed systems on the management of crop genetic diversity is also analysed.

Chapter 13 concludes the book by providing an overview of DNA-based screening of Brassica germplasm for sustainable and enhanced crop production. The Brassica genus contains many agriculturally important oilseed and vegetable crops. Brassica germplasm, including natural accessions and breeding populations, are maintained globally for sustainable management and enhancement of Brassica crop production which is critical to meet the demands of population growth and challenges of environmental stresses due to global climate change. DNA based markers, such as SNPs, are commonly used to screen large numbers of Brassica germplasm for conservation, genetic mapping and association studies. The chapter focuses on the application of SNP genotyping technologies for conservation of Brassica germplasm, uncovering the genetic basis of various biotic and abiotic stresses and screening for yield related traits and oil quality through marker-trait association studies.

Part 1

Importance and value of conservation and use of plant genetic diversity

Chapter 1

Plant genetic resources for food and agriculture for sustainable development[1]

Chikelu Mba, Seeds and Plant Genetic Resources Team – Food and Agriculture Organization of the United Nations, Italy; M. Ehsan Dulloo, Bioversity International, Italy; and Kent Nnadozie, Secretariat of the International Treaty on Plant Genetic Resources for Food and Agriculture – Food and Agriculture Organization of the United Nations, Italy

1 Introduction

In 2019, about 2 billion people across the world lacked regular access to safe, nutritious and sufficient food (FAO, IFAD, UNICEF, WFP and WHO, 2020). This dire statistic, which mirrored the earlier estimation of about 2.5 billion people who suffer from various forms of nutrient deficiencies and obesity globally (FAO, IFAD, UNICEF, WFP and WHO, 2018, 2019), underscored a progressively

1 The views expressed in this publication are those of the authors and do not necessarily reflect the views or policies of the Food and Agriculture Organization of the United Nations.

http://dx.doi.org/10.19103/AS.2020.0085.01
Published by Burleigh Dodds Science Publishing Limited, 2021.

worsening state of global food insecurity and malnutrition. With increasing global population expected to reach 9.7 billion by 2050 (Anon, 2020), whereby the steepest increases are in food insecure countries of the developing South, it is estimated that 50% more food, over the 2013 baseline, needs to be produced by 2050 in order to meet the ever-increasing demands (FAO, 2017a). Yet, with climate change imperiling crop production systems, there are scarcely any redundant arable lands and water resources to be further deployed for food production. It is against this challenging backdrop that the world, through the Sustainable Development Goals (SDGs), committed to a world free of hunger and malnutrition in the next 10 years (UN General Assembly, 2015; FAO, 2017a).

The prohibitive economic and environmental costs and dangers to human health make the use of additional external inputs (such as agrochemicals) non-viable options for the significantly increased food production needed for a hunger-free world. Therefore, it is certain that a business-as-usual mind-set is not an option. In the absence of additional agricultural natural resources base for increased deployment and the impracticality of unbridled increases in the use of external inputs, the inherent potentials of our crops must be tapped to enhance their productivities, that is, producing more yields with fewer inputs. It is in this vein that the Food and Agriculture Organization of the United Nations (FAO) called for a new paradigm of sustainable crop production intensification (SCPI) in which a genetically diverse portfolio of improved crop varieties that are best suited to agroecosystem and farming practices and more resilient to climate change should be promoted (FAO, 2011). Similarly, the further development of agriculture, and the world's continued food security, will depend on farmers and breeders continuing to have access to and use diverse plant genetic resources necessary to face new environmental and social challenges.

Our optimism regarding the prospects of this paradigm derives from the knowledge that the first wave of crop domestication over 10 000 years ago resulted in the current agricultural systems on which the global food security and nutrition depend (McCouch, 2004; Waines and Ehdaie, 2007). The repertoire of wild ancestors or related species of crops – known as crop wild relatives (CWR); non-related species which are donors of heritable traits used in plant breeding; farmers' varieties or landraces and modern crop varieties is referred to as plant genetic resources for food and agriculture (PGRFA) (FAO, 2009). In essence, PGRFA are plants used, or with potentials to be used, for food and other agricultural purposes and therefore encompass the diversity of the plant kingdom that underpin global food security and nutrition. The values of crops, which are cultivated and harvested for food or other uses, and wild and undomesticated plants that are similarly exploited as food, are obvious. However, perhaps more importantly, PGRFA also serve as a huge reservoir of heritable traits that enable crops to adapt to physical and biological stresses, for example, drought, heat, cold, pests and diseases – the direct consequences of climate change, and to enhance the nutritional attributes of crops (Mba et al., 2012).

The commitment of the international community to make PGRFA freely available, especially for research and development, through various normative processes and instruments is reviewed in the subsequent sections of this chapter. The actions that may be taken – through the conservation of PGRFA in situ and ex situ and the maintenance of on-farm diversity; their use in breeding progressively superior crop varieties; and delivery to farmers via effective seed systems to ensure that humanity benefits from these irreplaceable resources in ways that safeguard them for current and future generations are also highlighted. This chapter, in providing examples of these aspects of the conservation and sustainable use of PGRFA, also highlights the scientific and technological advances that make these possible. Finally, perspectives are shared on the imperative of the creation of the enabling environment and the fostering of winning partnerships for the sustainable management of PGRFA, especially with regard to unlocking their potentials through research and development.

2 Plant genetic resources for food and agriculture (PGRFA): a common concern for all countries

Today, cassava – *Manihot esculenta* Crantz – is a far more important food security crop in sub-Saharan Africa into which it was first introduced by Portuguese sailors in the 16th century than in the countries, Brazil and Bolivia, which adjoin the Amazon basin where it was originally domesticated several tens of millennia ago. In like manner, oil palm – *Elaeis guineensis* Jacq. – with the evidence of its use as food dating as far back as five millennia in West Africa, its primary centre of genetic diversity, is a far more major cash crop in Asian countries, Malaysia and Indonesia, into which it was introduced a few decades ago.

Other crops such as maize, potatoes, wheat, barley and rice are now so ubiquitous around the globe that it is not obvious to the average consumer that they were introduced into most of the places where they have become seemingly irreplaceable staples. Also, the pioneering collecting expeditions of the likes of N.I. Vavilov and Jack Harlan early in the 20th century are well chronicled along with the concerns they raised about the decline in the diversity of crops as evidenced by the increasingly diminishing cultivation of traditional crop varieties or landraces and their undomesticated relatives, that is, CWR on-farm and in nature, respectively (Vavilov, 1931; Harlan, 1957).

These probably explain why the international community has consistently, while demonstrating the critical importance of PGRFA to global food security and nutrition, also underscored the interdependence among nations as means to derive the most benefit from these resources. For instance, the FAO called for immediate action for the collection and conservation of landraces and CWR at the 10th session of its Conference in 1959. Following this, two major technical meetings in 1961 and 1967 took place. These were the Technical meeting on Plant Exploration and Introduction which was convened in 1961 to streamline germplasm conservation and distribution and established exploration centres in regions of greatest diversity (Scarascia-Mugnozza and Perrino, 2002). The 1967 International Technical Conference organized jointly by the FAO and the International Biological Programme (IBP) was a landmark conference that aimed to articulate a global strategy for the conservation of PGRFA. A Panel of Experts was constituted to develop the guidelines for the establishment of a global network for ex situ conservation and the associated plan of action (Frankel and Hawkes, 1975). The proposal of the Panel was presented at the 1973 FAO/IBP Technical Conference and subsequently was considered by the Technical Advisory Council of the CGIAR. This in turn led to the creation of a coordinating centre, the International Board on Plant Genetic Resources (IBPGR) in the FAO (which later became the CGIAR centre, the International Plant Genetic Resources Institute, later renamed Bioversity International and now the Alliance of Bioversity International and CIAT). The IBPGR was operated under the aegis of the FAO to promote and assist in the worldwide efforts to collect and conserve plant germplasm needed for future research and crop production (Scarascia-Mugnozza and Perrino, 2002).

Ever since, the international community, especially through the FAO's leadership, has been most active in emphasizing the interdependence of nations in both the responsibility of safeguarding PGRFA and the mutual benefits of using them. FAO's intergovernmental work on the conservation and use of PGRFA has been carried out through its Commission on Genetic Resources for Food and Agriculture (Commission) established in 1983 (FAO, 2010). Starting out as the Commission on Plant Genetic Resources and in 1995 taking on the responsibility for other components of agricultural biodiversity, it is an FAO forum for governments to discuss and negotiate matters relevant to genetic resources for food and agriculture. It is the only intergovernmental body that specifically deals with all components of biological diversity for food and agriculture. In this regard, it reviews and advises the FAO on policy matters, programmes and activities. It was in this vein that the Commission facilitated a Global System on Plant Genetic Resources for Food and Agriculture (Global System) which is a set of policy instruments and mechanisms to promote the safeguarding of PGRFA, their availability and sustainable use (FAO, 2010; Frison et al., 2011). The elements of the Global System and related international agreements are described in the following sections.

3 International agreements supporting the global system on PGRFA

The principal international agreements supporting the Global System on PGRFA are:

- International Undertaking on Plant Genetic Resources for Food and Agriculture;
- The Convention on Biological Diversity (CBD); and
- The International Treaty on Plant Genetic Resources for Food and Agriculture.

These are discussed in the following sections.

3.1 International undertaking on plant genetic resources for food and agriculture

Adopted by the FAO's Conference in 1983, the objective of this undertaking was 'to ensure that plant genetic resources of economic and/or social interest, particularly for agriculture, will be explored, preserved, evaluated and made available for plant breeding and scientific purposes. This undertaking is based on the universally accepted principle that plant genetic resources are a heritage of mankind and consequently should be available without restriction' (FAO, 1983). This set the tone for the ongoing recognition and treatment of PGRFA as a global commonwealth.

3.2 The convention on biological diversity (CBD)

The CBD is the international agreement for 'the conservation of biological diversity, the sustainable use of its components and the fair and equitable sharing of the benefits arising out of the utilization of genetic resources' (United Nations, 1993). Biodiversity encompasses PGRFA. In fact, the Strategic Plan for Biodiversity 2011–2020, including the Aichi Biodiversity targets, of the CBD is aimed at the promotion and the conservation and use, *inter alia*, of PGRFA. Aichi Target 13 specifically addresses the maintenance of the genetic diversity of cultivated plants, that is, crops, and their wild relatives. Also, Aichi Targets 7 and 13 address PGRFA explicitly. The Tenth Conference of Parties (or COP 10) of the CBD also endorsed the Global Strategy on Plant Conservation (GSPC) that addresses the challenges posed by threats to plant diversity. Target 9 of the GSPC aims at conserving 70% of genetic diversity of crops including their wild relatives. The post-2020 Biodiversity Framework, presently under negotiation, continues to highlight the contribution of genetic diversity to Nature's contribution to people and new targets for maintaining and sustainably utilizing genetic diversity of both wild and cultivated/domesticated species are being proposed. It is to be noted that the CBD specifically did not cover access to ex situ PGRFA collected before its entry into force, including genebank collections of the International Agriculture Research Centres of the Consultative Group on International Agricultural Research (CGIAR) as well as many national collections. The Nairobi Conference, where it was adopted, recognised the need to seek solutions to these and other outstanding matters concerning plant genetic resources within the Global System for the Conservation and Sustainable Use of Plant Genetic Resources for Food and Agriculture established by the FAO.

3.3 The international treaty on plant genetic resources for food and agriculture (treaty)

This Treaty is the result of the revision, in line with the CBD, of the International Undertaking. The Treaty was adopted by the Thirty-first session of the FAO Conference on 3 November 2001 and entered into force on 29 June 2004. The Treaty provides, in harmony with the CBD, for the special needs of plant genetic resources for food and agriculture and also provides for the governance of PGRFA outside of the CBD. In particular, it will allow for the conservation and continued flow across national boundaries of the plant genetic resources most important to sustain food security and on which all countries are interdependent. The Treaty provides for a multilateral approach by which countries agreed to establish an open multilateral system of conservation and exchange of PGRFA. In negotiating the Treaty, and in exercise of their sovereign rights, States chose to create this multilateral system for the world's key crops, based on their importance for food security and the degree to which countries are interdependent on them. This system is exemplified in the work and functioning of the CGIAR. The multilateral system is based on the understanding that plant genetic resources are a common concern of the international community and therefore must follow rules established multilaterally.

4 Global instruments supporting the global system on PGRFA

This section covers:

- rolling global plans;
- periodic reports; and
- the report on the State of the World's Biodiversity for Food and Agriculture.

These are discussed in the following sections.

Published by Burleigh Dodds Science Publishing Limited, 2021.

4.1 *The global plan for the conservation and sustainable utilization of plant genetic resources for food and agriculture (GPA)*

The GPA, the first in what was envisaged as a rolling plan to be continually updated, was adopted by representatives of 150 countries during the Fourth International Technical Conference on Plant Genetic Resources, Leipzig, Germany, from 17 June 1996 to 23 June 1996 (FAO, 1996a). It contains the priority actions for the conservation and use of PGRFA to be implemented at local, national, regional and international levels. At this conference, the Leipzig Declaration was also adopted; it focusses attention on the importance of plant genetic resources for world food security and commits countries to implement the plan.

4.2 *The second global plan of action for plant genetic resources for food and agriculture (second GPA)*

The Second GPA is a strategic framework for the conservation and sustainable use of the plant genetic diversity which, prepared under the aegis of the Commission, was adopted by FAO Council at its 143rd Session in November 2011 (FAO, 2012). The Second GPA is an update to the GPA (FAO, 1996a) based on the Second Report on the State of the World's Plant Genetic Resources for Food and Agriculture (FAO, 2010) and complementary inputs obtained from a series of regional consultations and worldwide experts.

4.3 *The state of the world's plant genetic resources for food and agriculture (first report)*

This was the first in the envisaged series of periodic reports on the global status of the conservation and use of PGRFA (FAO, 1997). It provided a snapshot of the then state of the diversity of PGRFA and capacities at the national, regional and global levels for managing them, that is, their conservation in situ and ex situ and utilization both directly by end users and in plant breeding.

4.4 *The second report on the state of the world's plant genetic resources for food and agriculture (FAO, 2010)*

This was an update to the First Report and reflected the changes and trends to the conservation and use of PGRFA since 1996. The identified gaps and needs fed into the development of the Second GPA.

4.5 The report on the state of the world's biodiversity for food and agriculture

This more recent report provided the results of the detailed assessment of the diversity and management worldwide of the repertoire of the domesticated plants and animals that are part of crop, livestock, forest or aquaculture systems, harvested forest and aquatic species, the wild relatives of domesticated species, and other wild species harvested for food and other products as well as the associated biodiversity, that is, the vast range of organisms that live in and around food and agricultural production systems which contribute to the sustenance and outputs of such systems (FAO, 2019c).

5 Global mechanisms supporting the global system on PGRFA

This section covers:

- The World Information and Early Warning System of Plant Genetic Resources for Food and Agriculture (WIEWS);
- The International Network of *Ex Situ* Collections; and
- Agreement between the FAO and the CBD.

These are discussed in the following sections.

5.1 The world information and early warning system of plant genetic resources for food and agriculture (WIEWS)

This is a web-based portal for the curation and retrieval of data on the implementation of the Second GPA (FAO, 2020). WIEWS is therefore an invaluable repository for the preparation of the periodic reports on the State of the World's PGRFA and, more recently, for the annual reporting on the plant component of Indicator 2.5.1 of the Sustainable Development Goals (SDG), that is, ex situ conservation of diversity. Additionally, the Treaty established, under Article 17, a global information system to facilitate the exchange of information, based on existing information systems, on scientific, technical and environmental matters related to PGRFA.

5.2 The international network of ex-situ collections

In 1994, 11 CGIAR centres signed agreements with the FAO to the effect that their ex situ germplasm collections were being held in trust for the international community thereby creating the International Network of *Ex Situ* Collections.

Published by Burleigh Dodds Science Publishing Limited, 2021.

With the Treaty coming into force, the centres would sign further agreements with the FAO in 2006, superseding the previous ones, which brought their germplasm holdings within the Multilateral System of the Treaty's Access and Benefit-sharing (ABS) mechanism. The import remained that PGRFA are global commonwealth.

5.3 Agreement between the FAO and the CBD

The Memorandum of Cooperation signed between the FAO and the CBD in 2006 paved the way for increased synergy between the two organizations in the conservation and use of biodiversity of relevance to food and agriculture. Each organization participates in the processes of the other. More recently, in 2016, the FAO Conference welcomed the initiative of FAO to be the Biodiversity Mainstreaming Platform (the Platform) in order 'to facilitate the integration of national, regional and international actions to support the conservation, sustainable use, management and restoration of biological diversity across the agricultural sectors' (FAO, 2018a, 2019a). Envisaged for implementation in collaboration with FAO partners, including the CBD and other United Nations organizations, the Platform facilitates the building of bridges across the agricultural and environmental sectors. This would ensure, importantly, that food and agricultural practices achieve the dual objectives of food security and nutrition on the one hand and the conservation of biological diversity on the other.

6 Codes of conduct and standards supporting the global system on PGRFA

This section covers:

- The International Code of Conduct for Plant Germplasm Collecting and Transfer; and
- The Genebank Standards for Plant Genetic Resources for Food and Agriculture.

These and other initiatives are discussed in the following sections.

6.1 The international code of conduct for plant germplasm collecting and transfer

Developed under the auspices of the Commission, adopted by the FAO Conference in November 1993 and based on the principle of national sovereignty over plant genetic resources, the Code 'aims to promote the rational collection and sustainable use of genetic resources, to prevent genetic erosion, and to protect the interests of both donors and collectors of germplasm' (FAO, 1993).

6.2 The genebank standards for plant genetic resources for food and agriculture

These Standards provide international standards for the ex situ conservation of PGRFA germplasm in seed banks, field genebanks and for in vitro and cryopreservation (FAO, 2013). The Genebank Standards are updates to the earlier FAO/IPGRI Genebank Standards (FAO, 1994).

6.3 Other initiatives

There are many more national, regional and international initiatives, usually multilateral in nature, in addition to the Global System under the aegis of FAO's Commission and the Treaty, (FAO, 2010), which demonstrate the extremely high premium that the international community places on PGRFA. Hodgkin et al. (2013) detailed the myriad of other international and regional agreements, regional networks and fora, including the crop- and thematic-based ones and information systems, such as:

- The GENESYS Platform (https://www.genesys-pgr.org/);
- funding mechanisms, for example, the Global Crop Diversity Trust (https://www.croptrust.org/; accessed on 27 November 2020) and the Global Environment Facility (GEF; https://www.thegef.org/; accessed on 27 November 2020); and
- international endeavours, such as the Svalbard Global Seed Vault (https://www.croptrust.org/our-work/svalbard-global-seed-vault/; accessed on 27 November 2020).

There are other support systems such as international non-governmental organizations and collaborative research efforts. In addition to these, Sonnino (2017) has also chronicled important PGRFA-related meetings and events from 1961 to 2016.

7 The conservation of PGRFA

Taken together, the 18 Priority Activities of the Second GPA stipulate what countries and other partners must do in order to both conserve PGRFA in perpetuity and derive the most benefits from them sustainably, that is, in ways that do not erode their diversity.

Published by Burleigh Dodds Science Publishing Limited, 2021.

PGRFA are safeguarded so as to preserve the available intra- and inter-specific diversities of the target plants. Typically, these are achieved by (a) safeguarding PGRFA in their natural habitats, that is, in situ conservation which targets both CWR and wild plants harvested for food; (b) in the case of domesticated plants, maintaining on-farm diversity, that is, on-farm management, through the cultivation of a wide spectrum of crops and their variants, including farmers' varieties and landraces; and (c) preserving the propagules of the varied samples of plants in genebanks, outside their natural habitats, that is, ex situ conservation (Dulloo et al., 2017). The in situ conservation and on farm management methods on the one hand and those for ex situ conservation were subjects of intense debates and negotiations in 1970s and 1980s (Scarascia-Mugnozza and Perrino, 2002), but it is now widely recognized that these two methods are complementary and actions for conservation of PGRFA should be taken in an integral manner (Dulloo et al., 2017).

7.1 In situ conservation

The requisite practices for safeguarding targeted PGRFA, such as CWR and other wild plants - some of which may be harvested for food, in nature are locating the species, describing the status of their conservation, and actively managing and monitoring the populations where they occur in nature (FAO, 2017b). This is particularly critical as the genetic diversity of PGRFA in nature is being eroded by various factors, that is, loss and fragmentation of habitats and extreme weather events that may be linked to climate change. In situations of threats to the continued existence of PGRFA in nature, it is imperative to collect and safeguard them in genebanks. However, it must be borne in mind that this truncates the evolution of adaptive traits, which the continuing natural selection in the wild engenders, and, therefore, as far as possible, should be implemented as a complement to in situ conservation. Efforts for the in situ conservation of CWR have been gaining grounds across the world over the past few decades and the FAO published voluntary guidelines to help countries develop strategies for the conservation and use of CWR (FAO, 2017b). This has been complemented by the development of an interactive tool for the planning of the in situ conservation of CWR (Holness et al., 2019; Magos Brehm et al., 2019; Ng'uni et al., 2019). Several countries, including Mauritius and Zambia, have prepared National Strategic Action Plans while others such as Indonesia and South Africa, have set conservation priorities for wild relatives for their countries (Dulloo and Maxted, 2019). At the global level, using species distribution modelling, Vincent et al. (2019) had recently identified 150 sites where 65.7% of the CWR of major crops are located and can be the foundation of a global network of genetic reserves for priority

CWR species. To facilitate the monitoring of in situ conservation of CWR in a consistent manner, a core set of descriptors has been published by Bioversity International (Thormann et al., 2013) and, more recently, the Secretariat of the Treaty reviewed this list to develop a globally accepted strategic short list of descriptors to facilitate the exchange of information on CWR that are conserved in situ (Alercia et al., 2020).

7.2 On-farm management of PGRFA

Farmer's fields, orchards and home gardens are the veritable repositories of a significant amount of crop diversity for which their utility has been demonstrated. These PGRFA include especially farmers' varieties/landraces (FAO, 2019b) which remain in cultivation on account of the preferences for their agronomic, culinary and other quality traits and for various sociocultural considerations. The provisions of support to community-based seed systems, the promotion of markets for niche products and various incentives, including subsidies, are some of the means for enhancing and/or maintaining the levels of inter- and intra-specific on-farm diversity. These support systems and incentives have become even more critically important as this diversity is continually threatened by various factors such as of the preponderance of monocropping and the cultivation of genetically uniform varieties (FAO, 2019b). The documentation of farmers' varieties/landraces and the complementary ex situ conservation are two essential means to safeguard them (FAO, 2010).

7.3 Ex situ conservation

A significant amount of crop diversity is safeguarded as orthodox seeds in seed genebanks. Species that produce non-orthodox seeds or are propagated vegetatively as live plants are preserved in field genebanks or under slow growth conditions as plantlets through various in vitro culture and cryopreservation methods (FAO, 2014). Routine genebank practices include the acquisition of germplasm accessions followed by their storage, characterization, evaluation, regeneration, safety duplication and documentation (FAO, 2014). There are about 650 genebanks worldwide which together hold about 5.3 million accessions under medium- and long-term conditions (FAO, 2020). It is imperative that germplasm accessions are safeguarded under the prescribed optimal conditions for their continued viability and, hence, availability – to plant breeders, other scientists and other end-users – to be assured (FAO, 2014). While significant progress has been made with safeguarding the germplasm accessions stored in international genebanks, especially those of the CGIAR centres and duplicates held at the Svalbard Global Seed Vault, many of the collections, especially in genebanks of developing countries, are vulnerable. Their major constraints relate to sub-optimal funding. This must be addressed so that the gains made from the previous significant investments in their collecting and conservation are not irredeemably lost.

A most potent justification for the investment of the usually scarce resources in the conservation of PGRFA is the demonstration of their utility - most effectively, in terms of improved food security, nutrition and livelihoods. That crop variety that outperforms all others in one or multiple critical criteria provides such a justification for the continued conservation of its progenitors just as that wild plant which is the source of a valuable nutrient or pharmaceutical product. But, because the need for traits continues to evolve, it is usually impossible to know which germplasm accession would be the source of the critical trait until after the fact. So, the conservation of PGRFA - that represents the widest possible spectrum of heritable variations - must be a continuing integral part of agricultural research and development.

8 Pre-breeding and conservation of PGRFA

The increasingly narrow genetic bases of cultivars, especially of the major crops (Nass and Paterniani, 2000; Martynov and Dobrotvorskaya, 2006; Tester and Langridge, 2010; Mba et al., 2012; Mba, 2013) make massive failures of food security crops inevitable. A season's cropping could be uniformly susceptible to the same, or combinations of, biotic stresses, for example, new variants and biotypes of more virulent pests and diseases, and abiotic stresses such as drought, flooding and variations in temperature. One means for combatting this genetic vulnerability is through the incorporation of novel traits, for example, resistance genes, often found in CWR and landraces/farmer varieties, into otherwise well-adapted varieties (Gur and Zamir, 2004). Pre-breeding, that is, 'all activities designed to identify desirable characteristics and/or genes from unadapted (exotic or semi-exotic) materials, including those that, although adapted have [not] been subjected to any kind of selection for improvement', is a means to introgress novel alleles into crop varieties. The enthusiastically adopted e-learning course on pre-breeding developed under the auspices of the then FAO-convened Global Partnership Initiative for Plant Breeding Capacity Building has contributed to make pre-breeding an integral part of crop improvement.

Novel traits have been introgressed from CWR and other non-adapted materials into cultivars of apple, banana, barley, cassava, chickpea, grapevine, groundnut, lettuce, maize, millet, oats, peanuts, pear, pigeonpea, potato, rice, sunflower, tomato and wheat (Rabinovich, 1998; Hajjar and Hodgkin, 2007; Maxted and Kell, 2008; Jung and Seo, 2014; Brozynska et al., 2015; CSA; Dempewolf et al., 2017). A comprehensive review by Hajjar and Hodgkin (2007), which focussed mostly on activities in CGIAR centres, reported that over 60 wild species had been used successfully for introgressing over 100 novel beneficial traits into 13 crops – banana, barley, cassava, chickpea, groundnut, lettuce, maize, millet, potato, rice, sunflower, tomato and wheat. Over 80% of the traits were related to pest and disease resistance while others were related to tolerance to abiotic stress, yield and quality attributes and male sterility or fertility restoration.

A major disincentive for pre-breeding is linkage drags whereby deleterious alleles are inherited along with desirable attributes in the offspring of crosses between non-adapted materials, such as CWR, and standard parental lines. To address this drawback, Dempewolf et al. (2017) reported the increasing use of marker-aided selection (MAS) made possible by the mapping of quantitative trait loci (QTL) and the applications of other means for associating traits with genome locations, such as genome-wide association studies (GWAS); multiparent advance generation intercross (MAGIC) and nested association mapping (NAM).

9 Plant breeding techniques to introduce desirable traits from PGRFA

Farmers need to grow 'a genetically diverse portfolio of improved crop varieties, suited to a range of agroecosystems and farming practices, and resilient to climate change' (FAO, 2011) as means to produce the significantly more food needed to feed an ever-growing population even as erratic extreme weather events imperil agricultural production systems. To attain this imperative, the potentials encoded into the genetic blueprint of PGRFA must be unlocked and harnessed into progressively superior crop varieties. This is known as plant breeding, the science of altering the heritable patterns of plants to increase their value (The Gale Encyclopedia of Science, 2020). In its simplest form, plant breeding involves the preferential selection of variants of a crop for cultivation – much like our hunter-gatherer forebears who began selecting off-type plants to grow as food some 10 000 years ago. Of course, advances in botany, genetics, biochemistry, physiology, pathology, statistics, and so on have increased the precisions with which promising parents for use in crosses could be identified and the values of their offspring predicted.

The relative abundance of food which is now taken for granted, especially in the richer and more developed global North, derives essentially from the superior higher-yielding crop varieties that plant breeders have been producing for over a century. By the early 20th century in the United States, plant breeding began generating significant numbers of newly improved high-yielding crop varieties, especially hybrids. This enhanced scale of the breeding, release and adoption of improved crop varieties would be replicated in other parts of the world only after the world war. However, the most consequential contribution of plant breeding to global development was the Green Revolution (Box 1). Confronted with these current consequential challenges, the time is

Published by Burleigh Dodds Science Publishing Limited, 2021.

ripe for leveraging the available substantial human and material resources and significant advances in the scientific and technical disciplines that drive crop improvement to recreate the success story of the Green Revolution for which plant breeding is credited with saving hundreds of millions of people from hunger and starvation.

Box 1 The Green Revolution and PGRFA

The Nobel Peace Prize (1970) was awarded to Norman E. Borlaug for a lifetime of work on increasing global food production, which resulted in the saving of up to 1 billion people worldwide from famine, starvation and death. His work, which consisted essentially of breeding and disseminating semi-dwarf wheat varieties, was made possible by the international exchange of wheat germplasm. Lumpkin (2015) and Dubin and Brennan (2009) are two of the several narrations of the improbable success story. A Japanese scientist, Gonjiro Inazuka, had in 1935, produced an improved semi-dwarf wheat variety, known as Norin 10, by crossing a semi-dwarf Japanese wheat landrace with two American varieties. A few years later, in the late 1940s, an American scientist working at a Washington State University - United States Department of Agriculture programme, Orville Vogel, used Norin 10 in crosses to produce high-yielding, semi-dwarf winter wheat varieties. Norman Borluag, who had been breeding wheat in Mexico through a Rockefeller Foundation-sponsored programme since 1944, obtained the seeds of these varieties in 1953. He crossed them with Mexican varieties to produce spring wheat varieties, which in addition to the initial breeding goal of rust resistance were short and stiff-strawed, tillered profusely, produced more grain per head, and were less likely to lodge than the check varieties. Some of these, especially Sonora 64 and Lerma Rojo 64, were subsequently introduced into India and Pakistan; the dramatic increases in yield led to the coining of the term, 'Green Revolution'. In fact, within 4 years, from 1963 to 1967, India's wheat production doubled to 20 million tonnes per year, catapulting the country from import dependency for the crop to self-sufficiency. Similarly, wheat production in Pakistan increased from 4.6 million in 1965 tons to 6.7 million tons in 1970.

That the extreme starvation projected for South Asia, that threatened to eclipse the Bengal famine of 1943, did not materialize is credited to these increased productions in the two countries. This success was made possible by the free exchange of plant germplasm that started some decades earlier. Parallel to the wheat breeding and dissemination endeavours, similar successes were achieved in rice breeding in the Philippines by the then unique International Rice Research Institute (IRRI), which was established in the country with funding by the Rockefeller and Ford Foundations. The development and subsequent adoption of the rice variety IR8, a cross between Dee-Geo-woo-gen and Peta, led to the annual rice production in the country more than doubling within two decades. The significant increases in the yields of these two cereals, rice and wheat, led to the Green Revolution. In effect, the Green Revolution was underpinned by the unlocking of the inherent potentials of PGRFA. But, in practical terms, these successful scientific feats - which provided the momentum for the establishment of the CGIAR - were made possible by international collaborations. The combined effects of free international exchange of germplasm and ideas and sustained funding and support by governments (in these instances, the Philippines, Mexico, India, Pakistan) and donor agencies, especially the Ford and Rockefeller Foundations and the United States Agency for International Development, and the Food and Agriculture Organization of the United Nations made all the difference.

The success of any need-based plant breeding depends essentially on the abilities of the breeders, germplasm curators and other associated scientists to understand the mechanisms for the inheritance of the desired trait, identify the sources of the heritable variation and integrate the desired trait into a suitable genetic background. In situations of ready access to well-characterized germplasm, this would be relatively straightforward. The processes would involve the use of quantitative genetics approaches to determine if the trait is, in fact, heritable - through the estimation of heritability; identify suitable parents - through the estimates of combining abilities; conduct crosses; isolate promising segregating progeny; and fix the trait in a suitable genetic background through selfing and/or backcrossing. In the best-case scenarios, these processes are laborious and time-consuming.

However, in real life, the desired trait is not always available in the genepool or if available, the particular germplasm accession is not accessible to the plant breeder - for a variety of reasons, including impediments engendered by ABS regimes. The constraint may also be biological, for example, the existence of crossing barriers between the selected parents. It may also be that the desired trait is available only in an undesirable genetic background, such as CWR, and resulting progeny co-inherit many deleterious alleles with the desired trait. Known as linkage drag, these unwanted attributes could be removed only through repeated cycles of backcrosses, which add to the tedium and duration for developing a variety.

9.1 Induced mutagenesis

Induced mutations have been used, in situations that the desirable traits were absent in accessible or usable germplasm, to generate heritable variations which were then subsequently exploited in breeding novel crop varieties. The Joint Division of the Food and Agriculture Organization of the United Nations and the International Atomic Energy Agency (Joint FAO/IAEA) in Vienna, Austria maintain a Mutant Varieties Database (MVD) for crop varieties developed through induced mutations (IAEA, 2020). Data on over 3300 crop varieties developed through induced mutagenesis - mostly using ionizing radiation - which have been released across the different continents of the world could be accessed from this database.

9.2 Tissue culture

Advances in cell biology, such as embryo rescue, have been used to bridge crossing barriers. A notable example is the so-called New Rice for Africa (NERICA; Box 2), a group of upland rice varieties developed for the sub-Saharan African agrooecologies by the AfricaRice Centre by crossing high-yielding Asian rice (*Oryza sativa* L.) with locally adapted African rice (*Oryza glaberrima* Steud.). With in vitro embryo rescue, mature fertile plants could be recovered from the wide crosses. However, the greatest enhancements in the efficiency of plant breeding have been through molecular genetics.

Box 2 NERICA: The best of African and Asian Rice PGRFA

Rice is major food security crop in Africa - with progressively increasing demands in many sub-Saharan African countries. By 1996, of the 11.6 million tonnes of milled rice consumed per year, 3.3 million tonnes (or 33.6%) were imported (FAO, 1996b), with significant foreign exchange burden. By 1999, West African countries spent about 850 million US dollars per year on rice importation (WARDA, 2001), an unsustainable annual cost that would surge to over 1 billion US dollars by 2002 (WARDA, 2002). It was against this sup-optimal food security and economic backdrop that the AfricaRice Centre (then West African Rice Development Association or WARDA), a CGIAR centre, sought to develop rice varieties suited to the largely low input production systems for the rainfed upland and lowland production ecosystems. The scientists successfully combined the hardiness (resistance to the prevailing local biotic and abiotic stresses) of the African rice species, *Oryza glaberrima,* with the productivity of the Asian rice, *O. sativa* in the NERICA (short for New Rice for Africa) family of varieties. The hybrid sterility was overcome using embryo rescue; embryos were harvested after fertilization and the interspecific hybrids subsequently regenerated on in vitro tissue culture media. While *O. glaberrima* is thought to have been domesticated in West Africa about 3500 years ago, *O. sativa* was introduced into the sub-region about 450 years ago. Though sub-Saharan African countries have not been able to achieve self-sufficiency in rice production and the increasing demands continue to be met through importation, the development of the NERICA series of varieties contributed to reducing the burden. In essence, this significant contribution to the quest for food security and nutrition was achieved through the deployment of appropriate science and technology to harness the potentials of PGRFA, including one obtained through germplasm exchange, demonstrating yet again the global interdependence on PGRFA.

9.3 Genetic modification

The 1953 description of the structure, that is, double helix, of deoxyribonucleic acid (DNA) spawned the discipline of molecular genetics and made possible such consequential milestones as the polymerase chain reaction, DNA sequencing, the Human Genome Project, and so on. Arguably one of the most important achievements of molecular biology in the 20th century was recombinant DNA technology - whereby hereditary materials from more than one source are brought together artificially in the laboratory in order to create *de novo* DNA sequences which may lead to the expressions of novel traits. One such technology, variously referred to as genetic engineering, genetic transformation or genetic modification (GM), is used to generate genetically modified organisms (or GMOs). Twenty-three years after the first commercialization of a GM event, 425 such events in 32 plants had been approved for commercialization, planting or importation as food or feed in a total of 70 countries. Of these, 26 countries planted the GM crops while the other 44 imported them as food and/or feeds (ISAAA, 2018). Controversies surrounding the release of GMOs into the environment and/or their use in food, livestock feeds, vaccine production, and so on are well documented. The international community, in recognizing the potentials of GM and at the same time acknowledging concerns about the safety of its products, agreed on the Cartagena Protocol on Biosafety to the Convention on Biological Diversity (Secretariat of the Convention on Biological Diversity, 2000) as means to exploit the technology safely. Adopted on 29 January 2000 and entered into force on 11 September 2003, it is an international agreement that aims to ensure the safe handling, transport and use of GMOs with respect to biological diversity and human health.

9.4 Genome editing

Genome editing (or gene editing) is the term for a relatively new set of technologies used to make precise changes to the genetic makeup of a living organism (Hua et al., 2019; Kim and Kim, 2019); these include zinc-finger nucleases (ZFNs); TAL (transcription-activator-like) effector nucleases (TALENs) and CRISPR (clustered regularly interspaced palindromic repeat) which are used to target specific DNA sequences (Gupta and Musunuru, 2014; Hsu et al., 2014; Trevino and Zhang, 2014; Jiang and Marraffini, 2015; Sternberg and Doudna, 2015; Langner et al., 2018). These changes which are heritable (that is, are mutations) include the insertion, deletion or substitution of one or more specific nucleotides at a predetermined site in the genome, such as a gene. Mutants therefore result and may be beneficial in terms of improvements in traits. Genome editing, especially its most commonly used technique, CRISPR/Cas9 and other CRISPR/Cas systems (Mao et al., 2019), has been demonstrated successfully in many plants, for example, the model plants - *Arabidopsis* and *Medicago truncatula* and other plants spanning a wide range of genera, such as rice (*Oryza* sp.), wheat (*Triticum aestivum*), maize (*Zea mays*), soybean (*Glycine max*), sorghum (*Sorghum bicolor*), cotton (*Gossypium hirsutum* L.), rapeseed (*Brassica napus* L.), barley (*Hordeum vulgare* L.), close relatives of tobacco - *Nicotiana benthamiana* and *N. attenuate*, tomato (*Solanum lycopersicum* L.), potato (*Solanum tuberosum*), sweet orange (*Citrus sinensis* L.), cucumber (*Cucumis sativus* L.), wild cabbage (*Brassica oleracea* L.), a wild legume - *Lotus japonicus* L., lettuce (*Lactuca sativa* L.), common liverwort (*Marchantia polymorpha*), tobacco (*Nicotiana tabacum* L.), petunia (*Petunia hybrida*), grape (Vitis vinifera L.), apple (Malus pumila), cassava (Manihot esculenta) and watermelon (Citrullus lanatus) (Waltz, 2016; Bomgardner, 2017; Manghwar et al., 2019). The modified phenotypes reported in the extensive review by Manghwar et al. (2019) included enhanced vigour and improved yield, herbicide tolerance, resistance to diseases, dwarf stature and altered quality attributes, including increased grain protein digestibility and reduced amylose content while some proofs of concept resulted in albinism, alteration of leaf morphology and changes to days to flowering.

Being relatively inexpensive to apply, the prognosis is for the continuing rapid expansion, including in developing countries, of this novel technique, especially its CRISPR variants. The award of the 2020 Nobel Prize in Chemistry to the pioneers of CRISPR, Emmanuelle Charpentier of the Max Planck Institute, Berlin, Germany and Jennifer Doudna of the University of California, Berkeley, USA (Ledford and Callaway, 2020), is probably the most eloquent testimony to the potentials of this genome editing technique.

The enabling policy environment and regulatory frameworks, however, have not kept pace with the rapid scientific and technological advances. For instance, countries differ in their opinions as to whether the products of genome editing are to be treated the same as GMOs or whether they should even be regulated at all. Interestingly, developed countries have been taking unilateral decisions as evidenced by the United States deciding that genome-edited crops 'could otherwise have been developed through traditional breeding techniques', and so not subject to any regulation (Waltz, 2016; USDA, 2018). In like manner, Australia would not regulate the application of gene editing in plants, animals and human cell lines so long as extraneous genetic materials are not introduced in the process (Mallapaty, 2019). On the contrary, for New Zealand, genome-edited organisms must be regulated just as GMOs (Fritsche et al., 2018). The apex court of the European Union (EU) ruled that genome-edited organisms should be treated in the same way as GMOs by its member countries (Callaway, 2018). Currently, aside from the EU and New Zealand - both of which regulate genome-edited products under their applicable biosafety regulations and both as a result of court rulings - no other country imposes any such regulations (Schmidt et al., 2020). However, the majority of developing countries do not seem to have reached any firm conclusions or, in many cases, even given the issue any systematic consideration.

9.5 Next-generation sequencing

The rapid advances in scientific and technological developments, institutional and human capacities and infrastructures for next-generation sequencing (NGS) and the associated bioinformatics and supercomputing have been continuing for the past 15 years. The effects have been the significantly increased throughput of DNA sequencing and the reductions in costs and durations for the processes which have resulted in massive amounts of sequence data, much of which are now publicly available. The genome assemblies of 323 flowering plants have been completed (Kersey, 2019) while the sequences of the transcriptomes, that is, gene coding regions, of almost 1200 green plants species have been generated (Leebens-Mack et al., 2019). Additionally, the representative genomes of 10 000 plants - encompassing every major clade of embryophytes, green algae and protists (excluding fungi) will be sequenced and characterized by 2023 under the auspices of the 10KP (10 000 plants) Genome Sequencing Project (Cheng et al., 2018).

9.6 Marker-aided selection

Leveraging these genomic resources which are increasingly available, the establishments of high-throughput NGS-based genotyping platforms which are used to develop comprehensive polymorphism datasets including single nucleotide polymorphisms (SNPs), InDels, structural variations and genomic rearrangement have therefore been made possible (Nguyen et al., 2019). These and comparable resources are enabling the accelerated mapping of QTL and gene discovery in crops including genome sequence assemblies, re-sequencing of several hundred lines, and, taking advantage of genome-wide methodologies, the development of HapMaps, high-density genetic maps, various marker genotyping platforms and the identification of markers associated with agronomic traits (Varshney et al., 2019). Consequently, the efficiencies of the characterization of germplasm and the development of improved crop varieties through MAS, marker-assisted recurrent selection (MARS) and, more recently, genome-wide or genomics selection (GS) and genomics-assisted breeding (GAB) are increasing rapidly in the order of magnitudes.

There is a paucity of information on the use of genomics resources in the plant breeding programmes of multinational companies. However, it is generally assumed that molecular breeding techniques, including MAS, MARS, GAB and GS, have become their methods of choice. In the recent past, national public and international agricultural research institutions, especially the CGIAR centres, are utilizing the publicly available genomics resources, some of which are generated by them, to improve the efficiencies of the genetic improvement of food security crops, for example, barley, chickpea, common bean, cowpea, groundnut, maize, pigeonpea, wheat, rice, sorghum and soybean (Kulwal et al., 2012). The African Orphan Crops Consortium plans to sequence the genomes of 101 under-researched crops in the continent (Ribaut and Ragot, 2019). The availability of these resources will increase significantly the scope for the use of MAS in the genetic improvement for less-studied and underfunded crops.

10 Seed delivery systems

Quality seeds and planting materials translate the potentials – for productivity and the attributes for enhanced nutritional quality and tolerance to biotic and abiotic stresses and other breeding objectives – which have been bred into improved crop varieties into the desired outcomes on the farmers' fields. Seed quality, defined in terms of genetic purity; germination percentage; vigour; appearance and the absences of contamination – for example, from weeds, other crops or varieties and debris; and freedom from disease-causing agents and pests, ensures that the propagule bears true fidelity to these potentials. The implication, therefore, is that farmers and by extension the society at large, can only benefit maximally from PGRFA in situations where quality seeds and planting materials of the most suitable improved crop varieties are used. Unfortunately, in many food-insecure countries of the world, this is not the case. In sub-Saharan Africa, for instance, where crop productivities are extremely low, the average adoption rates of improved crop varieties is about 30% (Walker et al., 2014) and up to 90% of seeds are sourced from a wide spectrum of insecure informal supply channels (McGuire and Sperling, 2016) which do not have any quality assurance mechanism.

These farmers are usually small-scale and resource-poor; they practice subsistence low-input production systems on very small family farms. Though they produce a significant amount of the global food supply, their production systems do not align with the business models of the large seed companies. To support them effectively and thereby mitigate this critical shortcoming, there must therefore be different forms and scales of seed enterprises, which would ideally be supported by community-based production and distribution systems. In addition, the mechanisms for quality assurance must accordingly be suitable for these circumstances. Aimed at supporting practitioners along the entire seed value chain, the six-module Seeds Toolkit, covering small-scale enterprises; processing; quality control; sector regulatory framework; marketing; and storage, respectively, (FAO, 2018b,c,d,e,f,g) could be valuable in strengthening institutional and human capacities. Similarly, the Voluntary

Guide for National Seed Policy Formulation (FAO, 2015) would be useful for articulating an effective enabling environment for seed delivery which caters to all farmers, irrespective of scale.

11 Conclusion

There is ample reason to be confident that the largely untapped potentials that are encoded into the genetic blueprint of PGRFA could be leveraged to address the generational challenge of feeding an ever-growing population without inflicting further damage on planet earth. This is evidenced by the recognition of international agreements (CBD, Treaty and SDGs) of the need to ensure that PGRFA are effectively conserved in situ to allow the diversity to continue to evolve and in ex situ facilities to allow the diversity to be accessible to users of germplasm. Most importantly, both the CBD and Treaty, through their respective Nagoya Protocol on Access and Benefit-sharing (Secretariat of the Convention on Biological Diversity, 2011) and the Multilateral System of Access and Benefit-sharing (FAO, 2009), provide internationally agreed mechanisms for the exchange of PGRFA, especially across national boundaries. The concept of ABS continues to be discussed by the parties to both instruments, especially in order to ensure that the intended aim of fairness and equity remains achievable in the face of advances in science and technology - such as digital sequence information and synthetic biology. However, these normative mechanisms provide clarity for the exchange of genetic resources and thereby enhance international cooperation in research and development.

Extraordinary advances have been made in various scientific and technological disciplines, including molecular biology, computing and high throughput genotyping and phenotyping platforms that will render the conservation and use of PGRFA more efficient and effective. Today, the ability to safeguard, characterize and evaluate germplasm for traits of interest efficiently, induce and/or introgress hereditary factors into desirable backgrounds and follow their inheritance in progeny in order to develop crop varieties has become routine. Though the North–South divide in the capabilities to benefit from these scientific advances remains evident, in general, a new crop variety can be bred in a fraction of the time that it took a few decades ago. The scientists who demonstrated that starvation did not have to be inevitable some five decades ago by breeding and disseminating the Green Revolution cereals could not call upon these advances in science and technology which are taken for granted today. However, compared to the current global order, the palpable goodwill and fraternity that were taken for granted then, more than compensated for whatever was lacking in scientific and technological knowhow and infrastructure.

In general, the aphorism of 'use it or lose it' continues to apply to PGRFA. In this regard, sustained funding for research and development and partnerships that span national boundaries are required to sustain the notion in the preamble of the Treaty that, 'PGRFA are a common concern of all countries, in that all countries depend very largely on PGRFA that originated elsewhere' (FAO, 2009). These made it possible for Japanese wheat germplasm – which had been enhanced both locally in Japan and in the United States – to be used in breeding wheat varieties in Mexico that arguably deflected the trajectory of human history, especially in Asian countries. Also, without this concept of PGRFA as a common concern of nations and funding by major donors, Asian rice germplasm and the human and material resources would not have been available to conduct the interspecific hybridization to create the NERICA varieties for use by farmers in Sub-Saharan Africa. In this context, the need for enhanced international cooperation and a world's network of genetic collections from which countries can access the PGRFA they need for research and development becomes more important. The Multilateral System established under the Treaty, which is currently the most systematic and largest global mechanism to ensure regular and facilitated access to PGRFA by breeders and researchers, exemplifies this.

Global coordination, such as the role that FAO and development partners have continued to play in fostering multilateral engagements for food security and nutrition, also remains critically important. Regarding PGRFA in particular, FAO's provision of the platform for international legal frameworks for the governance of the management and exchange of PGRFA, as well as the policy forum for decision-making should be sustained. This role of a neutral convener is probably even more important now than ever before. Leadership is required in many critical areas, especially with the hot potato issue of ABS, for both the products of value addition to PGRFA and their digital sequence information. Importantly also, there is a need for convening the international community for standard setting and consensus building for the powerful new and emerging biotechnologies such as genome editing and gene drive. These are certainly relevant for driving the foreseen knowledge-intensive and sustainable Green Revolution. However, the significant concerns, especially regarding ethics and equity, also demand attention. The more the development of internationally agreed enabling environment lags behind the pace of scientific and technological advancements, the more polarizing arguments fester, contrasting viewpoints calcify and the dimmer the hopes grow for leveraging these tools to attain sustainable food systems which everyone deserves in the 21st century. This is particularly the case for genome editing, for instance, for which the CBD, through the work of its Subsidiary Body on Scientific, Technical and Technological Advice (SBSTTA) on Synthetic Biology, provides the only intergovernmental forum for negotiations and consensus building. The pace of progress towards a global consensus on the regulation, or otherwise, of this powerful cutting edge tool, in spite of the trailblazing work of the pioneers being recognized through the 2020 Nobel Prize for Chemistry (Ledford and

Callaway, 2020), has been extremely slow. The elevated consciousness and momentum generated by the COVID-19 pandemic which has upended many precarious food systems and the need for attaining the lofty aims of the SDGs may well engender a sense of urgency that could sensitize the international community to the need to address above concerns.

12 Where to look for further information

Future research endeavours should focus on the most efficient and cost-effective means for harnessing the unique potentials of PGRFA for the attainment and sustenance of global food security and nutrition. Such efforts should take advantage of the most appropriate scientific and technological advances, most of which were highlighted in this Introduction, to safeguard variations in PGRFA – in genebanks, in nature and through enhanced intra- and inter-specific on-farm diversity – and to add value to them through their use in breeding progressively more productive, nutritious and resilient crop varieties. Efforts should address the constraints to seed security, especially for vulnerable populations. For impactful outcomes, there is a critical need for innovative approaches which complement research and development endeavours with requisite enabling legislations, policies and regulations. There is also a need for winning partnerships that bring together the public and private sectors and non-state actors in the academia, civil society and grassroots organizations in synergistic engagements.

The 'champions' of the conservation and use of PGRFA include principally:

- The FAO, especially through its Commission on Genetic Resources for Food and Agriculture, provides a permanent forum for the global community to achieve consensus on policies for the sustainable use and conservation of genetic resources for food and agriculture, including PGRFA (http://www.fao.org/cgrfa/overview/how-we-work/en/).
- The International Treaty on Plant Genetic Resources for Food and Agriculture, with its Secretariat hosted by the FAO, facilitates the conservation and sustainable use of PGRFA, in particular through its multilateral system of access and benefit sharing for the world's most important food crops (http://www.fao.org/plant-treaty/en/).
- The genebanks of 11 CGIAR centers safeguard and distribute the germplasm accessions of the crops that underpin global food security and nutrition. The centres also use these holdings to breed progressively superior crop varieties that are made available to countries usually royalty-free as public goods (https://www.cgiar.org/).
- The Global Crop Diversity Trust, in particular through its Endowment Fund, seeks to provide 'stable, predictable funding for a global system of crop

Published by Burleigh Dodds Science Publishing Limited, 2021.

collections, in perpetuity'. The collections consist of the holdings of the genebanks of 11 CGIAR centres and those of some regional and national programmes (https://www.croptrust.org/).

Examples of important reference materials that have not been cited in the chapter include:

- Engels, J. M. M., Ramanatha Rao, V., Brown, A. H. D. and Jackson, M. T. 2002. Managing plant genetic diversity. International Plant Genetic Resources Institute, Rome, Italy, CABI Publishing, 487 pp.
- Dulloo, M. E. 2019. Maintaining Diversity of Plant Genetic Resources as a Basis for Food Security. In: Ferranti, P., Berry, E. M., Anderson, J. R. (Eds.), *Encyclopedia of Food Security and Sustainability*, vol. 3, pp. 54–63. Elsevier.
- Maxted, N., Hunter, D., and Rios, R. O. 2020. Plant genetic conservation. Cambridge University Press, UK, 566 pp.

13 References

Alercia, A., López, F., Marsella, M. and Cerutti, A. L. 2020. Descriptors for Crop Wild Relatives conserved under *in situ* conditions (CWRI v.1). *International Treaty on Plant Genetic Resources for Food and Agriculture*. FAO, Rome.

Anon. 2020. World population worldometer. Available at: https://www.worldometers.info/world-population/ (Accessed on 16 June 2020).

Bomgardner, M. M. 2017. CRISPR: a new toolbox for better crops. *Chemical & Engineering News* 95(24), 30–4.

Brozynska, M., Furtado, A. and Henry, R. J. 2015. Genomics of crop wild relatives: expanding the gene pool for crop improvement. *Plant Biotechnology Journal* 14(4), 1070–85 [doi:10.1111/pbi.12454].

Callaway, E. 2018. EU law deals blow to CRISPR crops. *Nature* 560(7716), 16 [doi:10.1038/d41586-018-05814-6].

Cheng, S., Melkonian, M., Smith, S. A., Brockington, S., Archibald, J. M., Delaux, P. M., Li, F. W., Melkonian, B., Mavrodiev, E. V., Sun, W., Fu, Y., Yang, H., Soltis, D. E., Graham, S. W., Soltis, P. S., Liu, X., Xu, X.. and Wong, G. K.-S. 2018. 10KP: a phylodiverse genome sequencing plan. *GigaScience* 7(3), 1–9. doi:10.1093/gigascience/giy013.

Crop Trust n.d.a. Available at: https://www.croptrust.org/ (Accessed on 27 November 2020).

Crop Trust n.d.b. Svalbard global seed vault. Available at: https://www.croptrust.org/our-work/svalbard-global-seed-vault/ (Accessed on 27 November 2020).

Dempewolf, H., Baute, G., Anderson, J., Kilian, B., Smith, C. and Guarino, L. 2017. Past and future use of wild relatives in crop breeding. *Crop Science* 57(3), 1070–82 [doi:10.2135/cropsci2016.10.0885].

Dubin, H. J. and Brennan, J. P. 2009. Combating stem and leaf rust of wheat: historical perspective, impacts, and lessons learned. IFPRI Discussion Paper 00910. International Food Policy Research Institute, Washington DC, 55pp.

Dulloo, M. E. and Maxted, N. 2019. Editorial Special Issue Crop wild relative. *Plant Genetic Resources: Characterization and Utilization* 17(2), 101–2 [doi:10.1017/S1479262118000606].

Dulloo, M. E., Rege, J. E. O., Ramirez, M., Drucker, A. G., Padulosi, S., Maxted, N., Sthapit, B., Gauchan, D., Thormann, I., Gaisberger, H., Roux, N., Sardos, J., Ruas, M. and Rouard, M. 2017. Conserving agricultural biodiversity for use in sustainable food systems. In: Bioversity International. Mainstreaming Agrobiodiversity in Sustainable Food Systems: Scientific Foundations for an Agrobiodiversity, *Bioversity International*, Chapter 5, pp. 103–39.

FAO 1983. Resolution 8/83. International undertaking on plant genetic resources. FAO Conference, 23 November 1983. Food and Agriculture Organization of the United Nations, Rome, Italy.

FAO 1993. The international code of conduct for plant germplasm collecting and transfer. Available at: http://www.fao.org/fileadmin/templates/agphome/documents/PGR/GS/CCgermpE.pdf (Accessed on 18 May 2020). Food and Agriculture Organization of the United Nations, Rome, Italy, 11 pp.

FAO 1994. *GeneBank Standards*. Food and Agriculture Organization of the United Nations, Rome, International Plant Genetic Resources Institute, Rome, Italy. 13 pp. Available at: http://www.fao.org/tempref/docrep/fao/meeting/015/aj680e.pdf (Accessed on 18 May 2020).

FAO 1996a. *Global Plan of Action for the Conservation and Sustainable Utilization of Plant Genetic Resources for Food and Agriculture and the Leipzig Declaration*. Food and Agriculture Organization of the United Nations, Rome, Italy, 63 pp.

FAO 1996b. *Production Year Book* (Vol. 50). Food and Agriculture Organization of the United Nations, Rome, Italy.

FAO 1997. *The State of the World's Plant Genetic Resources for Food and Agriculture*. Food and Agriculture Organization of the United Nations, Rome, Italy, 511 pp.

FAO 2009. *International Treaty on Plant Genetic Resources for Food and Agriculture*. Food and Agriculture Organization of the United Nations, Rome, Italy. Available at: http://www.fao.org/plant-treaty/en/ (Accessed on 12 May 2020). Food and Agriculture Organization of the United Nations, Rome.

FAO 2010. *The Second Report on the State of the World's Plant Genetic Resources for Food and Agriculture*. Food and Agriculture Organization of the United Nations, Rome, Italy, 370 pp.

FAO 2011. *Save and Grow - A Policy Maker's Guide to the Sustainable Intensification of Smallholder Crop Production*. Food and Agriculture Organization of the United Nations, Rome, Italy, 102 pp.

FAO 2012. *Second Global Plan of Action for Plant Genetic Resources for Food and Agriculture*. Food and Agriculture Organization of the United Nations, Rome, Italy, 91 pp.

FAO. 2013. *Genebank Standards for Plant Genetic Resources for Food and Agriculture*. Food and Agriculture Organization of the United Nations, Rome, Italy, 166 pp.

FAO 2014. *GeneBank Standards for Plant Genetic Resources for Food and Agriculture*. Food and Agriculture Organization of the United Nations, Rome, Italy, 166 pp.

FAO 2015. *Voluntary Guide for National Seed Policy Formulation*. Food and Agriculture Organization of the United Nations, Rome, Italy, 60 pp.

FAO 2017a. *The Future of Food and Agriculture – Trends and Challenges*. Food and Agriculture Organization of the United Nations, Rome, Italy, 163 pp.

Published by Burleigh Dodds Science Publishing Limited, 2021.

FAO 2017b. *Voluntary Guidelines for the Conservation and Sustainable Use of Crop Wild Relatives and Wild Food Plants*. Food and Agriculture Organization of the United Nations, Rome, Italy, 92 pp.

FAO 2018a. *FAO Biodiversity Mainstreaming Platform*. Food and Agriculture Organization of the United Nations, Rome, Italy, 4 pp.

FAO. 2018b. *Seeds Toolkit: Module 1: Development of Small-scale Seed Enterprises*. The Food and Agriculture Organization of the United Nations and AfricaSeeds, Rome, Italy, 109 pp.

FAO. 2018c. *Seeds Toolkit: Module 2: Seed Processing: Principles, Equipment and Practice*. The Food and Agriculture Organization of the United Nations and AfricaSeeds, Rome, Italy, 79 pp.

FAO. 2018d. *Seeds Toolkit: Module 3: Seed Quality Assurance*. The Food and Agriculture Organization of the United Nations and AfricaSeeds, Rome, Italy, 111 pp.

FAO. 2018e. *Seeds Toolkit: Module 4: Seed Sector Regulatory Framework*. The Food and Agriculture Organization of the United Nations and AfricaSeeds, Rome, Italy, 64 pp.

FAO. 2018f. *Seeds Toolkit: Module 5: Seed Marketing*. The Food and Agriculture Organization of the United Nations and AfricaSeeds, Rome, Italy, 97 pp.

FAO. 2018g. *Seeds Toolkit: Module 6: Seed Storage*. The Food and Agriculture Organization of the United Nations and AfricaSeeds, Rome, Italy, 102 pp.

FAO 2019a. *FAO Strategy on Mainstreaming Biodiversity across Agricultural Sectors. Hundred and Sixty-Third Session. Rome, 2-6 December 2019*. Food and Agriculture Organization of the United Nations, Rome, Italy, 11 pp.

FAO 2019b. *Voluntary Guidelines for the Conservation and Sustainable Use of Farmers' Varieties/Landraces*. Food and Agriculture Organization of the United Nations, Rome, Italy, 135 pp.

FAO 2019c. The state of the world's biodiversity for food and agriculture. In: Bélanger, J. and Pilling, D. (Eds). *FAO Commission on Genetic Resources for Food and Agriculture Assessments*. Food and Agriculture Organization of the United Nations, Rome, Italy, 572 pp.

FAO 2020. World information and early warning system on plant genetic resources for food and agriculture (WIEWS). Available at: http://www.fao.org/wiews/en/ (Accessed on 18 May 2020).

FAO, IFAD, UNICEF, WFP and WHO 2018. *The State of Food Security and Nutrition in the World 2018. Building Climate Resilience for Food Security and Nutrition*. Food and Agriculture Organization of the United Nations, Rome, Italy, 181 pp.

FAO, IFAD, UNICEF, WFP and WHO 2019. *The State of Food Security and Nutrition in the World 2019. Safeguarding against Economic Slowdowns and Downturns*. Food and Agriculture Organization of the United Nations, Rome, Italy, 212 pp.

FAO, IFAD, UNICEF, WFP and WHO 2020. *The State of Food Security and Nutrition in the World: Transforming Food Systems for Affordable Healthy Diets*. Food and Agriculture Organization of the United Nations, Rome, Italy, 287 pp.

Frankel, O. H. and Hawkes, J. 1975. *Crop Genetic Resources for Today Tomorrow*. IBS Series (Vol. 2). Cambridge University Press, Cambridge, UK.

Frison, C., Lopez, F. and Esquinas-Alcazar, J. T. (Eds) 2011. *Plant Genetic Resources and: Stakeholder Perspectives on International Treaty on Plant Genetic Resources Agriculture*. Food and Agriculture Organization of the United Nations, Bioversity International and Earthscan, Rome, Italy, 321 pp.

Fritsche, S., Poovaiah, C., MacRae, E. and Thorlby, G. 2018. A New Zealand perspective on the application and regulation of gene editing. *Frontiers in Plant Science* 9, 1323. doi:10.3389/fpls.2018.01323.

Global Environment Facility n.d. Available at: https://www.thegef.org/ (Accessed on 27 November 2020).

Gupta, R. M. and Musunuru, K. 2014. Expanding the genetic editing tool kit: ZFNs, TALENs, CRISPR-Cas9. *Journal of Clinical Investigation* 124(10), 4154–61 [doi:10.1172/JCI72992].

Gur, A. and Zamir, D. 2004. Unused natural variation can lift yield barriers in plant breeding. *PLoS Biology* 2(10), e245 [doi:10.1371/journal.pbio.0020245].

Hajjar, R. and Hodgkin, T. 2007. The use of wild relatives in crop improvement: a survey of developments over the last 20 years. *Euphytica* 156(1–2), 1–13 [doi:10.1007/s10681-007-9363-0].

Harlan, H. V. 1957. *One Man's Life with Barley*. Exposition Press, New York.

Hodgkin, T., Demers, N. R. and Frison, E. 2013. The evolving global system of conservation and use of plant genetic resources for food and agriculture: what is it, and where does the Treaty fit it? In: Halewood, M., López Noriega, I. and Louafi, S. (Eds), *Crop Genetic Resources as a Global Commons*. Bioversity International, Italy, pp. 344–67.

Holness, S., Hamer, M., Magos Brehm, J. and Raimondo, D. 2019. Priority areas for the in situ conservation of crop wild relatives in South Africa. *Plant Genetic Resources: Characterization and Utilization* 17(2), 115–27 [doi:10.1017/S1479262118000503].

Hsu, P. D., Lander, E. S. and Zhang, F. 2014. Development and applications of CRISPR-Cas9 for genome engineering. *Cell* 157(6), 1262–78 [doi:10.1016/j.cell.2014.05.010].

Hua, K., Zhang, J., Botella, J. R., Ma, C., Kong, F., Liu, B. and Zhu, J.-K. 2019. Perspectives on the application of genome-editing technologies in crop breeding. *Molecular Plant* 12, 1047–59.

IAEA 2020. Mutant Varieties Database of the Joint FAO/IAEA Division of Nuclear Techniques in Food and Agriculture. Available at: https://mvd.iaea.org/#!Home (Accessed on 28 May 2020).

ISAAA 2018. *Global Status of Commercialized Biotech/GM Crops in 2018: Biotech Crops Continue to Help Meet the Challenges of Increased Population and Climate Change. ISAAA Brief 54 Executive Summary*. ISAAA S.E. Asia Center, Los Baños, The Philippines.

Jiang, W. and Marraffini, L. A. 2015. CRISPR-cas: new tools for genetic manipulations from bacterial immunity systems. *Annual Review of Microbiology* 69, 209–28 [doi:10.1146/annurev-micro-091014-104441].

Jung, W. J. and Seo, Y. W. 2014. Employment of wheat-rye translocation in wheat improvement and broadening its genetic basis. *Journal of Crop Science and Biotechnology* 17(4), 305–13 [doi:10.1007/s12892-014-0086-1].

Kersey, P. J. 2019. Plant genome sequences: past, present, future. *Current Opinion in Plant Biology* 48, 1–8 [doi:10.1016/j.pbi.2018.11.001].

Kim, J.-I. and Kim, J.-Y. 2019. New era of precision plant breeding using genome editing. *Plant Biotechnology Reports* 13(5), 419–21 [doi:10.1007/s11816-019-00581-w].

Kulwal, P. L., Thudi, M. and Varshney, R. K. 2012. Crop breeding for sustainable agriculture, genomics interventions. In: Meyers, R. A. (Ed.), *Encyclopedia of Sustainability Science and Technology*. Springer, New York, NY.

Langner, T., Kamoun, S. and Belhaj, K. 2018. CRISPR crops: plant genome editing toward disease resistance. *Annual Review of Phytopathology* 56, 479–512 [doi:10.1146/annurev-phyto-080417-050158].

Ledford, H. and Callaway, E. 2020. Pioneers of revolutionary CRISPR gene editing win chemistry Nobel. *Nature* 586(7829), 346–7 [doi:10.1038/d41586-020-02765-9].

Leebens-Mack, J. H., Barker, M. S., Carpenter, E. J., et al. 2019. One thousand plant transcriptomes and the phylogenomics of green plants. *Nature* 574, 679–85.

Lumpkin, T. A. 2015. How a gene from Japan revolutionized the world of wheat: CIMMYT's quest for combining genes to mitigate threats to global food security. In: Ogihara, Y., Takumi, S. and Handa, H. (Eds). *Advances in Wheat Genetics: From Genome to Field*. Proceedings of the 12th International Wheat Genetics Symposium, Springer, Tokyo, Heidelberg, New York, Dordrecht, London, pp. 13–20.

Magos Brehm, J., Kell, S., Thormann, I., Gaisberger, H., Dulloo, M. E. and Maxted, N. 2019. New tools for crop wild relative conservation planning. *Plant Genetic Resources: Characterization and Utilization* 17(2), 208–12. doi:10.1017/S1479262118000527.

Mallapaty, S. 2019. Australian gene-editing rules adopt 'middle ground'. *Nature* [doi:10.1038/d41586-019-01282-8].

Manghwar, H., Lindsey, K., Zhang, X. and Jin, S. 2019. CRISPR/Cas system: recent advances and future prospects for genome editing. *Trends in Plant Science* 24(12), 1102–25 [doi:10.1016/j.tplants.2019.09.006].

Mao, Y., Botella, J. R., Liu, Y. and Zhu, J.-K. 2019. Gene editing in plants: progress and challenges. *National Science Review* 6(3), 421–37 [doi:10.1093/nsr/nwz005].

Martynov, S. P. and Dobrotvorskaya, T. V. 2006. Genealogical analysis of diversity of Russian winter wheat cultivars (*Triticum aestivum* L.). *Genetic Resources and Crop Evolution* 53(2), 379–86 [doi:10.1007/s10722-004-0572-2].

Maxted, N. and Kell, S. P. 2008. Linking in situ and ex situ conservation with use of crop wild relatives. Maxted, N., FordLloyd, B. V., Kell, S. P., Iriondo, J. M., Dulloo, M. E. and Turok, J. (Eds). *Crop Wild Relative Conservation and Use*. CABI Publishing, Wallingford, pp. 450–67.

Mba, C. 2013 Induced mutations unleash the potentials of plant genetic resources for food and agriculture. *Agronomy* 3(1), 200–31 [doi:10.3390/agronomy3010200].

Mba, C., Guimaraes, E. P. and Ghosh, K. 2012. Re-orienting crop improvement for the changing climatic conditions of the 21st century. *Agriculture and Food Security* 1, 7 [doi:10.1186/2048-7010-1-7].

McCouch, S. 2004. Diversifying selection in plant breeding. *PLoS Biology* 2(10), e347 [doi:10.1371/journal.pbio.0020347].

McGuire, S. and Sperling, L. 2016. Seed systems smallholder farmers use. *Food Security* 8(1), 179–95 [doi:10.1007/s12571-015-0528-8].

Nass, L. L. and Paterniani, E. 2000. Pre-breeding: a link between genetic resources and maize breeding. *Scientia Agricola* 57(3), 581–7 [doi:10.1590/S0103-90162000000300035].

Ng'uni, D., Munkombwe, G., Mwila, G., Gaisberger, H., Magos Brehm, J. M., Maxted, N., Kell, S. and Thormann, I. 2019. Spatial analyses of occurrence data of crop wild relatives (CWR) taxa as tools for selection of sites for conservation of priority CWR in Zambia. *Plant Genetic Resources: Characterization and Utilization* 17(2), 103–14 [doi:10.1017/S1479262118000497].

Nguyen, K. L., Grondin, A., Courtois, B. and Gantet, P. 2019. Next-generation sequencing accelerates crop gene discovery. *Trends in Plant Science* 24(3), 263–74 [doi:10.1016/j.tplants.2018.11.008].

Rabinovich, S. V. 1998. Importance of wheat-rye translocations for breeding modern cultivar of *Triticum aestivum* L. *Euphytica* 100(1/3), 323–40 [doi:10.1023/A:1018361819215].

Ribaut, J. M. and Ragot, M. 2019. Modernising breeding for orphan crops: tools, methodologies, and beyond. *Planta* 250(3), 971-7 [doi:10.1007/s00425-019-03200-8].

Schmidt, S. M., Belisle, M. and Frommer, W. B. 2020. The evolving landscape around genome editing in agriculture. *EMBO Reports* 21(6), e50680 [doi:10.15252/embr.202050680].

Secretariat of the Convention on Biological Diversity 2000. *Cartagena Protocol on Biosafety to the Convention on Biological Diversity: Text and Annexes*. Secretariat of the Convention on Biological Diversity, Montreal, 30 pp.

Secretariat of the Convention on Biological Diversity 2011. *Nagoya Protocol on Access to Genetic Resources and the Fair and Equitable Sharing of Benefits Arising from Their Utilization to the Convention on Biological Diversity: Text and Annex*. Secretariat of the Convention on Biological Diversity, Montreal, 25 pp.

Scarascia-Mugnozza, G. T. and Perrino, P. 2002. The history of ex situ conservation and use of plant genetic resources. In: Engels, J. M. M., Rao, V. R., Brown, A. H. D. and Jackson, M. (Eds), *Managing Plant Genetic Diversity*. CABI Publishing, Wallingford; New York, pp. 1-22.

Sonnino, A. 2017. International instruments for conservation and sustainable use of plant genetic resources for food and agriculture: an historical appraisal. *Diversity* 9(4), 50. doi:10.3390/d9040050.

Sternberg, S. H. and Doudna, J. A. 2015. Expanding the biologist's toolkit with CRISPR-Cas9. *Molecular Cell* 58(4), 568-74 [doi:10.1016/j.molcel.2015.02.032].

Tester, M. and Langridge, P. 2010. Breeding technologies to increase crop production in a changing world. *Science* 327(5967), 818-22 [doi:10.1126/science.1183700].

The Gale Encyclopedia of Science 2020. "Plant breeding." The gale encyclopedia of science. . Available at: https://www.encyclopedia.com/science/encyclopedias-almanacs-transcripts-and-maps/plant-breeding-0 (Retrieved on October 16, 2020).

Thormann, I., Alercia, A. and Dulloo, M. E. 2013. *Core Descriptors for In Situ Conservation of Crop Wild Relatives* (Vol.1). Bioversity International, Rome, Italy.

Trevino, A. E. and Zhang, F. 2014. Genome editing using Cas9 nickases. *Methods in Enzymology* 546, 161-74 [doi:10.1016/B978-0-12-801185-0.00008-8].

UN General Assembly, Transforming our world: the 2030 Agenda for Sustainable Development, 21 October 2015, A/RES/70/1, available at: https://www.refworld.org/docid/57b6e3e44.html [Accessed on 21 January 2021].

United Nations 1993. Convention on Biological Diversity (with annexes). Concluded at Rio de Janeiro on 5 June 1992; entered into force 29 December 1993. *Treaty Series* 1760(30619), 79-307.

USDA 2018 Press Release No. 0070.18. Available at: https://www.usda.gov/media/press-releases/2018/03/28/secretary-perdue-issues-usda-statement-plant-breeding-innovation (Accessed on 26 November 2020).

Varshney, R. K., Pandey, M. K., Bohra, A., Singh, V. K., Thudi, M. and Saxena, R. K. 2019. Toward the sequence-based breeding in legumes in the post-genome sequencing era. *Theoretical and Applied Genetics* 132, 797-816.

Vavilov, N. I. 1931. The problem of the origin of world agriculture in the light of the latest investigations. In: Bukharin, N. I. *Science at the Crossroads*. Kniga, London, pp. 1-20. (Reprint New York 1971).

Vincent, H., Amri, A., Castañeda-Álvarez, N. P., Dempewolf, H., Dulloo, E., Guarino, L., Hole, D., Mba, C., Toledo, A. and Maxted, N. 2019. Modeling of crop wild relative species identifies areas globally for *in situ* conservation. *Communication Biology* 2, 1-8 [doi:10.1038/s42003-019-0372-z].

Walker, T., Alene, A., Ndjeunga, J., Labarta, R., Yigezu, Y., Diagne, A., Andrade, R., Muthoni Andriatsitohaina, R., De Groote, H., Mausch, K., Yirga, C., Simtowe, F., Katungi, E., Jogo, W., Jaleta, M. and Pandey, S. 2014. Measuring the effectiveness of crop improvement research in sub-Saharan Africa from the perspectives of varietal output, adoption, and change: 20 crops, 30 countries, and 1150 cultivars in farmers' fields. Report of the Standing Panel on Impact Assessment (SPIA), CGIAR Independent Science and Partnership Council (ISPC) Secretariat: Rome. Italy, 83 pp.

Waines, J. G. and Ehdaie, B. 2007. Domestication and crop physiology: roots of green revolution wheat. *Annals of Botany* 100(5), 991–8 [doi:10.1093/aob/mcm180].

Waltz, E. 2016. Gene-edited CRISPR mushroom escapes US regulation. *Nature* 532(7599), 293 [doi:10.1038/nature.2016.19754].

WARDA 2001. *NERICA: Rice for Life*. West African Rice Development Association, Bouaké 01, Côte d'Ivoire, 8 pp.

WARDA 2002. *NERICA on the Move: A Symbol of Hope for Rice Farmers in Africa*. West African Rice Development Association, Bouaké 01, Côte d'Ivoire, 2 pp.

Chapter 2

Valuing plant genetic resources in genebanks: Past, present and future

Melinda Smale, Michigan State University, USA; and Nelissa Jamora and Luigi Guarino[1], Global Crop Diversity Trust, Germany

1 Introduction

During the 1960s, the specter of famine in India drove the development of short-statured crop varieties which responded to fertilizer with dramatically higher yields, when moisture conditions were adequate (Lele and Goldsmith, 1989). Less than a decade later, as he witnessed the rapid replacement of landraces by such improved varieties across the Asian landscape, Harlan challenged the scientific community to heed the disastrous 'genetic erosion' of 'some of the most vital resources for human survival' (1972, p. 212). They did and initiated the massive collections of crop genetic resources that are now stored in genebanks, including in CGIAR Centers. Around the world today there are some 1750 genebanks, housing over 7 million accessions, 2 million of which are estimated to be unique (Fu, 2017). Typically, genebanks are publicly funded, because they are public goods intended to serve the welfare of society as a whole. Accordingly, Target 2.5 of the Sustainable Development Goal (SDG) of the United Nations calls for 'soundly managed and diversified seed and plant banks at national, regional and international scales' in order to end global hunger and malnutrition.

Though the intrinsic value to society of plant genetic resources conserved in genebanks may be recognized, assigning a dollar value to them remains elusive.

1 The authors gratefully acknowledge the continued intellectual contributions of Doug Gollin, Kelly Day-Rubenstein, Paul Heisey, and Charlotte Lusty to this chapter.

http://dx.doi.org/10.19103/AS.2020.0085.02

Placing a monetary value on resources such as these is often considered to be human-centric and presumptuous (Wilson, 1988). Because crops are the result of domestication by humans, rather than endangered species of wild animals, most economists have emphasized the value-in-use of crop genetic resources by humans rather than their intrinsic value, which is potentially limiting from both ecosystem and ethical viewpoints (Gollin, 2020). Practically speaking, whenever resources are constrained, we need some measure of relative magnitude of value to help guide investments, and we need investments to fund genebanks.

Assigning a cardinal value to genebanks is also difficult because of the public good characteristics of the resources they maintain and the multiple dimensions of their economic value. Most of the total economic value embodied in the crop genetic resources conserved by genebanks, as well as the services genebanks provide[2], have no market prices. Modern improved varieties, which may be sold as seed for a market price, embody the immeasurable genetic contributions of numerous and varied ancestors. The more highly bred the improved variety, the more internationally distributed and complex the ancestry, and perhaps, the more likely that important segments remain unknown. Arguably, the most 'valuable' plant genetic resources conserved in the genebank are the rarest and most genetically distinct from others, but these have unknown probabilities of future use and payoff. Examples of these include crop wild relatives in the three genepools as defined by Harlan and De Wet (1971). Referring to species preservation, Krutilla (1967) suggested that when we know little about the cardinal value of benefits, we might employ our best scientific estimates to rank candidates for conservation.

However, perhaps it is not the value of the genebank that matters most for investment decisions. A library is one analogy (Brown, 1990). Rather than valuing the books we have already read and are likely to read again, genebank managers must choose which 'book' to keep from the many not yet read, taking into account preconceived notions of society's future reading habits. Another criterion to consider is cost: if two accessions are expected to have the same value, and resources are limited, the least-cost option should be conserved. The time value of money also means that those used tomorrow will have greater value than those only likely to be used in the distant future. Choosing which of the remaining genetic resources in the three genepools to collect from farmers or wild places is another strategic decision with cost implications. Logistically, this is not as easy as it was, for example, Iltis (1988, p. 103)

2 Other than the provision of germplasm, services typically provided by the CGIAR genebanks include: in-depth advice on selection, use, or management of germplasm; conservation or distribution of breeders' and researchers' materials; hosting safety duplicates on behalf of partners or colleagues; provision of planting material for specific accessions on a large scale (outside of normal germplasm distribution); provision of germplasm data; taxonomy verification; and multiplication, cleaning, processing of materials other than genebank accessions (CGIAR Genebank Platform, 2019). CGIAR genebanks also follow a quality management system to address eight standards of practice established by the FAO (2014) regarding acquisition, conservation, regeneration, characterization, safety duplication, distribution, information management, and collecting.

described the 'serendipity' of finding *Lycopersicon chmielewskii* – 'tiny, slimy seeds of a useless, ugly weed stuck to an old newspaper'– while on a collecting expedition for potato in Peru in 1962. The marginal cost of collecting this wild relative of tomato was only $42, but it generated an estimated commercial value of $8 million per year (in 1987 dollars, when originally calculated). Today, collecting missions must respect a potentially daunting tangle of regulations designed to ensure the fair and equitable sharing of benefits arising from the use of genetic resources.

In this chapter, we first summarize the main messages of the small applied economics literature on genebank values that began in the late 1990s. We then report the central findings from a set of recent studies supported by the CGIAR Genebank Platform (www.genebanks.org) and the Global Crop Diversity Trust (Crop Trust) (www.croptrust.org) that explore various ways of valuing genetic resources held by the international genebanks. We supplement this literature with what is known about the value of national genebanks. The essay closes with a discussion of the needs for future applied economics research on this topic.

2 Past research on genebanks

Two decades ago, applied economics research on genebank values was motivated by concerns that despite the increase in numbers, capacity, and breadth of stored genetic materials from the 1970s, plant breeders rarely used genebanks as a source of material from which to transfer diversity into improved varieties for farmers. Raeburn (1995) described genebanks as 'seed morgues'. Wright (1997, p. 110) called for 'a comprehensive review of the extent of use of major genebanks, how they are used, what materials are used, what they are used for, and the nature of the users.' Kingwell et al. (2001, p.265) cited 'considerable economic waste in germplasm maintenance' exacerbated by funding discontinuity. If genebanks were not being used, why continue to support them?

Evenson and Gollin (1997) examined the pedigrees of improved rice varieties released by the International Rice Research Institute (IRRI) and national rice research programs. In the first study of its kind, they used an econometric model to correlate changes in the number of accessions in genebanks with productivity. They found that adding 1000 accessions to the genebank catalogue was associated with the release of 5.8 additional varieties, generating a present value income stream (in 1990 dollars) of $325 million. They reported that the present value of a landrace added to the genebank (that had not yet been used in a released variety) could be as high as $50 million, justifying the continued operation of the genebank, and indeed its expansion.

Evenson and Gollin (1997) recognized that their result should be considered with caution because they were not able to link causally the acquisition of accessions and the release of varieties. Seeking to better understand how plant breeders 'use' genebanks, Gollin then led a study at the International Maize and Wheat Improvement Center (CIMMYT). The model proposed by Gollin et al. (2000) related the search for useful traits to the size of the collection. Applying the model to actual search costs and using plausible probability distributions for the levels of resistance to Russian wheat aphid and *Septoria* leaf blotch, the authors showed that the value of large collections is derived from the increased chance of finding rare traits of economic importance. In fact, a single successful search for resistance to a devastating plant disease or pest can more than justify the cost of maintaining a huge collection. Further, they demonstrated how knowing 'where to search' in a collection can tremendously enhance benefits by reducing the time to success attesting to the value of scientific knowledge about the collection. A third example illustrated the point that, under the existing technologies, the relatively high cost of incorporating traits from certain accessions, such as wild relatives, was an impediment to their use by plant breeders. Gollin et al. (2000) concluded that the observation that genebanks were 'used' infrequently by plant breeders in no way implied that they had no, or even low, value.

While Gollin et al. (2000) provided key insights into the value of genebanks, their analysis did not enable them to draw conclusions about the optimal size of a genebank or the value of adding an accession to the collection. In order to do so, their model could be applied to more comprehensive data, like that now contained in large-scale databases such as GRIN-Global, Genesys or the Plant Treaty's Global Information System (GLIS). However, given current technologies, the process of searching for useful traits might also be modeled differently.

The expected value of an additional genebank accession continues to be pertinent for the decision to expand a collection, discard an accession, or restrict the further use of an accession through imposition of property rights. Most work by economists suggests that while the process of shuffling numerous plant genetic resources in plant breeding generates huge economic values, and some individual genetic resources do prove to have great value when they confer specific traits, the marginal contribution of the last resource used is expected to be slight.

The main reason for this is that excess capacity, redundancy, and the availability of close substitutes, generally drives the expected marginal value of an individual genetic resource toward zero (Simpson et al., 1996). Economically important traits are not evenly distributed across plant genetic resources, with varying likelihood of encountering useful levels. The traits demanded by society, such as resistance to plant pests and diseases, and quality attributes preferred by consumers, also change frequently in response to biotic and environmental stresses and socioeconomic conditions, keeping plant breeders on a treadmill to surpass past accomplishments and to achieve new goals. Breeding products (crop varieties) contain many 'ingredients' that are also genetic resources and these products are in turn combined with others to produce the next variety. For this reason, Gollin and Smale (1998) cautioned against 'the myth of enormous value' – the mistaken notion that huge benefits could be generated by bioprospecting.

The study by Zohrabian et al. (2003) built on the same theoretic search framework of Gollin et al. (2000) to estimate the marginal value of an accession and compare it to the costs of collecting and conservation. Considering the example of a search for resistance to a soybean cyst nematode in the US soybean collection, they estimated a benefit-cost ratio ranging from 36 to 61 times the costs of collecting and conserving an additional accession. This implied that expansion of the US soybean collection was justified, even for a single trait.

Wright (1997) also recognized that 'it might be much easier to judge whether the marginal benefits are likely to exceed measurable marginal costs of conservation than to accurately calculate those benefits.' He argued that better understanding of cost structures could facilitate the choice of conservation technologies, and overcome claims of poor performance in the areas of documentation and characterization. A set of studies conducted with CGIAR genebank managers (Koo et al., 2004) used the cost minimization approach to confirm the relatively low costs of maintaining large collections of some major crops, while also providing projections for the size of an endowment to be managed by the Crop Trust to support globally important collections in perpetuity. The main message of this body of work was that, if the costs of conserving an accession are lower than any sensible lower-bound estimate of benefits, undertaking the challenging exercise of benefits estimation is not necessary to justify its conservation. Zohrabian et al. (2003) reiterated this message.

The CGIAR cost analysis also showed that the present value of conserving and distributing an accession in perpetuity varied significantly by crop, reflecting in part its reproductive biology. Other cost studies of national genebanks in Kenya, India, and Colombia were published in a volume edited by Virchow (2003). As with Koo et al. (2004), these studies reveal variations in conservation cost by method of conservation, crop, and genebank location. Garming et al. (2010) also noted the large share of collecting costs in the national conservation effort of nations located in centers of origin and diversity. Schreinemachers et al. (2014) analyzed the cost structure of the World Vegetable Center (AVRDC) collection. They found that regeneration, seed processing, characterization, and germplasm distribution were the four most expensive operations of the genebank.

Turning again to Wright's (1997) challenge regarding 'use', Smale and Day-Rubenstein (2002) summarized the findings from a 2000 survey of requestors of germplasm samples from the US National Plant Germplasm System (US NPGS). Within the brief 5-year period covered by the survey, 11% of germplasm samples received had already been incorporated into breeding programs. Another 43% were still under evaluation, and 19% were reported as useful in other ways, leaving 28% categorized as 'not useful'. Day-Rubenstein et al. (2006) tested the effect of accompanying data on the share of accessions received that were reported to be 'useful'. Their analysis confirmed that accompanying data improved the chances that a sample received from the US NPGS was immediately incorporated into a breeding program, evaluated, or used in other ways. Breeders surveyed by Brennan et al. (1999) in Australia reported that they had made an average of five requests per year from the Winter Cereals Collection over the preceding five years. Breeders in general judged the material they obtained as 'useful', 'very useful', or 'essential'. Of course, rates calculated over such short time frames understate overall use.

The global interdependence of genebank use has been documented in increasingly sophisticated ways (as databases expand) over the decades by analyzing the flows of genetic resources among countries (Evenson and Gollin, 1997; Fowler et al., 2001; Johnson et al., 2003; Galluzzi et al., 2016).

3 Current research on genebanks

The study by Gollin et al. (2000) hinted at the option and insurance values of genebanks when comparing short-term against long-term benefits of conservation, but to our knowledge, no published research estimated these values until recently. Taking the Greek national genebank as an example, Xepapadeas et al. (2014) confirmed that insurance values alone considerably exceeded operating costs.

Another recent contribution has been the assessment of public willingness-to-pay for the conservation of plant genetic resources. Tyack and Ščasný (2018) applied an on-line survey to estimate the willingness-to-pay (WTP) to conserve additional crop varieties in the Czech national genebank for 10 years. A major finding of their research is that Czechs who are provided with information about the importance of crop diversity would be willing to pay a total of $68 million for crop conservation during the coming decade. This amount represents about 4.5 times more than the current genebank conservation costs; once again, estimated benefits far outweigh costs.

A recent set of studies funded by the CGIAR Genebank Platform and the Crop Trust adds to this earlier research, expanding the evidence of the value-in-use of genebanks (Smale, 2020; Smale and Jamora, 2020). These studies contributed new insights by tracing the contributions of genebank accessions to the ancestry of widely adopted crop varieties and demonstrating the value-in-use of crop varieties derived from landraces and wild crop relatives. End-user surveys and cost analyses were enhanced by new approaches.

At the IRRI, Villanueva et al. (2020) built on the earlier study by Evenson and Gollin (1997). They were able to combine detailed data from a survey of 8640 farms in four states of Eastern India with digitized pedigree information to fully characterize the contribution of the International Rice Genebank (IRG) to the ancestry of widely grown rice varieties. The authors quantified the genetic contribution of materials from the IRG to the mean and variance of rice yields in the econometric framework of a yield response function, controlling for the effects of other farm inputs and environmental factors. They found that 45–77% of the genetic composition of the rice varieties grown by farmers they surveyed was derived from IRG accessions, depending on whether the accession was identified by its genebank identifier (definite contribution) or simply by name (possible contribution). With a high degree of statistical significance, a 10% increase in the genetic contribution of these accessions was associated with a 27% increase in rice yields on farms, but it was neutral for yield variance and kurtosis. The authors also found high genealogical diversity among the 10 most widely adopted varieties, as indicated by a low coefficient of parentage, and ancestry tracing back to 19 countries of origin.

Bernal-Galeano et al. (2020) applied an economic surplus model to demonstrate, as Gollin et al. (2000) had suggested in their simulations, that the estimated gross benefits from a single improved potato variety ('Victoria') grown in Uganda ($1.04 billion in 2016 dollars) exceeds the operating costs of the entire International Potato Center (CIP) genebank over its entire lifetime thus far. An estimated 72% of these benefits correspond to the genebank's contribution. The fact that 'Victoria' includes a wild relative and a Peruvian landrace (6.3% each) in its ancestry underscored the importance of maintaining wide diversity in collections. This analysis extended a study by Robinson and Srinivasan (2013), who were able to link economic benefits from a very successful potato variety, C-88, to the CIP genebank, but did not apportion benefits based on pedigree analysis.

Despite the challenges of incorporating them into plant breeding programs, the potential of crop wild relatives is increasingly recognized, especially as traits such as tolerance of drought and salinity increase in economic importance. Tyack and Dempewolf (2015) summarized the findings of a number of studies that estimated the economic value of crop wild relatives in crop improvement. Annual benefit estimates ranged from $8 million to $165 billion (in 2012 dollars). Yet, the authors note that the reported estimates were often based on 'back-of-the-envelope' calculations in which the entire value of the increase in the yield or quality of the finished variety was attributed to the wild crop relative with the known trait. They recommend a study on wild coffee by Hein and

Gatzweiler (2006), who took costs into consideration. Tyack and Dempewolf (2015) propose a new conceptual framework for analysis which considers not only the production value of introgression of wild genetic material, but also reductions in input costs, and environmental and health externalities.

Aberkane et al. (2020) traced the impact pathway of improved wheat varieties derived from wild relatives. This starts from the genebanks at CIMMYT and at the International Center for Agricultural Research in the Dry Areas (ICARDA), continues with the use of accessions of 'goat grass' (*Aegilops tauschii*) in pre-breeding of synthetic hexaploid wheat (SHW) that serves as a bridge to transfer useful traits directly into breeding lines, moves on to the distribution of SHW-derived lines in international nurseries, and culminates with the adoption of 86 varieties released in 20 countries. Survey estimates show the greatest scale of adoption so far in southwest China, at 34% of the reported wheat area, and in India at 7%. The analysis highlights the importance of pre-breeding to expand allelic variation and to incorporate beneficial traits, including resistance to pests and pathogens, tolerance to abiotic stresses, agronomic performance and quality.

Tyack and Ščasný (2020) discuss the Czech national genebank's collection of numerous species of wild wheat relatives, including *Aegilops*. The authors applied stated preference methods not only to the conservation of accessions, but to characterization, evaluation and pre-breeding. They find that the Czech Republic is willing to support substantial investments in characterization and evaluation as well as further conservation of wheat wild relatives, and, to a lesser extent, pre-breeding. WTP estimates for non-core genebank activities are lower than they had estimated in the earlier study (Tyack and Ščasný 2018). A major contribution of this article is to consider the valuation of genebank activities other than conservation.

Sellitti et al. (2020) also trace the 'journey' of an accession, in this case from its introduction to the genebank at the International Center for Tropical Agriculture (CIAT) to its use by farmers in Rwanda. They show that the scope of the genetic diversity housed in the CIAT collection underpinned the development of iron-biofortified beans to address the high rates of anemia in dispersed, rural populations with little means to address iron deficiency other than their own farm production. Applying farm adoption and impact models to detailed household survey data collected in 2015 by HarvestPlus, they found significant positive effects of the new iron-biofortified beans on yields. Iron-biofortified beans enhance diet quality through the pathway of raising production among farm households who consume an important share of the beans they grow, as is the case in Rwanda.

A study by Kitonga et al. (2020) is distinctive in its treatment of the use and benefits of tree germplasm of *Calliandra* and *Gliricidia* requested from the World Agroforestry genebank. Kitonga's phone survey targeted farmers in

Kenya, representing 95% of requests for samples of the two species, the most popular among smallholders. Farmers identify improved food security, higher incomes, increased milk production, reduced vulnerability to drought, reduced soil erosion and enhanced soil fertility as benefits of using the germplasm they received. Findings demonstrate the role of the genebank in delivering ecosystem and livelihoods services.

One of the lacunae in the applied economics literature is the linkage between research on in situ and on-farm conservation on the one hand and ex situ conservation in genebanks on the other. The concept of complementary conservation is by no means novel, but valuation studies tend to address one or the other types of conservation separately. A recent study by McLean-Rodríguez et al. (2019) examines the abandonment of maize landraces over the last 50 years in Morelos, Mexico, by tracing a collection of 93 maize landrace accessions stored in the CIMMYT genebank back to the original 66 donor families and examining the causes for abandonment or conservation of seed lots. By 2017, only 13 families had conserved a total of 14 seed lots directly descended from the 1967 collection. Interrelated changes in maize cultivation practices, markets for maize and other crops, national policies affecting land use, urbanization and cultural preferences were disincentives for landrace conservation on farms. Processes also differed by location and landrace.

Ocampo-Giraldo et al. (2020) lay out the history of one of the major landraces of maize in Mexico (*Jala*), and describe a strategy for dynamic conservation being developed and tested through the rematriation approach. Dynamic conservation is defined as a circular process in which relatively static conservation ex situ is complemented by evolutionary change in situ, leading to germplasm exchange involving both utilization and conservation. As applied by the Instituto Nacional de Investigaciones Forestales, Agrícolas y Pecuarias (INIFAP) and the CIMMYT genebank, rematriation is a 'co-creative process' with *Jala*-growing communities, involving the return of conserved *Jala* germplasm back to its place of origin for further development, and its subsequent conservation both on farm and in the genebank.

Returning to Wright's (1997) concern for the 'usefulness' of materials stored in genebanks, the distribution of conserved germplasm provides a rough indicator of the demand for genetic resources. In 2018, USDA's Economic Research Service administered another study of requestors of the U.S. National Plant Germplasm System, focusing on the same 10 crops covered in 2000. Preliminary results suggest anticipated future use of NPGS germplasm was very similar over the two periods, with more users in both periods expecting increased or unchanged use than anticipating decreased use. However, in 2018 a greater percentage of NPGS samples were found to be 'useful' than in the 2000 study (K. Day Rubenstein and P. Heisey, personal communication).

Over the last 10 years, the CGIAR genebanks have distributed more than 1 million germplasm samples to recipients both within and outside the CGIAR. From 2017 to 2019, 63% of the materials requested were traditional cultivars (or landraces) and wild relatives of cultivated crops. Another 30% were breeding or research materials and advanced or improved cultivars (Fig. 1). About 80% of the materials sent outside the CGIAR were distributed to national agricultural research systems and advanced research institutes (Fig. 2). Only 9% were distributed to the commercial sector. An even smaller percentage (5%) was distributed directly to farmers or nongovernmental organizations, which suggests a potential opportunity for national genebanks.

In a 2018 survey of users of the IRRI genebank (n=244), 83% of responses confirmed the usefulness of IRRI rice germplasm for research (65% extremely useful and 18% moderately useful) (CGIAR Genebank Platform, Crop Trust,

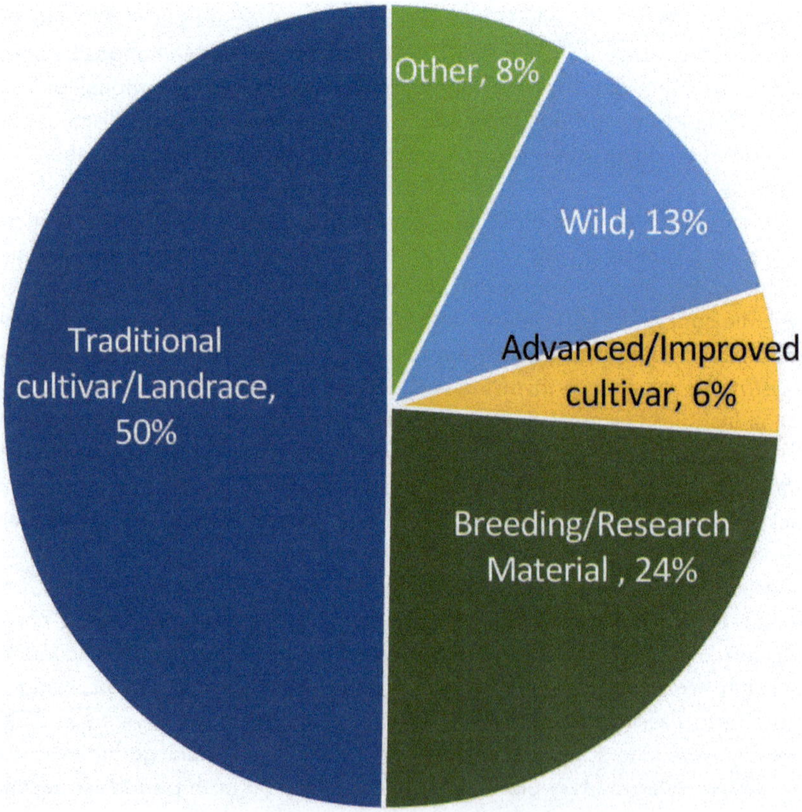

Figure 1 Types of materials distributed, % by number of samples distributed from 2017 to 2019. Source: N. Jamora, based on data from Genebank Online Reporting Tool (ORT). CGIAR Genebank Platform and the Crop Trust.

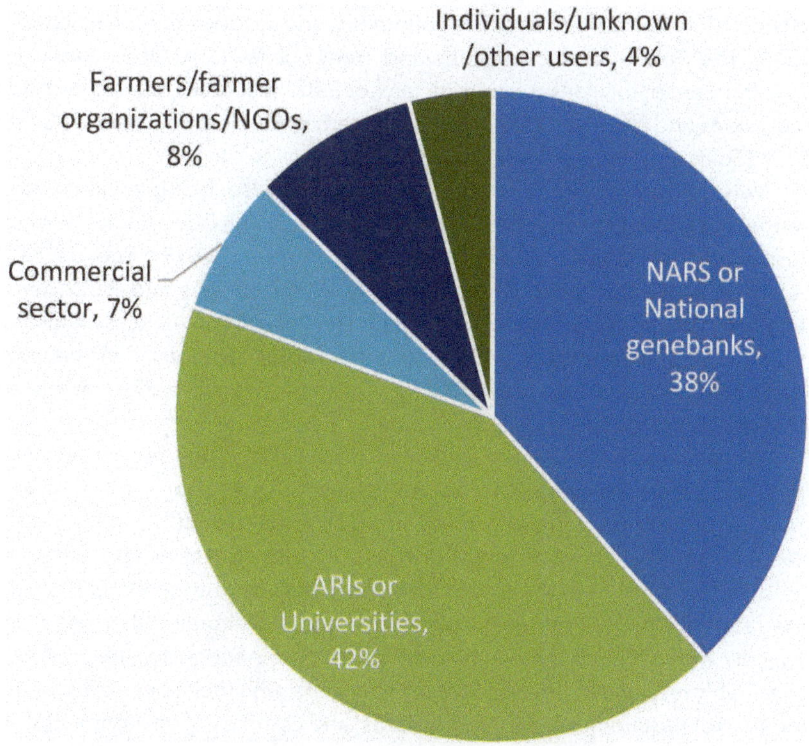

Figure 2 Types of recipients, % by number of accessions distributed to external recipients from 2017 to 2019. Source: N. Jamora, based on data from Genebank Online Reporting Tool (ORT). CGIAR Genebank Platform and the Crop Trust.

and IRRI, 2021). When asked to elaborate on the usefulness of materials, 95 respondents (43%) reported scientific publications as major outputs resulting from the use of IRRI materials. About 35% of the respondents have generated characterization and evaluation data, 23% produced genetic markers, and another 35% identified useful traits for potential incorporation in crop improvement programs. Further, more than half of the respondents (52%) expect an increase in their demand for IRRI rice germplasm in the next 10 years. When asked about the types of germplasm they are likely to request, there was significant increased interest for wild species and genetic stocks, but the majority of the respondents will also continue to source traditional cultivars or landraces from IRRI. A similar survey of users of the ICARDA genebank in 2019 found similar trends (CGIAR Genebank Platform, Crop Trust and ICARDA, 2019).

The World Vegetable Center, one of the largest international public genebanks outside the CGIAR system, distributes around 7000 accessions and breeding lines annually to about 70 countries worldwide (Schreinemachers et al., 2014). The Centre for Pacific Crops and Trees (CePaCT), which is considered a global center for conservation of taro genetic resources, distributed 3059 accessions and close to 22 000 samples of taro germplasm between 2004 and 2017 (Ebert and Waqainabete, 2018).

Very large collections of crop diversity are also being maintained by national genebanks in countries such as Australia, Brazil, China, India, Germany, Russia, and the United States (Bramel and Krishnan, 2019). The US NPGS has over 500000 accessions, representing 15977 species and 2539 genera (https://npgsweb.ars-grin.gov/gringlobal). The N.I. Vavilov All Russian Institute of Plant Genetic Resources (VIR), one of the oldest genebanks in the world, holds 346666 accessions representing 64 botanical families, 376 genera, and 2169 species (Dzyubenko, 2018).

While the CGIAR Centers, together, have built up the largest and most frequently accessed network of global collections of major staples, national genebanks hold unique diversity of crops and varieties of local importance. National genebanks work as repositories of local diversity and are major conduits for new diversity introduced from outside the country. They also interface with the full range of institutions and stakeholders that need crop diversity within the country. In-country capacity is needed to effectively and sustainably respond to the needs of local farmers, with their specific adaptation and market requirements, farming systems, and complex of environmental and climatic challenges.

With adequate resourcing and infrastructure upgrades, national genebanks could spearhead a much-needed change in the global food system and broaden the genetic basis of agricultural production. However, the future of critical crop collections in national genebanks is often threatened by deteriorating infrastructure and unpredictable public funding even in higher-income countries. Díez et al. (2018, p. 1) in their assessment of crop collections in Spain acknowledged that significant increase in public funding is unlikely and listed several shortcomings, including the 'unsatisfactory rate of use by end users, existence of detected and undetected duplicates, insufficient rate of regeneration, and insufficient phenotyping, genotyping, and epiphenotyping'.[3]

Further, many of these national genebanks maintain sparse data on distributions and costs of operations. Limited information not only hinders their pursuit of efficiency, but also efforts to raise funds for investment in their maintenance and expansion.

3 Epiphenotyping includes data about the relationship between the plant and the environment, including ethnobotanical aspects (Díez et al., 2018, p. 1).

4 Future trends in research

Past and current studies of the value of genebanks and the plant genetic resources they maintain, while welcome, offer no surprises for those tracking debates in crop diversity conservation. We highlight three recurring, salient points:

1. **The value of information**. Access to data on conserved germplasm, about demand for crop diversity, and on innovations in technology are needed to meet evolving challenges in crop diversity conservation. Significant investments to support the work of genebanks will depend on both better describing the genetic diversity they maintain and using it more effectively (Bramel and Krishnan, 2019). Germplasm users increasingly require access to digital information associated with each accession, as much as to the physical material itself, though this is proving contentious (Halewood et al., 2018). Innovations in technology also promise gains in efficiency and effectiveness in the conservation of plant genetic resources. Advances in genomics and informatics are revolutionizing the breeding model, which now allows users of germplasm to choose parents and select progeny based directly on genotype rather than phenotype, much faster and more cheaply (Gollin et al., 2000). However, it is clear that genebanks need to change in many and significant ways to make full use of new technologies, or they revert to being 'seed morgues'. Though the genomic and informatics revolutions have the capacity to dramatically change the range of activities of a genebank, there needs to be a concomitant paradigm shift in the way genebanks are perceived, organized, managed, and funded (Wambugu et al., 2018).
2. **A diversity of values**. No single model or approach can capture the multiple kinds of market and non-market values of plant genetic resources in general and genebanks in particular in a simple, objective manner (Guarino et al., 2019). A diversity of approaches is needed to document the diversity of ways that genebanks benefit society through the diversity they manage and make available. This cannot be adequately or usefully summarized in a single dollar value. Beyond the provision of germplasm for use in breeding and research, genebanks also add values to the conserved materials through characterization, evaluation, documentation, and health testing and cleaning. In contrast with financial banks, plant genebanks maintain resources that are alive and need constant attention. The preservation of germplasm requires methods and conditions that support longevity and regular rejuvenation. Seed must be carefully re-sowed to produce new seed that maintains

the genetic nature of the original stock, before that stock loses its ability to germinate. Plantlets kept in vitro must be subcultured, and plants in field genebanks must be carefully tended. The consequence of not doing this effectively is the loss of often irreplaceable diversity. There is value inherent in the sheer fact of conservation, as much as in the use of the conserved material.

3. **Establishing priorities for collecting and conservation based on cost-effectiveness.** Gollin (2020) highlighted the need to focus on the efficient management of genetic resources. However, this requires information that is not readily available. For example, it is essential to have a much better idea of which resources are unique in international and national genebanks, which are still only to be found in farmer's fields, and which remain in the wild. Taking regular stock of where different crop genepools stand in terms of conservation and access reveals the gaps and helps to identify opportunities. For example, there are currently 26 existing conservation strategies used by the Crop Trust to support ex situ conservation around the world. To stay relevant, the strategies need to be continuously updated with new information.

Prioritization and rationalization of activities could be a double-edged sword. While we continue to demonstrate the use of conserved germplasm, we must face the reality that funding for the work of genebanks will be increasingly difficult. Competing funding priorities is a perpetual challenge and many stakeholders are primarily interested in the economic potential of genetic materials, overlooking the core activities of the genebank to sustainably manage and conserve germplasm for future use. Due to budget constraints, many genebanks have eliminated the element of supportive conservation research for genebanks (Fu, 2017). Improvements in genebank efficiency will depend on new knowledge and innovative technologies. These can only come from comprehensive and integrated research programs on seed storage, regeneration, and germplasm health.

5 Conclusion

We discuss the past, present, and future of genebank valuation by applied economists. Our review indicates that the multifaceted benefits of genebanks, even when assessed only one facet at a time, have typically far outweighed their costs. Why, then, are we obliged to continue making the case for genebanks? Because the future is uncertain and we have not ended global hunger (SDG 2). While there is no silver bullet to achieve this goal, no effort can be sustained, or ultimately successful, without continued and expanded access to plant genetic resources. Genebanks must remain relevant or else they revert to 'seed morgues' and we lose the future options of crop diversity as raw material for the development of new, improved varieties on which we depend as societies and climates change.

A background paper prepared as part of the 2020 System Level Review of Genebank Costs and Operations (GCO) Review (CGIAR and the Crop Trust, 2020) puts forward a vision of the genebank of the future (Sackville Hamilton, 2020). The future genebank must conserve and deliver the right resources, in the right way, under future scenarios which are increasingly uncertain except insofar as that they involve massive change in multiple dimensions. In economic terms, this means the future genebank should aim for higher levels of efficiency (the right resources) and effectiveness (in the right way). Conserving and delivering 'the right resources' will require carefully targeted investments in characterization, evaluation, and genotyping, along with management of the resulting data, and its use to develop packages of breeder- and researcher-oriented materials. Conserving and delivering resources 'in the right way' will require investment in supportive research and various organizational optimization processes, including automation. None of these investments will be easy to obtain without an understanding of the evolving values of crop diversity that is more accurate but also more comprehensive of the diversity and complexity of genebanks.

6 Where to look for further information

For future research on this topic, consult the work of the Genebank Platform and Global Crop Diversity Trust at: https://www.genebanks.org/resources.

7 References

Aberkane, H., Payne, T., Kishi, M., Smale, M., Amri, A. and Jamora, N. (2020). Transferring diversity of goat grass to farmers' fields through the development of synthetic hexaploid wheat. *Food Security* 12(5), 1017–1033. https://doi.org/10.1007/s12571-020-01051-w [Accessed 25 Aug 2020].

Bernal-Galeano, V., Norton, G., Ellis, D., Anglin, N. L., Hareau, G., Smale, M., Jamora, N., Alwang, J. and Pradel, W. (2020). Andean potato diversity conserved in the International Potato Center GeneBank helps develop agriculture in Uganda: the example of the variety 'Victoria'. *Food Security* 12(5), 959–973. https://doi.org/10.1007/s12571-020-01037-8 [Accessed 25 Aug 2020].

Bramel, P. and Krishnan, S. (2019). Systematic assessment for conservation and utilisation of crop genetic resources. In: Zeigler, R. (Ed.), *Sustaining Global Food Security: The Nexus of Science and Policy*. Clayton: CSIRO Publishing, pp. 3–14.

Brennan, J. P., Godden, D., Smale, M. and Meng, E. (1999). Breeder demand for and utilisation of wheat genetic resources in Australia. *Plant Varieties and Seeds* 12, 113–127.

Brown, G. M., Jr. (1990). Valuation of genetic resources. *The Preservation and Valuation of Biological Resources*, 6(3), 256–259, Seattle: University of Washington Press.

CGIAR Genebank Platform (2019). Annual Report. Bonn: Global Crop Diversity Trust.

CGIAR and the Crop Trust (2020). System Level Review of Genebank Costs and Operations (GCO). Bonn: Global Crop Diversity Trust [CGIAR and the Global Crop Diversity Trust].

CGIAR Genebank Platform, Crop Trust and IRRI (2021). *Results from the IRRI Genebank User Survey*. Genebank Impacts Working Paper No. 12.

CGIAR Genebank Platform, Crop Trust and ICARDA (2019). *Results from the ICARDA Genebank User Survey*. Internal report.

Day-Rubenstein, K. D., Smale, M. and Widrlechner, M. P. (2006). Demand for genetic resources and the U.S. National plant germplasm system. *Crop Science* 46(3), 1021–1031.

Díez, M. J., De la Rosa, L., Martín, I., Guasch, L., Cartea, M. E., Mallor, C., Casals, J., Simó, J., Rivera, A., Anastasio, G., Prohens, J., Soler, S., Blanca, J., Valcárcel, J. V. and Casañas, F. (2018). Plant genebanks: present situation and proposals for their improvement. The case of the Spanish Network. *Frontiers in Plant Science* 9, 1794. https://doi.org/10.3389/fpls.2018.01794 [Accessed 25 Aug 2020].

Dzyubenko, N. I. (2018). Vavilov's collection of worldwide crop genetic resources in the 21st century. *Biopreservation and Biobanking* 16(5), 377–383. https://doi.org/10.1089/bio.2018.0045 [Accessed 25 Aug 2020].

Ebert, A. W. and Waqainabete, L. M. (2018). Conserving and sharing taro genetic resources for the benefit of global taro cultivation: a core contribution of the Centre for Pacific Crops and Trees. *Biopreservation and Biobanking* 16(5), 361–367. http://doi.org/10.1089/bio.2018.0017 [Accessed 25 Aug 2020].

Evenson, R. E. and Gollin, D. (1997). Genetic resources, international organizations, and improvement in Rice varieties. *Economic Development and Cultural Change* 45(3), 471–500.

FAO (2014). *GeneBank Standards for Plant Genetic Resources for Food and Agriculture*. Rome: Food and Agriculture Organization.

Fowler, C., Smale, M. and Gaiji, S. (2001). Unequal exchange? recent transfers of agricultural resources and their implications for developing countries. *Development Policy Review* 19(2), 181–204.

Fu, Y.-B. (2017). The vulnerability of plant genetic resources conserved ex situ. *Crop Science* 57(5), 2314–2328. http://doi.org/10.2135/cropsci2017.01.0014 [Accessed 25 Aug 2020].

Galluzzi, G., Halewood, M., Noriega, I. L. and Vernooy, R. (2016). Twenty-five years of international exchanges of plant genetic resources facilitated by the CGIAR genebanks: a case study on global interdependence. *Biodiversity and Conservation* 25(8), 1421–1446.

Garming, H., Van den Houwe, I.. and Roux, N. (2010). *The Impact of the Musa International Transit Centre (ITC), Belgium: A Review of Its Services and Cost-Effectiveness, and Recommendations for Rationalization of Its Operations*. Rome: Bioversity International.

Gepts, P. (2006). Plant genetic resources conservation and utilization: the accomplishments and future of a societal insurance policy. *Crop Science* 46(5), 2278–2292. http://doi.org/10.2135/cropsci2006.03.0169gas [Accessed 25 Aug 2020].

Gollin, D. (2020). Conserving genetic resources for agriculture: economic implications of emerging science. *Food Security* 12(5), 919–927. https://doi.org/10.1007/s12571-020-01035-w [Accessed 25 Aug 2020].

Gollin, D. and Smale, M. (1998). Valuing genetic diversity: crop plants and agroecosystems. In: Collins, W. W. and Qualset, C. O. (Eds), *Biodiversity in Agroecosystems*. London: CRC Press, pp. 237–265.

Gollin, D., Smale, M. and Skovmand, B. (2000). Searching and ex situ collection of wheat genetic resources. *American Journal of Agricultural Economics* 82(4), 812–827.

Guarino, L., Spielman, D. and Wiebe, K. (2019). How Do We Estimate the Consequences of Global Inaction on Genetic Diversity Conservation, Exchange, and Use? Washington, DC: IFPRI. https://www.ifpri.org/blog/how-do-we-estimate-consequences-global-inaction-genetic-diversity-conservation-exchange-and-use [Accessed 25 Aug 2020].

Halewood, M., Chiurugwi, T., Sackville Hamilton, R., Kurtz, B., Marden, E., Welch, E., Michiels, F., Mozafari, J., Sabran, M., Patron, N., Kersey, P., Bastow, R., Dorius, S., Dias, S., McCouch, S. and Powell, W. (2018). Plant genetic resources for food and agriculture: opportunities and challenges emerging from the science and information technology revolution. *New Phytologist* 217(4), 1407–1419. https://doi.org/10.1111/nph.14993 [Accessed 25 Aug 2020].

Harlan, J. R. (1972). Genetics of disaster. *Journal of Environmental Quality* 1(3), 212–215.

Harlan, J. R. and De Wet, J. M. J. (1971). Toward a rational classification of cultivated plants. *Taxon* 20(4), 509–517.

Hein, L. and Gatzweiler, H. (2006). The economic value of coffee genetic resources. *Ecological Economics* 60, 176–185.

Iltis, H. H. (1988). Serendipity in the exploration of biodiversity. What good are weedy tomatoes? In: Wilson, E. O. (Ed), *Biodiversity*. Washington, DC: National Academy Press, pp. 98–105.

Johnson, N. L., Pachico, D. and Voysest, O. (2003). The distribution of benefits from public international germplasm banks: the case of beans in Latin America. *Agricultural Economics* 29(3), 277–286.

Kingwell, R., Godden, D., Kambuou, R. and Jackson, G. (2001). Managing and funding germplasm preservation in Papua New Guinea–for improved indigenous foods. *Food Policy* 26(3), 265–280.

Kitonga, K., Jamora, N., Smale, M. and Muchugi, A. (2020). Use and benefits of tree germplasm from the World Agroforestry GeneBank for smallholder farmers in Kenya. *Food Security* 12(5), 993–1003. https://doi.org/10.1007/s12571-020-01047-6 [Accessed 25 Aug 2020].

Koo, B., Pardey, P. G. and Wright, B. D. (2004). *Saving Seeds: The Economics of Conserving Crop Genetic Resources Ex Situ in the Future Harvest Centres of the CGIAR*. Wallingford: CABI Publishing.

Krutilla, J. (1967). Conservation reconsidered. *The American Economic Review* 57, 777–786.

Lele, U. and Goldsmith, A. A. (1989). The development of national agricultural research capacity: India's experience with the Rockefeller Foundation and its significance for Africa. *Economic Development and Cultural Change* 37(2), 305–343.

McLean-Rodríguez, F. D., Camacho-Villa, T. C., Almekinders, C. J. M., Pè, M. E., Dell'Acqua, M. and Costich, D. E. (2019) The abandonment of maize landraces over the last 50 years in Morelos, Mexico: a tracing study using a multi-level perspective. *Agriculture and Human Values* 36(4), 651–668. https://doi.org/10.1007/s10460-019-09932-3.

Ocampo-Giraldo, V., Camacho-Villa, C., Costich, D. E., Vidal Martínez, V. A., Smale, M. and Jamora, N. (2020). Dynamic conservation of genetic resources: rematriation of the maize landrace *Jala*. *Food Security* 12(5), 945–958. https://doi.org/10.1007/s12571-020-01054-7 [Accessed 25 Aug 2020].

Raeburn, P. (1995). *The Last Harvest: The Genetic Gamble That Threatens to Destroy American Agriculture*. New York: Simon and Schuster.

Robinson, J. and Srinivasan, C. S. (2013). *Case Studies on the Impact of Germplasm Collection, Conservation, Characterization and Evaluation in the CGIAR*. Rome: CGIAR Standing Panel on Impact Assessment.

Sackville Hamilton, R. (2020). Scope and roles of the CGIAR genebanks: 2030 vision. Background paper No. 4 prepared for the 2020 System Level Review of Genebank Costs and Operations (GCO). CGIAR and the Global Crop Diversity Trust. Bonn: Global Crop Diversity Trust.

Schreinemachers, P., Ebert, A. W. and Wu, M.-H. (2014). Costing the ex situ conservation of plant genetic resources at AVRDC-the World Vegetable Center. *Genetic Resources and Crop Evolution* 61(4), 757–773.

Sellitti, S., Vaiknoras, K., Smale, M., Jamora, N., Andrade, R., Wenzl, P. and Labarta, R. (2020). The contribution of the CIAT GeneBank to the development of iron-biofortified bean varieties and well-being of farm households in Rwanda. *Food Security* 12(5), 975–991. https://doi.org/10.1007/s12571-020-01038-7.

Simpson, R. D., Sedjo, R. A. and Reid, J. W. (1996). Valuing biodiversity for use in pharmaceutical research. *Journal of Political Economy* 104(1), 163–185.

Smale, M. (2020). *A Sourcebook: Economics of Conserving Plant Genetic Resources in Genebanks*. Genebank Impact Working Paper 1. Bonn: CGIAR Genebank Platform and the Global Crop Diversity Trust. https://www.genebanks.org/resources/impacts/working-papers/ [Accessed 25 Aug 2020].

Smale, M. and Day-Rubenstein, K. (2002). The demand for crop genetic resources: international use of the US National Plant Germplasm System. *World Development* 30(9), 1639–1655.

Smale, M. and Jamora, N. (2020). Valuing genebanks. *Food Security* 12(5), 905–918. https://doi.org/10.1007/s12571-020-01034-x [Accessed 25 Aug 2020].

Tyack, N. and Dempewolf, H. (2015). The economics of crop wild relatives Under climate change. In: Redern, R., Yadav, S. S., Maxted, N., Dulloo, M. E., Guarino, L. and Smith, P. (Eds), *Crop Wild Relatives and Climate Change*. Hoboken, New Jersey: John Wiley & Sons, pp. 279–289.

Tyack, N. and Ščasný, M. (2018). Social valuation of GeneBank activities: assessing public demand for genetic resource conservation in the Czech Republic. *Sustainability* 10(11), 3997. https://doi.org/10.3390/su10113997 [Accessed 25 Aug 2020].

Tyack, N. and Ščasný, M. (2020). 'Warehouse' or research centre? Analyzing public preferences for conservation, pre-breeding and characterization activities at the Czech genebank? *Food Security* 12(5), 1035–1046. https://doi.org/10.1007/s12571-020-01040-z [Accessed 25 Aug 2020].

Villanueva, D., Smale, M., Jamora, N., Capilit, G. L. and Sackville Hamilton, R. S. (2020). The contribution of the International Rice GeneBank to varietal improvement and crop productivity in Eastern India. *Food Security* 12(5), 929–943. https://doi.org/10.1007/s12571-020-01036-9 [Accessed 25 Aug 2020].

Virchow, D. (Eds.) (2003). *Efficient Conservation of Crop Genetic Diversity. Theoretical Approaches and Empirical Studies*. Berlin: Springer Verlag.

Wambugu, P. W., Ndjiondjop, M. N. and Henry, R. J. (2018). Role of genomics in promoting the utilization of plant genetic resources in genebanks. *Briefings in Functional*

Genomics 17(3), 198–206. https://doi.org/10.1093/bfgp/ely014 [Accessed 25 Aug 2020].

Wilson, E. O. (Eds.) (1988). *Biodiversity*. Washington, DC: National Academy Press, pp. 98–105.

Wright, B. D. (1997). Crop genetic resource policy: the role of ex situ genebanks. *The Australian Journal of Agricultural and Resource Economics* 41(1), 81–115.

Xepapadeas, A., Ralli, P., Kougea, E., Spyrou, S., Stavropoulos, N., Tsiaousi, V. and Tsivelikas, A. (2014). Valuing insurance services emerging From a gene bank: the case of the Greek Gene Bank. *Ecological Economics* 97, 140–149.

Zohrabian, A., Traxler, G., Caudill, S. and Smale, M. (2003). Valuing pre-commercial genetic resources: a maximum entropy approach. *American Journal of Agricultural Economics* 85(2), 429–436.

Chapter 3

Monitoring plant genetic resources for food and agriculture

M. Ehsan Dulloo, Bioversity International, Italy; Prishnee Bissessur, Bioversity International, Mauritius; and Jai Rana, Bioversity International, India

1 Introduction

Genetic resources are the building blocks of what make up biodiversity, the total variation found within all living organisms, along with the ecological complexes they inhabit, and encompass diversity at the ecosystem, community, population, species, variety and genetic levels of biological organization (Wilson, 1992). The part of the overall biodiversity that plays a critical role either directly or indirectly for food and agriculture includes crops, livestock, forestry and fisheries and has been defined as agricultural biodiversity (Bioversity International, 2017). It comprises the diversity of genetic resources (varieties, breeds) and species used for food, fodder, fibre, fuel and pharmaceuticals. It also includes the diversity of non-harvested species that supports production (soil micro-organisms, predators, pollinators) and those in the wider environment that support agroecosystems (agricultural, pastoral, forest and aquatic) as well as the diversity of the agroecosystems (Bioversity International, 2017). Plant genetic resources for food and agriculture (PGRFA) is an important part of the spectrum of agricultural biodiversity that provides the essential raw materials for sustaining food production and making production system and livelihoods more resilient to shocks and stresses, including the impact of climate change (Bélanger and Pilling, 2019). IPGRI (1993) defined PGR as 'Genetic material for

http://dx.doi.org/10.19103/AS.2020.0085.03

plants which is of value as a resource for the present and future generations of people', but the term has been defined in several ways (see Dulloo (2019) for a review). In essence, PGRFA consists of the plant species that are cultivated for food and agriculture and includes varieties, improved varieties, cultivars, landraces as well as their wild relatives. In this chapter, we focus on the PGRFA component of agricultural biodiversity.

Despite their significant value in meeting the global needs for food and nutrition of the growing number of the human population, estimated to reach 9.7 billion by the 2050 (Anon, 2020), the diversity of plant genetic resources is fast declining and is thus reducing our options in coping with the major threats for increasing food production, such as pest and diseases, climate change, changes in environmental conditions leading to more drought in some areas or floods in other areas, increasing salinity, as well as the globalization and harmonization of food diets and farming systems (FAO, 2010, 2015; Redden et al., 2015; IPES-Food International Panel of Experts on Sustainable Food Systems, 2016).

The most recent global assessment of the state of biodiversity for food and agriculture (Bélanger and Pilling, 2019) states that many of the key components of biodiversity for food and agriculture are in decline. Specifically, regarding PGRFA, the report mentions that evidence suggests that overall the crop diversity present on farmers' field has declined and that threats to diversity are getting stronger. It further acknowledges the difficulty in precisely quantifying the extent of genetic diversity of within-crop species, especially in farmers' field (Bélanger and Pilling, 2019). Similarly, one of the key messages (A6) of the recent summary report of the global assessment of Biodiversity and Ecosystem Services for policy makers by Intergovernmental Science-Policy Platform on Biodiversity and Ecosystem Services (IPBES, 2019) states that 'Globally, local varieties and breeds of domesticated plants and animals are disappearing. This loss of diversity, including genetic diversity, poses a serious risk to global food security by undermining the resilience of many agricultural systems to threats such as pests, pathogens and climate change'. Both these global assessments on the state of biodiversity point to the decline in plant genetic diversity and raise concerns that this may have a drastic impact on global food and nutrition security. As expressed in these assessments and by other authors (Dulloo et al., 2016, 2017; Thormann and Engels, 2015), monitoring the changes in genetic diversity of crop plants over time at national, regional and global levels is proving to be very challenging. As yet there is no systematic global monitoring system for the status of genetic diversity in place to do that (Bélanger and Pilling, 2019; Bruford et al., 2017; Dulloo et al., 2016) and those that have tentatively been set up, such as FAO Second Global Plan of Action and CBD Strategic Plan for Biodiversity 2011–2020 (and now being replaced by post2020 Global Biodiversity framework currently in negotiation), have not yielded the desired results. Other indicators, indices and frameworks for monitoring plant genetic diversity have been proposed but are limited in implementation on a sustained basis.

Generally, biodiversity can be measured at different levels of the biological organization–ecosystems, communities, populations, species, varieties and genetic–and at different scales–local, national, regional and global. For PGRFA, it is more common to assess diversity at the former four levels (i.e. species, varieties, populations and genetic levels). The ability to adequately monitor plant genetic diversity is highly dependent on the data and knowledge we have about the amount of diversity and their distribution across their range and the rate at which they are changing over time. Unfortunately, our knowledge on the extent of diversity of PGRFA is limited by the data available (see Dulloo et al. (2017) for a full discussion on the challenges in monitoring the status of plant genetic diversity). As mentioned earlier, there are no global databases that comprehensively and systematically collect plant genetic data that can be used for monitoring purposes. While it may be easier to document the diversity of crop species that are cultivated in countries using national agricultural statistics, the greater challenge lies in the measurement and monitoring of varietal and genetic diversity across their populations and species, as such information is neither collected in national statistics nor at a global level. Another important factor to consider in the assessment of diversity and monitoring is where the diversity occurs–space and time. Genetic resources are conserved within three broad, interconnected realms (Dulloo et al., 2017), namely (a) on farm in farmers' fields (managed by farmers on farm and thus allowing responses to natural and human selection); (b) in the wild (occurring in natural habitats, in situ, that are under selective forces of nature) and (c) in ex situ collections (diversity that has been collected, conserved and managed in offsite facilities, e.g. genebanks). While it is easier to document the diversity of the genetic materials conserved in genebanks, it is much more difficult to know how much diversity lies in natural areas and on farmers' fields.

This chapter reviews the methodologies developed to assess and monitor the diversity of PGRFA, at species, varietal and genetic levels and discusses their strengths and challenges at different scales.

2 Monitoring species diversity

In the context of PGRFA, the focus on species diversity is on the so-called crop wild relatives (CWR), defined as wild species closely related to cultivated crops and that provide a broad reservoir of genetic traits that can be used in plant breeding programmes (Heywood et al., 2007; Maxted et al., 2010). However, like most wild plant species, these wild relatives are not only poorly conserved worldwide (Heywood and Dulloo, 2005) and undervalued (Ford-Lloyd et al., 2011) but also equally threatened by various drivers that may lead them to extinction (Castañeda-Álvarez et al., 2016). For example, in the European region, about 80% of their flora are known to be CWRs (Kell et al., 2008), and the assessment of 572 CWR species related to 25 economically important crops showed that at least 8–50% are threatened (Kell et al., 2012). The main anthropogenic threats are infrastructural development, tourism and recreation and agricultural intensification (Kell et al., 2012). This implies that these wild plants are subjected to constant changes in their immediate environment which might adversely affect their population trends and dynamics, as well as interactions with other congeners and mutualists.

Climate change is also a significant threat which will have repercussions on agricultural biodiversity (Kotschi, 2007; Maxted et al., 2013). Bioclimatic modelling provides evidence suggesting that climate change could induce substantial contraction in the distribution ranges of CWRs. Some examples include wild populations of peanut (*Arachis* spp.), potato (*Solanum* spp.) and cowpea (*Vigna* spp.) where studies reported that 16–22% of these species may become extinct by 2055, with most species possibly halving their range size (Jarvis et al., 2008). Given the dynamic nature of the environment, changes are expected in the plant populations, species diversity and species richness within a given region or area, species distribution range and abundance through space and time and thus need to be monitored (De Carvalho et al., 2016). However, often due to a lack of standard baseline information, fixed monitoring sites, standardized methods and partnerships for systematic monitoring of the conservation status of crop genetic diversity (de Hanne et al., 2016), the changes can go undetected and might eventually lead to unforeseen major reductions and fragmentation in population sizes, local extirpations or even extinctions altogether and thus loss of genetic diversity in nature. Therefore, the establishment of systematic monitoring, both accounting for spatial and temporal changes, is crucial to document the loss of agrobiodiversity as early as possible (Hunter and Heywood, 2011). This would allow the identification of gaps in our knowledge about the species diversity and richness present in an area and thus enhance our ability to conserve them.

Surveillance monitoring and targeted monitoring are the two broad categories which can be used (Nichols and Williams, 2006). Surveillance monitoring aims at gathering as much information as possible about the biodiversity over time, while targeted monitoring aims at addressing the impact of specific drivers on biodiversity change (Pereira et al., 2017). The efficiency of the monitoring techniques should be ensured to generate reproducible, reliable and robust datasets (Jetz et al., 2019). Thus, when devising a long-term monitoring plan for the regular assessment of species diversity, abundance, richness, geographical distribution and threat status, the scope of monitoring, sampling size, frequency, practicality, cost and its sustainability must be given due consideration (Hunter and Heywood, 2011). For example, the more threatened a population is, the more frequently it may need to be monitored. However, it should be ensured that the monitoring does not impact the target species over time (Pereira et al., 2017)

Monitoring is essentially defined as a repeated observation to create a time series which enables the detection of change. For the implementation of a systematic and long-term monitoring approach for species diversity of CWRs, the first step in the monitoring would be to identify and delimit the scale (local, national, regional or global) at which it will be conducted. Once delimited, the species and their habitat diversity can be regularly evaluated using a set of indicators and using the CWR checklist and inventory usually derived from the flora, literature, herbaria and field surveys (Maxted et al., 2007, 2010) as a baseline dataset to estimate the geographical diversity (e.g. at country level–China (Kell et al., 2015), Finland (Korpelainen et al., 2008), Mauritius (Bissessur et al., 2019), at regional level (Kell et al., 2008) and at the global level (Castañeda-Álvarez et al., 2016)) and detect eventual changes in species diversity. Essentially, the main indicators for in situ monitoring may include changes in (i) geographical distribution of the species, (ii) habitat diversity, (iii) number of species in the targeted area (species richness), (iv) species abundance, (v) phenology and biotic interactions and (vi) threat status (Mihoub et al., 2017; Pereira et al., 2017; Jetz et al., 2019).

2.1 Geographical distribution and habitat diversity

Monitoring changes in the geographical distribution of the CWR can be done through different methods, namely through (i) in situ mapping, i.e. distribution maps of the different CWR populations are generated based on occurrence data from literature, herbarium specimens and in situ field surveys (Jongman et al., 2017) and (ii) remote sensing which analyses satellite imagery (Borre et al., 2011). These would allow the detection of new populations as well as their loss. Their new occurrences should be recorded in the existing databases such as Global Biodiversity Information Facility (GBIF; https://www.gbif.org/) for future monitoring of population trends. Climatic conditions like temperature, rainfall, humidity, topography, soil nutrient content often define the geographical distribution of the CWR species through habitat heterogeneity. These parameters require regular monitoring to assess their impact on the population dynamics (e.g. Jewitt et al., 2017; von Wettberg et al., 2018). The climate data gathered can be used for bioclimatic modelling to project future changes such as contraction or increase in the population distribution of the species (Pecl et al., 2017).

2.2 Species richness, abundance and demographics

Species richness (total number of species in an area) and species abundance (total number of individual plants per species in an area) (Colwell, 2009) can change through time and affect the population structure and genetic flow within and among populations. Hence, plant populations must systematically be monitored through the collection of data over time to assess population viability, detect changes in population age structure and trends and estimate extinction risk. Several methods are applicable for species richness and abundance, given that plants are sessile, field-based methods such as quadrats, transects, cover, counts, mapping of vegetation cover and environmental DNA can be used (for more information, see Pereira et al., 2017) to estimate diversity indices (e.g. Shannon and Simpson diversities) (Colwell, 2009). For plant population demographics, assessment of population structure and size may include the stratified sampling of the number of seedlings, juveniles and adults. Population trend can be determined by assessing the survival rate, growth rate, reproduction (fertility rate) and mortality of plants in a specific population. These repeated measures can be done annually or every five years, depending on the type of wild relatives (annual or perennial) being monitored. Often, CWR populations may indicate demographically unstable populations which would call for specific conservation actions.

2.3 Phenology and biotic interactions

Leafing, flowering and fruiting phenology are among the species traits that are recommended as one of the essential biodiversity indicators for monitoring (Pereira et al., 2013). Although this can be done while conducting the population counts, newer methods include citizen science to report flowering or fruiting events through mobile applications (e.g. phenology programs such as iNaturalist, Nature's Calendar, ClimateWatch, Project Budburst) (Chandler et al., 2017), or automated repeat digital photography can also be used (e.g. Benton, 2009). Moreover, CWR species depend on pollinators and seed dispersers to maintain the gene flow within and among populations. Thus, regular monitoring and assessment of target species should be carried out. This can be done by collating information through field surveys or non-intrusive methods such as camera trapping to quantify the presence/absence of flower visitors, pollinators and seed dispersers; their relative abundance; interaction strength; and frequency of visits (Lortie et al., 2012; Krauss et al., 2018). Monitoring would also detect the changes in pest and diseases and any rise or decrease in the abundance of potential herbivores or predators. Modelling the biotic interactions in terms of pollinator abundance or presence of predators can also prove useful in the prediction of the CWR species' spatial distributions (range contraction or shift) and their responses to changing climate (e.g. Bale et al., 2002; Jaramillo et al., 2011).

2.4 Threat status

The conservation status of plants provides a useful indicator for the assessment of the survival and long-term persistence of the species within a given area. IUCN Red List of Threatened Species (see www.iucnredlist.org) is the best global source of information on the conservation status of plants. The Red List evaluates the relative risk of extinction, especially to identify those taxa that are more at risk of extinction and set different threat categories (CR, EN, VU, NT, LC) based on different criteria (Walter and Gillet, 1998), including the identification of anthropogenic drivers such as accidental or intentional introductions of alien invasive species, deforestation or active logging of habitats, pollution, climate change and other human-induced disturbances (Ford-Lloyd et al., 2011). Regular monitoring of these different threats at each site would help determine their impacts on the individual plant species population sizes, wild plant communities and interactions at the overall scale and to re-evaluate the degree of endangerment of the different species over time and to assess the effectiveness of conservation measures in place (Frances et al., 2018).

The goal of monitoring is to allow conservation practitioners to review and revisit on-going conservation and management actions by providing evidence on which practices work and which do not and tailoring them according to the site's requirements. Applying these different methods of monitoring will therefore contribute to more knowledge regarding different aspects of the wild plant diversity. Changes in the species abundance, diversity, richness, range contraction, rate of change in geographical distribution, increase or decrease in plant populations, composition and structure and possible homogenization of landscape would be detected. Moreover, measuring the potential impact of invasive alien species (e.g. Bale et al., 2002; Jaramillo et al., 2011), landscape changes (e.g. Jarvis et al., 2015), possible movement of species across altitudinal gradients (e.g. Skarbø and VanderMolen, 2016; Pecl et al., 2017) will be possible. Monitoring is scale dependent (Hunter and Heywood, 2011) and can sometimes be tedious and therefore requires the participation of various stakeholders who can conduct the monitoring on a regular basis. Employing the network approach is recently being promoted such that there is collaboration and institutional partnerships to ensure that protected area managers, farmers, breeders and other stakeholders work together to carry out monitoring as a regular activity under the conservation planning (e.g. Maxted et al., 2012, 2015; Fitzgerald et al., 2019) and supports judicious resource allocation, be it manpower or monetary.

3 Monitoring varietal diversity

Over the millennia, many of 7000 species of plants known to have been used for food by humans (Wilson, 1992) have been domesticated and improved by a process of natural selection and selection by early farmers for desirable traits. These selection processes gave rise to a great diversity of morphologically recognizable varieties that have been given specific names by local farmers. Ever since the agricultural revolution, breeders have further accelerated the development of improved crop varieties using rapidly advancing biotechnologies (Lenaerts et al., 2019). As a result of these advances, the number of varieties, at least for the major crops, is now counted in the thousands. For example, Delêtre et al. (2013) mentioned an estimated 200 000 or more landraces of rice (*Oryza sativa* L.) and wheat (*Triticum aestivum* L. subsp. *aestivum*) worldwide. According to the FAO State of World report on Plant Genetic Resources for Food and Agriculture (FAO, 1998), there are about 47 000 varieties of sorghum (*Sorghum bicolor* (L.) Moench); 30 000 varieties of common bean (*Phaseolus vulgaris* L.), chickpea (*Cicer arietinum* L.) and maize (*Zea mays* L.); ca. 20 000 varieties of pearl millet (*Pennisetum glaucum* (L.) R.Br.); 15 000 varieties of peanut (*Arachis hypogaea* L.); and between 7000 and 9000 varieties of manioc (*Manihot esculenta* Crantz).

Various terms have been used in the literature to describe the diversity of different varieties of crops, for example landraces, local varieties, farmer's varieties, traditional varieties, heirloom varieties and varietal diversity (Widawsky and Rozelle, 1998; Carolina et al., 2005; Vetelainen et al., 2009; Van de Wouw et al., 2009). This diversity of crop varieties is fundamentally important for the sustainability of crop production and livelihood of poor farmers around the world. They are characterized by having specific adaptation to the environmental conditions of their area of cultivation and closely associated with the knowledge and culture of farming systems where they have been developed (Vetelainen et al., 2009; Carolina et al., 2005). It is widely recognized that this varietal diversity is continually evolving and producing new populations over time (Döring et al., 2011), such that they are no longer the same genetically or morphologically to their predecessors. It should also be borne in mind that the evolutionary processes and selection forces that help shape diversity in agroecosystems are extremely dynamic and take place at a much higher rate than in natural ecosystems, thus making an estimate of varietal diversity even more complicated (Dulloo et al., 2014).

The question now is how the diversity of varieties on farm is assessed. The simplest measures would be the number of distinct varieties of each crop that is cultivated and managed as the identifiable unit of diversity on the farm and the proportion of the total farm area devoted to the different varieties (Brown and Hodgkin, 2007; Sadiki et al., 2007). Jarvis et al. (2008) defined three concepts of diversity to estimate the levels of crop genetic diversity on farm. These include the number of traditional varieties on farm (richness), the frequencies of different variants being grown on farm (evenness) and the probability that two randomly chosen households within the same community are growing different varieties (divergence). Using these measures, a good estimate of varietal diversity that is maintained on farm can be derived. However, for an accurate estimate of varietal diversity, a correct naming of local varieties is needed. Often there are inconsistencies in the naming of local varieties. Brown and Hodgkin (2007) and Sadiki et al. (2007) discuss how farmers recognize distinct local varieties based on agro-morphological, ecological-adaptive and use qualities and how farmers name their varieties. Sometimes, the same farmer-named local varieties are not genetically distinct, and in contrast different farmer-named varieties are genetically similar.

To assess varietal diversity, a community-level methodology called four-cell analysis (FCA) can be used. The method was developed in Nepal based upon the local assessment of richness (area planted to a crop or variety) and evenness (number of farmers growing them) at the village level (Sthapit et al., 2006). The method involves carrying out site-level focus group discussion with the most knowledgeable members of the local community to understand the status of local varieties in the village. A square divided into four quadrats is drawn on a large sheet of paper or on the ground, and farmers are asked to list the varieties for each crop that is grown in the community and place the name of the local variety on one of the four quadrats depending on whether the variety is grown by few or many farmers and on small or large areas (based on average areas) (Fig. 1). Case study A provides a concrete example of the application of the four-cell analysis in nine states in India, as part of a United Nation Environment implanted GEF project 'Mainstreaming agricultural biodiversity conservation and utilization in agricultural sector to ensure ecosystem services and reduce vulnerability in India' being executed jointly by the Alliance of Bioversity

Figure 1 Four-cell analysis of varietal diversity in Assam, India (Photo credit: Rajib).

International and the International Centre for Tropical Agriculture (CIAT) and the Indian Council of Agricultural Research (ICAR) (Bisht et al., 2020).

Case Study A: Assessment of local crop diversity in nine states in India

India is a recognized mega-diversity centre which possesses unique crop diversity, including several crops that have long been naturalized here. The crop diversity that exists in several pockets around India continues to be a major natural asset and represents an essential element in the livelihood strategies of much of the rural population. Traditional crops and varieties conserved by the Indian farmers over generations have been utilized in crop improvement, the value of which has increased further in the context of the new global challenge such as climate change and COVID-19 pandemic. The crop diversity available with Indian farmers now constitutes an essential resource to deal with the new challenges that agriculture is facing. Nevertheless, this diversity remains under threat from the continuing adoption of new varieties, changes in land use and agricultural practices, social trends, national policy to promote new varieties, weak seed system and climate change. In order to promote conservation and use of native crop varieties, Bioversity International and Indian Council of Agriculture Research have executed a UN Environment implemented project named 'Mainstreaming agrobiodiversity conservation and utilization in agricultural sector to ensure ecosystem services and reduce vulnerability' in nine states of India. The project focuses on five major areas which include (a) seed system and crop diversity of target crops, (b) farm livelihoods, income and market linkages, (c) household consumption and nutrition, (d) resilience to climate change impacts and related ecosystem services and (e) awareness about policy frameworks related to crop diversity. This will help in developing local community-based approaches, together with the necessary national framework that will enable the conservation and use of crop diversity to be mainstreamed into India's agricultural production and environmental management strategies.

In this project, 20–25 most knowledgeable male and female farmers, who know most about target crops, from 63 villages of nine states of India, were invited to participate in focus group discussion using a participative four-cell analysis method. A separate discussion was also held in these villages with women and men groups to understand the differences in knowledge between genders. Data was gathered about the crops that are grown in each village, varieties of each crop (both landraces and improved varieties), the number of farmers growing each variety and the areas planted. Farmers were also asked to specify if the cultivation of each variety has changed over the past 10 years and if any varieties that have been lost and why. The uses of each of the varieties are also discussed with the farmers to understand the important features of each variety and their traditional knowledge. The analysis of these data provides ground-level information on the extent of varietal diversity that can be used to monitor changes at the village and state levels and helps define policies and strategies for the conservation and use of native varietal diversity.

4 Monitoring genetic diversity

The highly diverse heterogeneous local varieties are rapidly disappearing, being replaced by a fewer more productive, higher yield improved varieties which are being grown over large expanses of land in monocultures. With it, we are also losing genetic diversity and gene complexes that constitute the insurance for future breeding and adaptation. Genetic erosion has also been defined as the loss of genetic diversity, in a particular location and over a particular period of time, including the loss of individual genes and the loss of particular combinations of genes such as those manifested in landraces or varieties (Mathur, 2011). At the ecosystem level, wild and on-farm populations are being lost due to various threats, and this has an impact on the genetic diversity. In fact, the population is the key unit at which evolution and adaptation take place, and genetic diversity within and among populations is the primary determinant for ensuring resilience and survival of the species. The capacity of populations in the wild and on farm to respond to environmental change and to be resilient depends on the breadth of the genetic diversity and traits contained within the populations that allows them to evolve and adapt to environmental and climatic changes. The large-scale genetic erosion of crop genetic resources has been reviewed by many authors (Mathur, 2011; Thormann and Engels, 2015; Bélanger and Pilling, 2019) who provide ample evidence that the world is losing valuable genetic resources and its capacity to respond to current and future changes in the global agricultural landscape. In fact, genetic diversity underpins other levels of biodiversity, including functional traits, populations, species and ecosystems (Bruford et al., 2017; Hoban et al., 2020). Rockström et al. (2009) have identified that the loss of biodiversity as one of the key variables for understanding whether society is within a 'safe operating space' for sustainability has already been exceeded. Maintaining and preventing the loss of genetic diversity should therefore be a key objective of new post2020 Global Biodiversity framework. Laikre et al. (2020) suggested that the post2020 Global Biodiversity framework should explicitly commit to maintain genetic diversity within all species and to implement strategies to halt genetic erosion and preserve the adaptive potential of populations of both wild and domesticated species.

Genetic diversity is the diversity within species which allows species to adapt (Hoban et al., 2020). Different metrics of genetic diversity have been used to measure genetic diversity (see Carvalho et al., 2019; Hughes et al., 2008). Examples include:

- Allelic diversity–number of alleles;
- Allelic richness–average number alleles per locus;

- Nucleotide diversity (π) –average number of nucleotide differences per site between two random individuals selected from a population;
- Genotypic richness–number of genotypes per population;
- Heterozygosity–average proportion of loci that carry two different alleles at a single locus within an individual. Observed heterozygosity (Ho) can be estimated with codominant molecular markers, but estimates are biased by the number of individuals sampled within a population. Expected heterozygosity (He) can be estimated with both dominant and codominant markers when assumptions are made about the mode of inheritance, as well as the size and structure of populations; and
- Nei FST indices.

Genetic diversity has conventionally been assessed using different kinds of DNA markers including biochemical markers and molecular markers (Spooner et al., 2005). Biochemical markers such as allozyme (allelic variants of enzymes) have a direct genetic link and have been used in the past to provide a measure of the gene and genotypic frequencies within populations. Molecular markers, on the other hand, measure diversity and the genetic distances at the DNA level. Molecular markers can be divided into non-polymerase chain reaction (PCR)-based techniques, such as restriction length polymorphisms (RFLPs) and minisatellites analysis (also known as a variable number of tandem repeats (VNTR) analysis) or PCR-based techniques such as DNA sequencing and sequence-tagged sites (STS) including simple sequence repeats (SSR)), random amplified polymorphic DNA (RAPD) analysis and amplified fragment length polymorphism (AFLP) (Spooner et al., 2005). But these techniques are now largely superseded by more advanced genomic techniques including microsatellites, mitochondrial and chloroplast DNA, single nucleotide polymorphism (SNP), DNA sequencing and new-generation sequencing (NGS) (see Bruford et al., 2017 for a review of the advantages and disadvantages of these different techniques). In plants, chloroplast DNA (cpDNA) markers predominate, especially in ecological studies (Carvalho et al., 2019). With advances in genomics, technology and computing power the monitoring of genetic diversity is now getting more affordable and more robust than it used to be, but also getting more complex as a multitude of techniques are becoming available to measure within-species genetic diversity (see Bruford et al., 2017 for a review of genetic monitoring). Genomics data are most often now produced using high-throughput sequencing technologies like NGS technologies that can sequence millions of DNA fragments across the genome (Flanagan et al., 2018) and thus increases our ability to quickly and cheaply assess genetic diversity.

Even though these DNA-based technologies are fast developing and becoming cheaper, the genomic data are not always available to allow monitoring of genetic diversity at desired scales and frequencies for most species. Even if they are available, often it requires specialized skills for data interpretation which may be costly. In the absence of knowledge of genetic diversity per se, other surrogate measures such as ecogeographical information (Parra-Quijano et al., 2012) or geographic occurrences (Khoury et al., 2019) can be used to monitor genetic diversity. It is assumed that the ecogeographical diversity across the species range is correlated with genetic diversity because the genetic variation within a wild population of species is influenced by natural selection and isolation (Parra-Quijano et al., 2012). This requires that in genetic diversity monitoring process, individuals and populations are sampled across the geographic range of the species and samples are collected from greater than 50 plants per population and at least from five populations (Hoban et al., 2020).

5 Targets and indicators framework for monitoring genetic diversity

Monitoring genetic diversity of crops in production systems over time at a global level is a very challenging exercise, and indicators are required to do so. The FAO developed a core set of indicators and higher-level composite indicators for monitoring the progress in the implementation of 20 priority areas of Second Global Plan of Action for plant genetic resources for food and agriculture (FAO, 2014). The reporting on the implementation of the second GPA is a country-driven process whereby countries report back regularly to the Commission on Genetic Resources for Food and Agriculture through a reporting tool called WIEWS (World Information and Early Warning System on Plant Genetic Resources for Food and Agriculture, http://www.fao.org/wiews/en/). The results are then used for preparing a periodic global assessment of the status of PGRFA. The FAO also uses one of the indicators to report to the target 2.5 of the Sustainable Development Goal on Zero Hunger–'SDG target 2.5: by 2020 maintain genetic diversity of seeds, cultivated plants, farmed and domesticated animals and their related wild species, including through soundly managed and diversified seed and plant banks at national, regional and international levels, and ensure access to and fair and equitable sharing of benefits arising from the utilization of genetic resources and associated traditional knowledge as internationally agreed'.

The CBD Biodiversity Strategic Plan 2011–2020 provided an overarching biodiversity framework for halting the loss of biodiversity globally to ensure that by 2020 ecosystems are resilient and continue to provide essential services, thereby securing the planet's variety of life and contributing to human well-being and poverty eradication. The plan sets out 20 Aichi targets organized in five strategic goals with accompanying headline indicators to monitor progress towards the achievement of the targets. Aichi Target 13 is devoted to the genetic diversity of cultivated plants and farmed and domesticated animals and of wild relatives, including other socio-economically as well as culturally valuable species. It calls for the genetic diversity of these species to be maintained, and strategies developed and implemented for minimizing genetic erosion and safeguarding their genetic diversity. As 2020 is coming to an end soon, it is becoming evident that this target, among others, will not be met. Negotiations to develop a post2020 global biodiversity framework are currently underway to define new goals and targets, focusing on the different levels of biodiversity viz., ecosystems, populations and genetic diversity within species. While genetic diversity features prominently as a high-level goal in the zero draft of the post2020 global biodiversity framework as discussed by at the open-ended working group meetings held in Rome in February 2020, the CBD stakeholders are still discussing the feasibility in properly monitoring it at a global level. However, a team of experts was convened in Davos, Switzerland in February 2020 by the Earth Commission in collaboration with the Future Earth and the CBD Secretariat to prepare an information document synthesizing scientific evidence relevant to the development of the post2020 global biodiversity framework (Díaz et al., 2020a). The report considers that the monitoring of genetic diversity in both wild and domesticated species will be crucial for achieving the CBD 2050 vision of 'Living in harmony with Nature', as genetic diversity supports species survival and adaptation, ecosystem stability and the provision of nature's contributions to people (Laikre et al., 2020). Further in a recent publication in Science, Diaz et al. (2020b) argue that only by setting highest level of ambition for each goal of the different level of biodiversity and implementing all goals in an integrated manner, will give a realistic chance in achieving the CBD 2050 vision. To monitor the progress to achieve the genetic diversity goals, Hoban et al. (2020) proposed three indicators– (i) number of populations within species with effective population size (Ne) above 500 versus those with Ne below 500; (ii) proportion of population maintained within species and (iii) the number of species and population in which genetic diversity is being monitored using DNA-based methods. It is also essential to improve agricultural ecosystems to alleviate poverty and ensure food security in a sustainable fashion (Brown and Hodgkin, 2007).

Many indicators have been proposed for monitoring genetic diversity, but none have been taken up widely due to difficulties of collating data on a wide scale. The availability and access of data or the ease of measuring them cost-effectively often determine the indicators which can usefully be applied. The most sensitive measure of genetic diversity would be allelic diversity measured at the DNA level with molecular tools (Laikre et al., 2020; Bruford et al., 2017; Hoban et al., 2014; Pereira et al., 2013). However, there is no globally acceptable framework in place for it to be implemented. Dulloo et al. (2016) reviewed the development of indicators for monitoring crop genetic diversity (see the summary in Table 1).

Khoury et al. (2019) propose an indicator called 'comprehensiveness of conservation of useful wild plants', as a pragmatic proxy to measure the genetic diversity of wild plants species using eco-geographic methods, although with potential applications to cultivated crops and possibly to animals as well. The Alliance of Bioversity International and CIAT has also developed an Agrobiodiversity Index that measures the status and progress of agrobiodiversity use and conservation in healthy diets, sustainable agriculture and genetic resource management for current and future use options so as to identify concrete actions to help achieve diverse, sustainable and resilient food systems (Bioversity International, 2018).

As mentioned above, the FAO set of indicators (FAO, 2014) remains the globally accepted monitoring mechanism for monitoring genetic diversity of plant genetic resources for food and agriculture. The indicators for monitoring the crop genetic diversity in SDG target 2.5 only consider the number of accessions and the number of genebanks ex situ, which is insufficient for measuring the status of genetic diversity in the natural habitats and where it matters most for reasons described above.

6 Conclusion and future trends in research

To summarize, systematic and long-term monitoring of the wild plant diversity of a given site via the establishment of a baseline dataset reporting the status of vascular plants within that area can be demanding and time-consuming (Hunter and Heywood, 2011). However, it allows to (i) detect the geographic areas and habitats which are required to be conserved to prevent extinctions and maintain an adequate conservation status of the vascular plants within, (ii) identify the major threats and to propose measures and conservation actions to address and curtail them and finally (iii) reinforce the network of experts who are involved in plant conservation (botanists, herbaria specialists, ecologists, conservation managers, breeders and others) to regularly update the assessment information and use their expertise to target and address the highest conservation priorities (Heywood and Dulloo, 2005; Maxted et al., 2012; Vincent et al., 2013).

One main challenge for monitoring crop genetic diversity is the availability of and access to data over space and time, as discussed above. However, monitoring of genetic diversity needs to be promoted as it will be the only way that data can be produced for determining the trends in genetic diversity. Many crops and their wild relatives that form the basis of food security in the world are under-conserved and underutilized and neglected by science (Ford-Lloyd et al., 2011; Khoury et al., 2019). For these species, there is a lack of primary data on their global occurrence, which is however improving with time. For example, many countries are now conducting surveys and inventories to identify the wild relatives in their territory and devising conservation strategies

Table 1 Summary of major indicators (or sets of indicators) proposed for monitoring crop genetic diversity

Proponents	Characteristics of the indicators	Year	Reference
Organization for Economic Co-operation and Development (OECD)	Number of varieties or breeds registered and certified for marketing; Share of key crop or breed varieties in total marketed production for individual crops or breeds	2001	OECD (2001)
Food and Agriculture Organization of the United Nations (FAO)	83 indicators and 68 complementary indicators for monitoring progress in the implementation of the 20 priority areas of the First Global Plan of Action PA (FAO, 2002)	2002	FAO (2002)
FAO and IPGRI–joint FAO/IPGRI workshop Rome, 11–14 September 2002. FAO, Rome	22 potential indicators for genetic diversity, genetic erosion and genetic vulnerability	2002	GDEV (2002)
OECD Expert meeting Zurich, Switzerland, November 2001	Expanded list of indicators based on the previous OECD list developed in 2001 (OECD, 2001)	2003	Wetterich (2003)
Streamlining European Biodiversity Indicators (SEBI)	Set of biodiversity indicators for Europe to measure progress towards the target of halting the loss of biodiversity in Europe	2005	Biala et al. (2012)
Centre for Genetic Resources (CGN/DLO), The Netherlands	A extended set of indicators based on pressure–state-response framework from OECD (2001) and Wetterich (2003) and a restricted list of indicators including areas of varieties in landscape/production areas, in low production/high biodiversity agriculture, percentage of seeds originating on farm and number of varieties in genebank	2006	Eaton et al. (2006)
Thematic Background study–FAO	22 indicators for genetic diversity and erosion both in situ and ex situ for cultivated and wild plant species	2008	Brown (2008)
Food and Agriculture Organization of the United Nations (FAO)	Revised core set of 63 indicators and 3 targets established and adopted by the FAO Fourteenth Regular Session of Commission on Genetic resources for Food and Agriculture for monitoring progress of the implementation of second Global Plan of Action	2012	FAO (2014)

Proponents	Characteristics of the indicators	Year	Reference
Bonneuil et al (2012)	Integrative index (HT*) for cultivated species, varietal richness and spatial evenness, and between- and within-variety genetic diversity	2012	Bonneuil et al. (2012)
International Partnership for the Satoyama Initiative (IPSI)	20 ecological and socio-cultural factors for Socio-Ecological Production Landscapes (SEPLs)	2013	Bergamini et al. (2013)
Last et al. (2014)	Five indicators for the estimation of genetic diversity of crop accession or breed diversity at farm level, including 'crop-species richness', 'crop-cultivar diversity', 'type of crop accessions', 'livestock-species richness' and 'breed diversity'	2014	Last et al. (2014)
Convention on Biological Diversity Strategic plan 2011–2020	Indicators for Aichi target 13: Number of plant and animal genetic resources for food and agriculture secured in either medium- or long-term conservation facilities (indicator for SDG target 2.5)	2016	CBD (2016)
Alliance of Bioversity International and CIAT	Agrobiodiversity index : measure agrobiodiversity use and conservation in healthy diets, sustainable agriculture and genetic resource management for current and future use options It includes 16 indicators: 3 indicators to monitor commitments to use and conserve agrobiodiversity; 4 indicators to monitor evidence of actions taken to use and conserve agrobiodiversity, and; 9 indicators to monitor the status of agrobiodiversity	2018	Bioversity International (2018)
Khoury et al. (2019)	Indicator on comprehensiveness of conservation of useful socio-economically as well as culturally valuable species of wild plants including crop wild relatives, wild utilized plants, any other wild species (ex situ and in situ)	2019	Khoury et al. (2019)
Laikre et al. (2020)	Indicators measuring genetically effective population size, number of species or populations in which genetic diversity is monitored by national agencies or universities using DNA markers and rate of loss genetically distinct population within species	2020	Laikre et al. (2020)

See Dulloo et al. (2016) for details.

to ensure their long-term protection and sustainable use. These primary data are however often not publicly available or have limited licensing, which makes the implementation of a global monitoring system for wild relatives challenging. For a global monitoring system, it would require much efforts in terms of data harmonization and coming up with global indicators that all stakeholders involved in monitoring can use (Hoban et al., 2020; Jetz et al., 2019; Pereira et al., 2017).

Establishing a global operative long-term in situ monitoring system on genetic diversity for plant genetic resources for food and agriculture remains a high priority to enable a reliable assessment of temporal and spatial dynamics of in situ conserved genetic diversity. These are prerequisites for the generation of evidence about the extent to which diversity is being lost, to assess the ongoing evolutionary and adaptation processes in farmers' fields and to create evidence about ecosystem services. To-date, monitoring for genetic diversity relies on practical proxy indicators. There is an urgent need to develop robust, practical methodologies and indicators for measuring the status of genetic diversity of PGRFA per se. This calls for a multi-scale network in fixed localities that are hotspots, using semi-standardized methodologies that allow systematic comparisons of species over time and space.

7 Where to look for further information

Relevant information about this topic may be found at the following websites:

- FAO Commission on Genetic Resources for Food and Agriculture: http://www.fao.org/cgrfa/en/.
- International Treaty on Plant Genetic Resources: http://www.fao.org/plant-treaty/en/.
- Convention on Biological Diversity: https://www.cbd.int/.
- Biodiversity Indicators Partnerships: https://www.bipindicators.net/.
- Agrobiodiversity Index : https://www.bioversityinternational.org/abd-index/.
- Crop wild relative portal: http://www.cropwildrelatives.org/.

Some key further reading materials would include:

- Fowler, C. and Mooney, P. (1990) *Shattering Food, Politics, and the Loss of Genetic Diversity*. The University of Arizona Press, Tucson, AZ.
- Maxted, N., Dulloo, M. E. and Ford Lloyd, B. (eds) (2016) *Enhancing Crop Genepool Use: Capturing Wild Relative and Landrace Diversity for Crop Improvement*. CABI Publishing, Wallingford, UK. ISBN 9781780646138.
- Walters, M. and Scholes, R. J. (Eds). *The GEO Handbook on Biodiversity Observation Networks*, DOI: 10.1007/978-3-319-27288-7_5.

8 References

Anon (2020). World population worldometer. Available at: https://www.worldometers.info/world-population/ (accessed 22 May 2020).

Bale, J. S., Masters, G. J., Hodkinson, I. D., Awmack, C., Bezemer, T. M., Brown, V. K., Butterfield, J., Buse, A., Coulson, J. C., Farrar, J., Good, J. E. G., Harrington, R., Hartley, S., Jones, T. H., Lindroth, R. L., Press, M. C., Symrnioudis, I., Watt, A. D. and Whittaker, J. B. (2002). Herbivory in global climate change research: direct effects of rising temperature on insect herbivores, *Glob. Chang. Biol.* 8(1), 1–16.

Bélanger, J. and Pilling, D. (2019). *The State of the World's Biodiversity for Food and Agriculture*, Rome: FAO Commission on Genetic Resources for Food and Agriculture Assessments. Available at: http://www.fao.org/3/CA3129EN/CA3129EN.pdf.

Benton, L. M. L. (2009). Automated repeat digital photography for continuous phenological monitoring: an analysis of flowering in a semiarid shrubland, Thesis (PhD), University of Arizona.

Bergamini, N., Blasiak, R., Eyzaguirre, P., Ichikawa, K., Mijatovic, D., Nakao, F. and Subramanian, S. M. (2013). *Indicators of Resilience in Socio-Ecological Production Landscapes (SEPLs)*, Yokohama, Japan: United Nations University Institute of Advanced Studies (UNU-IAS).

Biala, K., Condé, J., Delbaere, B., Jones-Walters, L. and Torre-Marín A. (2012). *Streamlining European biodiversity Indicators* 2020: Building a future on lessons learnt from the SEBI 2010 process. European Environment agency Technical Report No11/2012.

Bioversity International (2017). *Mainstreaming Agrobiodiversity in Sustainable Food Systems: Scientific Foundations for an Agrobiodiversity Index*, Rome: Bioversity International.

Bioversity International (2018). *The Agrobiodiversity Index Methodology Report v.1.0.*, Rome: Bioversity International, 88 p.

Bisht, I. S., Rana, J. C. and Ahlawat, S. P. (2020). The future of sustainability farming in India: some sustainability considerations, *Sustainability* 12, 3751. doi:10.3390/su12093751.

Bissessur, P., Baider, C., Boodia, N., Badaloo, M. G. H., Bégué, J. A., Jhumka, Z., Meunier, A., Mungroo, Y., Gopal, V., Kell, S. P., Magos Brehm, J., Thormann, I. and Jaufeerally-Fakim, Y. (2019). Crop wild relative diversity and conservation planning in two isolated oceanic islands of a biodiversity hotspot (Mauritius and Rodrigues), *Plant Genet. Resour.* 17(2), 174–184.

Bonneuil, C., Goffaux, R., Bonnin, I., Montalent, P., Hamon, C., Balfourier, F. and Goldringer, I. (2012). A new integrative indicator to assess crop genetic diversity, *Ecol. Indic.* 23, 280–289.

Borre, J. V., Paelinckx, D., Mücher, C. A., Kooistra, L., Haest, B., De Blust, G. and Schmidt, A. M. (2011). Integrating remote sensing in natura 2000 habitat monitoring: prospects on the way forward, *J. Nat. Conserv.* 19(2), 116–125.

Brown, A. H. D. and Hodgkin, T. (2007). Measuring, managing and maintaining crop genetic diversity on farm. In: Jarvis, D., Padoch, C. and Cooper, H. D. (Eds), *Managing Biodiversity in Agricultural Ecosystem*, New York: Bioversity International and Columbia University Press, pp. 13–33.

Brown, A. H. D. (2008). *Indicators of Genetic Diversity, Genetic Erosion and Genetic Vulnerability for Plant Genetic Resources for Food and Agriculture*, Rome: Thematic Background Study, FAO.

Bruford, M. W., Davies, N., Dulloo, M. E., Faith, D. P. and Walters, M. (2017). Monitoring changes in genetic diversity. In: Walters, M. and Scholes, R. J. (Eds), *The GEO Handbook on Biodiversity Observation Networks*, Switzerland: Springer, pp. 107–128.

Carolina, T., Villa, C., Maxted, N., Scholten, M., Ford-Lloyd, B. (2005). Defining and identifying crop landraces, *Plant Genet. Resour.* 3(3), 373–384.

Carvalho, Y. G. S., Vitorino, L. C., Souza, U. J. Bd and Bessa, L. A. (2019). Recent trends in research on the genetic diversity of plants: implications for conservation, *Diversity* 11(4), 62. doi:10.3390/d11040062.

Castañeda-Álvarez, N. P., Khoury, C. K., Achicanoy, H. A., Bernau, V., Dempewolf, H., Eastwood, R. J., Guarino, L., Harker, R. H., Jarvis, A., Maxted, N., Müller, J. V., Ramirez-Villegas, J., Sosa, C. C., Struik, P. C., Vincent, H. and Toll, J. (2016). Global conservation priorities for crop wild relatives, *Nat. Plants* 2(4), 16022.

Chandler, M., See, L., Copas, K., Bonde, A. M. Z., López, B. C., Danielsen, F., Legind, J. K., Masinde, S., Miller-Rushing, A. J., Newman, G., Rosemartin, A. and Turak, E. (2017). Contribution of citizen science towards international biodiversity monitoring, *Biol. Conserv.* 213, 280–294.

Colwell, R. K. (2009). Biodiversity: concepts, patterns and measurement. In: Levin, Simon A. (Ed.), *The Princeton Guide to Ecology*, Princeton, NJ: Princeton University Press, pp. 257–263.

Convention on Biological Diversity (2016). Decision taken by the conference of the parties to the Convention on Biological Diversity XIII/28. Indicators for the Strategic Plan for Biodiversity 2011-2020 and the Aichi Biodiversity Targets. CBD/COP/DEC/XIII/28.

De Carvalho, M. Â. A. P., Bebeli, P. J., da Silva, A. M. B., Bettencourt, E., Slaski, J. J. and Dias, S. (2016). Agrobiodiversity: the importance of inventories in the assessment of crop diversity and its time and spatial changes. In: Ahuja, M. R. and Jain, S. M. (Eds), *Genetic Diversity and Erosion in Plants*, Switzerland: Springer, pp. 307–335.

De Haan, S., Polreich, S., Rodriguez, F., Juarez, H., Plasencia, F., Ccanto, R., Alvarez, C., Otondo, A., Sainz, H., Venegas, C. and Kalazich, J. (2016). A long-term systematic monitoring framework for on-farm conserved potato landrace diversity. In: Maxted, N., Dulloo M. E. and Ford-Lloyd, B. V. (Eds), *Enhancing Crop Genepool Use: capturing wild relative and landrace diversity for crop improvement*, CABI Publishers, Wallingford, pp. 289–297.

Delêtre, M., Gaisberger, H. and Arnaud, E. (2013). Agrobiodiversity in perspective–a review of questions, tools, concepts and methodologies. Report prepared for Bioversity International and IRD-Sud Expert Plantes, p. 78.

Díaz, S., Broadgate, W., DeClerk, F., Dobrota, S., Krug, C., Moersberg, H., Obura, D., Spehn, E., Tewksbury, J., Verburg, P., Zafra-Calvo, N., Bellon, M., Burgess, N., Cariño, J., Castañeda-Alvarez, N., Cavender-Bares, J., Chaplin-Kramer, R., De Meester, L., Dulloo, E., Fernández-Palacios, J. M., Garibaldi, L. A., Hill, S., Isbell, F., Leadley, P., Liu, J., Mace, G. M., Maron, M., Martín-López, B., McGowan, P., Pereira, H., Purvis, A., Reyes García, V., Rocha, J., Rondinini, C., Shannon, L., Shaw, R., Shin, Y. J., Snelgrove, P., Strassburg, B., Subramanian, S., Visconti, P., Watson, J., Zanne, A., Bruford, M., Colli, L., de Dornelas, M. A., Bascompte, J., Forest, F., Hoban, S., Johnson, M., Jones, S., Jordano, P., Kassen, R., Khoury, C., Laikre, L., Maxted, N., Miloslavich, P., Moreno-Mateos, D., Ogden, R., Segelbacher, G., Souffreau, C., Svenning, J. C. and Vázquez, E. (2020a). Synthesizing the scientific evidence to inform the development of the post-2020 Global Framework on biodiversity. Earth Commission Meeting Report to the Convention on Biological Diversity. Subsidiary Body on Scientific, Technical and Technological Advice, Convention on Biological Diversity, CBD/SBSTTA/24/inf/9.

Díaz, S., Zafra-Calvo, N., Purvis, A., Verburg, P. H., Obura, O., Leadley, P., Chaplin-Kramer, R., De Meester, L., Dulloo, E., Martín-López, E., Shaw, M. B., Visconti, P., Broadgate, W., Bruford, M. W., Burgess, N. D., Cavender-Bares, J., DeClerck, F., Fernádez-Palacios, J. M., Garibaldi, L. A., Hill, S. L. L., Isbell, F., Khoury, C. K., Krug, C. B., Liu, J., Maron, M., McGowan, P. J. K., Pereira, H. M., Reyes-García, V., Rocha, J., Rondinini, C., Shannon, L., Shin, Y. J., Snelgrove, P. V. R., Spehn, E. M., Strassburg, B., Subramanian, S. M., Tewksbury, J. J., Watson, J. E. M. and Zanne, A. E. (2020b). Setting ambitious goal for biodiversity and sustainability. *Science* 370 (6515), 411–413.

Döring, T. F., Knapp, S., Kovacs, G., Murphy, K. and Wolfe, M. S. (2011). Evolutionary plant breeding in cereals–into a new era, *Sustainability* 3(10), 1944–1971. doi:10.3390/su3101944.

Dulloo, M. E., Hunter, D. and Leaman, D. (2014). Plant diversity in addressing food, nutrition and medicinal needs. In: Gurib-Fakim, A. (Ed.), *Novel Plant Bioresources: Applications in Food, Medicine and Cosmetics*, Chichester, UK: John Wiley & Sons Ltd, pp. 3–22. doi:10.1002/9781118460566.

Dulloo, M. E., Drucker, A. and Thormann, I. (2016). What do we have to lose, monitoring genetic diversity. In: Maxted, N., Dulloo, M. E. and Ford Lloyd, B. (Eds), *Enhancing Crop Genepool Use: Capturing Wild Relative and Landrace Diversity for Crop Improvement*, Wallingford: CABI Publishing, pp. 421–436. http://eu.wiley.com/WileyCDA/Section/id-302479.html?query=Mohammad+Ehsan+Dulloo.

Dulloo, M. E., Rege, J. E. O., Ramirez, M., Drucker, A. G., Padulosi, S. Maxted, N., Sthapit, B., Gauchan, D., Thormann, I., Gaisberger, H., Roux, N., Sardos, J., Ruas, M. and Rouard, M. (2017). Conserving agricultural biodiversity for use in sustainable food systems. In: *Bioversity International. Mainstreaming Agrobiodiversity in Sustainable Food Systems: Scientific Foundations for an Agrobiodiversity*, Rome: Bioversity International, pp. 103–139.

Dulloo, M. E. (2019). Maintaining diversity of plant genetic resources as a basis for food security. In: Ferranti, P., Berry, E. M. and Anderson, J. R. (Eds), *Encyclopedia of Food Security and Sustainability*, Elsevier, pp. 54–63.

Eaton, D., Windig, J. J., Hiemstra, S. J., van Veller, M., Trach, N. X., Hao, P. X., Doan, D. H. and Hu, R. (2006). *Indicators for Livestock and Crop Biodiversity*, Wageningen, the Netherlands: CGN/DLO Foundation.

FAO (1998). *The State of the World's Plant Genetic Resources for Food and Agriculture*, Rome: Food and Agriculture Organization of the United Nations, p.510.

FAO (2002). *Indicators and Reporting Format for Monitoring the Implementation of the Global Plan of Action for the Conservation and Sustainable Utilization of Plant Genetic Resources for Food and Agriculture. Ninth Session of the Commission on Genetic Resources for Food and Agriculture Document CGRFA-9/02/Inf2*, Rome: Food and Agriculture Organization.

FAO (2010). The second report on the state of the world's plant genetic resources for food and agriculture, Rome. Available at: http://www.fao.org/docrep/013/i1500e/i1500e.pdf.

FAO (2014). *Targets and Indicators for Plant Genetic Resources for Food and Agriculture. Fifteenth Regular Session of the Commission on Genetic Resources for Food and Agriculture*, Document CGRFA-15/15/4.1, Rome: FAO.

FAO (2015). Coping with climate change–the roles of genetic resources for food and agriculture, Rome. Available at: http://www.fao.org/3/a-i3866e.pdf.

Fitzgerald, H., Palmé, A., Asdal, Å, Endresen, D., Kiviharju, E., Lund, B., Rasmussen, M., Thorbjörnsson, H. and Weibull, J. (2019). A regional approach to Nordic crop wild relative *in situ* conservation planning, *Plant Genet. Resour.* 17(2), 196–207.

Flanagan, S. P., Forester, B. R., Latch, E. K., Aitken, S. N. and Hoban, S. (2018). Guidelines for planning genomic assessment and monitoring of locally adaptive variation to inform species conservation, *Evol. Appl.* 11(7), 1035–1052.

Ford-Lloyd, B. V., Schmidt, M., Armstrong, S. J., Barazani, O., Engels, J., Hadas, R., Hammer, K., Kell, S. P., Kang, D., Khoshbakht, K., Li, Y., Long, C., Lu, B., Ma, K., Nguyen, V. T., Qiu, L., Ge, S., Wei, W., Zhang, Z. and Maxted, N. (2011). Crop wild relatives–undervalued, underutilized and under threat?, *BioScience* 61(7), 559–565.

Frances, A. L., Smith, A. B. and Khoury, C. K. (2018). Conservation status and threat assessments for North American crop wild relatives. In: Greene, S., Williams, K., Khoury, C., Kantar, M. and Marek, L. (Eds), *North American Crop Wild Relatives*, Cham: Springer, pp. 189–208.

GDEV (2002). Review and development of indicators for genetic diversity, genetic erosion and genetic vulnerability (GDEV). Summary report of a joint FAO/IPGRI workshop, Rome.

Heywood, V. H. and Dulloo, M. E. (2005). In situ conservation of wild plant species–a critical global review of good practices. IPGRI Technical Bulletin, 11, Rome: International Plant Genetic Resources Institute.

Heywood, V. H., Casas, A., Ford-Lloyd, B., Kell, S. P. and Maxted, N. (2007). Conservation and sustainable use of crop wild relatives, *Agric. Ecosyst. Environ.* 121(3), 245–255.

Hoban, S., Arntzen, J. A., Bruford, M. W., Godoy, J. A., Hoelzel, A. R., Segelbacher, G., Vilà, C. and Bertorelle, G. (2014). Comparative evaluation of potential indicators and temporal sampling protocols for monitoring genetic erosion, *Evol. Appl.* 7(9), 984–998.

Hoban, S., Bruford, M., D'Urban Jackson, J., Lopes-Fernandes, M., Heuertz, M., Hohenlohe, P. A., Paz-Vinas, I., Sjögren-Gulve, P., Segelbacher, G., Vernesi, C., Aitken, S., Bertola, L. D., Bloomer, P., Breed, M., Rodríguez-Correa, H., Funk, W. C., Grueber, C. E., Hunter, M. E., Jaffe, R., Liggins, L., Mergeay, J., Moharrek, F., O'Brien, D., Ogden, R., Palma-Silva, C., Pierson, J., Ramakrishnan, U., Simo-Droissart, M., Tani, N., Waits, L. and Laikre, L. (2020). Genetic diversity targets and indicators in the CBD post2020 Global biodiversity framework must be improved, *Biol. Conserv.* 248: 108654. Available at: https://www.sciencedirect.com/science/article/pii/S0006320720307126.

Hughes, A. R., Inouye, B. D., Johnson, M. T., Underwood, N. and Vellend, M. (2008). Ecological consequences of genetic diversity, *Ecol. Lett.* 11(6), 609–623.

Hunter, D. and Heywood, V. (2011). *Crop Wild Relatives: A Manual of In Situ Conservation*,London, UK and Washington DC: Earthscan Ltd.

IPBES (2019). *Summary for Policymakers of the Global Assessment Report on Biodiversity and Ecosystem Services of the Intergovernmental Science-Policy Platform on Biodiversity and Ecosystem Services*, Bonn, Germany: IPBES.

IPES-Food International Panel of Experts on Sustainable Food Systems (2016). From uniformity to diversity: a paradigm shift from Industrial Agriculture to Diversified Agroecological Systems.

IPGRI (1993). *Diversity for Development*, Rome: The Strategy of the International Plant Genetic Resources Institute.

Jaramillo, J., Muchugu, E., Vega, F. E., Davis, A., Borgemeister, C. and Chabi-Olaye, A. (2011). Some like it hot: the influence and implications of climate change on coffee berry borer (*Hypothenemus hampei*) and coffee production in East Africa, *PLoS ONE* 6(9), e24528.

Jarvis, D. I., Brown, A. H. D., Cuong, P. H., Collado-Panduro, L., Latournerie-Moreno, L., Gyawali, S., Tanto, T., Sawadogo, M., Mar, I., Sadiki, M., Hue, N. T., Arias-Reyes, L., Balma, D., Bajracharya, J., Castillo, F., Rijal, D., Belqadi, L., Rana, R., Saidi, S., Ouedraogo, J., Zangre, R., Rhrib, K., Chavez, J. L., Schoen, D., Sthapit, B., De Santis, P., Fadda, C. and Hodgkin, T. (2008). A global perspective of the richness and evenness of traditional crop-variety diversity maintained by farming communities, *Proc. Natl Acad. Sci. U. S. A.* 105(14), 5326–5331.

Jarvis, S., Fielder, H., Hopkins, J., Maxted, N. and Smart, S. (2015). Distribution of crop wild relatives of conservation priority in the UK landscape, *Biol. Conserv.* 191, 444–451.

Jetz, W., McGeoch, M. A., Guralnick, R., Ferrier, S., Beck, J., Costello, M. J., Fernandez, M., Geller, G. N., Keil, P., Merow, C., Meyer, C., Muller-Karger, F. E., Pereira, H. M., Regan, E. C., Schmeller, D. S. and Turak, E. (2019). Essential biodiversity variables for mapping and monitoring species populations, *Nat. Ecol. Evol.* 3(4), 539–551.

Jewitt, D., Goodman, P. S., Erasmus, B. F. N., O'Connor, T. G. and Witkowski, E. T. F. (2017). Planning for the maintenance of floristic diversity in the face of land cover and climate change, *Environ. Manag.* 59(5), 792–806.

Jongman, R. H. G., Skidmore, A. K., Mücher, C. A., Bunce, R. G. H. and Metzger, M. J. (2017). Global terrestrial ecosystem observations: why, where, what and how? In: Walters, M. and Scholes, R. J. (Eds), *The GEO Handbook on Biodiversity Observation Networks*, Cham, Switzerland: Springer, pp. 19–38.

Kell, S. P., Jury, S. L., Knüpffer, H., Ford-Lloyd, B. V. and Maxted, N. (2008). Crops and wild relatives of the Euro-Mediterranean region: making and using a conservation catalogue. In: Maxted, N. (Ed.), *Crop Wild Relative Conservation and Use*, Wallingford: CABI Publishing, pp. 69–109.

Kell, S. P., Maxted, N. and Bilz, M. (2012). European crop wild relative threat assessment: knowledge gained, and lessons learnt. In: Maxted, N., Dulloo, M. E., Ford-Lloyd, B. V., Frese, L., Iriondo, J. M. and de Carvalho, M. A. A. P. (Eds), *Agrobiodiversity Conservation Securing the Diversity of Crop Wild Relatives and Landraces*, Wallingford: CABI Publishing, pp. 218–242.

Kell, S. P., Qin, H., Chen, B., Ford-Lloyd, B. V., Wei, W., Kang, D. and Maxted, N. (2015). China's crop wild relatives: diversity for agriculture and food security, *Agric. Ecosyst. Environ.* 209, 138–154.

Khoury, C. K., Amariles, D., Soto, J. S., Diaz, M. V., Sotelo, S., Sosa, C. C., Ramírez-Villegas, J., Achicanoy, H. A., Velásquez-Tibatá, J., Guarino, L., León, B., Navarro-Racines, C., Castañeda-Álvarez, N. P., Dempewolf, H., Wiersema, J. H. and Jarvis, A. (2019). Comprehensiveness of conservation of useful wild plants: an operational indicator for biodiversity and sustainable development targets, *Ecol. Indic.* 98, 420–429.

Korpelainen, H., Takaluoma, S., Pohjamo, M. and Helenius, J. (2008). Diversity and conservation needs of crop wild relatives in Finland. In: Maxted, N. (Ed.), *Crop Wild Relative Conservation and Use*, Wallingford: CABI Publishing, pp. 152–164.

Kotschi, J. (2007). Agricultural biodiversity is essential for adapting to climate change, *GAIA Ecol. Perspect. Sci. Soc.* 16(2), 98–101.

Krauss, S. L., Roberts, D. G., Phillips, R. D. and Edwards, C. (2018). Effectiveness of camera traps for quantifying daytime and nighttime visitation by vertebrate pollinators, *Ecol. Evol.* 8(18), 9304–9314.

Laikre, L., Hoban, S., Bruford, M. W., Segelbacher, G., Allendorf, F. W., Gajardo, G., Rodríguez, A. G., Hedrick, P. W., Heuertz, M., Hohenlohe, P. A., Jaffé, R., Johannesson, K., Liggins, L., MacDonald, A. J., OrozcoterWengel, P., Reusch, T. B. H., Rodríguez-Correa, H., Russo, I. M., Ryman, N. and Vernesi, C. (2020). Post2020 overlook genetic diversity, *Science* 367(6482), 1083–1085.

Last, L., Arndorfer, M., Balázs, K., Dennis, P., Dyman, T., et al. (2014). Indicators for the on-farm assessment of crop cultivar and livestock breed diversity: a survey-based participatory approach. *Biodiversity and Conservation* 23, 3051–3071.

Lenaerts, B., Collard, B. C. Y. and Demont, M. (2019). Review: improving global food security through accelerated plant breeding, *Plant Sci.* 287, 110207.

Lortie, C. J., Budden, A. E. and Reid, A. M. (2012). From birds to bees: applying video observation techniques to invertebrate pollinators, *J. Pollinat. Ecol.* 6(0), 125–128.

Mathur, P. N. (2011). Assessing the threat of genetic erosion. In: Guarino, L., Rao, V. R. and Goldberg, E. (Eds), *Collecting Plant Genetic Diversity Technical Guidelines–2011 Update*, Rome: Bioversity International, pp. 1–7.

Maxted, N., Kell, S. P., Toledo, Á, Dulloo, E., Heywood, V., Hodgkin, T., Hunter, D., Guarino, L., Jarvis, A. and Ford-Lloyd, B. (2010). A global approach to crop wild relative conservation: securing the gene pool for food and agriculture, *Kew Bull.* 65(4), 561–576.

Maxted, N., Kell, S. P., Ford-Lloyd, B. V., Dulloo, E. and Toledo, Á (2012). Toward the systematic conservation of global Crop wild relative diversity, *Crop Sci.* 52(2), 774–785.

Maxted, N., Kell, S. P. and Magos Brehm, J. (2013). Crop wild relatives and climate change. In: Jackson, M., Ford-Lloyd, B. V. and Parry, M. (Eds), *Plant Genetic Resources and Climate Change*, Wallingford, UK: CABI Publishing, pp. 114–136.

Maxted, N., Scholten, M., Codd, R. and Ford-Lloyd, B. V. (2007). Creation and use of a national inventory of crop wild relatives, *Biol. Conserv.* 140(1–2), 142–159.

Maxted, N., Avagyan, A., Frese, L., Iriondo, J. M., Magos Brehm, J., Singer, A. and Kell, S. P. (2015). *'ECPGR Concept for In Situ Conservation of Crop Wild Relatives in Europe,' Wild Species Conservation in Genetic Reserves Working Group*, Rome, Italy: European Cooperative Programme for Plant Genetic Resources, pp. 1–20.

Mihoub, J.-B., Henle, K., Titeux, N., Brotons, L., Brummitt, N. A. and Schmeller, D. S. (2017). Setting temporal baselines for biodiversity: the limits of available monitoring data for capturing the full impact of anthropogenic pressures, *Sci. Rep.* 7, 41591.

Nichols, J. D. and Williams, B. K. (2006). Monitoring for conservation, *Trends Ecol. Evol.* 21(12), 668–673.

OECD (2001). *Environmental Indicators for Agriculture*, Vol 3. Paris: Organization for Economic Co-operation and Development.

Parra-Quijano, M., Iriondo, J. M., Frese, L. and Torres, E. (2012). Spatial and ecogeographical approaches for selecting genetic reserves in Europe. In: Maxted, N., Dulloo, M. E., Ford-Lloyd, B. V., Frese, L., Iriondo, J. M. and Pinheiro de Carvalho, M. A. A. (Eds), *Agrobiodiversity Conservation: Securing the Diversity of Crop Wild Relatives and Landraces*, Wallingford, UK: CAB International, pp. 20–28.

Pecl, G. T., Araújo, M. B., Bell, J. D., Blanchard, J., Bonebrake, T. C., Chen, I. C., Clark, T. D., Colwell, R. K., Danielsen, F., Evengård, B., Falconi, L., Ferrier, S., Frusher, S., Garcia,

R. A., Griffis, R. B., Hobday, A. J., Janion-Scheepers, C., Jarzyna, M. A., Jennings, S., Lenoir, J., Linnetved, H. I., Martin, V. Y., McCormack, P. C., McDonald, J., Mitchell, N. J., Mustonen, T., Pandolfi, J. M., Pettorelli, N., Popova, E., Robinson, S. A., Scheffers, B. R., Shaw, J. D., Sorte, C. J., Strugnell, J. M., Sunday, J. M., Tuanmu, M. N., Vergés, A., Villanueva, C., Wernberg, T., Wapstra, E. and Williams, S. E. (2017). Biodiversity redistribution under climate change: impacts on ecosystems and human well-being, *Science* 355(6332), eaai9214.

Pereira, H. M., Ferrier, S., Walters, M., Geller, G. N., Jongman, R. H. G., Scholes, R. J., Bruford, M. W., Brummitt, N., Butchart, S. H., Cardoso, A. C., Coops, N. C., Dulloo, E., Faith, D. P., Freyhof, J., Gregory, R. D., Heip, C., Höft, R., Hurtt, G., Jetz, W., Karp, D. S., McGeoch, M. A., Obura, D., Onoda, Y., Pettorelli, N., Reyers, B., Sayre, R., Scharlemann, J. P., Stuart, S. N., Turak, E., Walpole, M. and Wegmann, M. (2013). Essential biodiversity variables, *Science* 339(6117), 277–278.

Pereira, H. M., Belnap, J., Böhm, M., Brummitt, N., Garcia-Moreno, J., Gregory, R., Martin, L., Peng, C., Proenca, V., Schmeller, D. and van Swaay, C. (2017). Monitoring essential biodiversity variables at the species level. In: Walters, M. and Scholes, R. J. (Eds), *The GEO Handbook on Biodiversity Observation Networks*, Switzerland: Springer, pp. 79–105.

Redden, R. J., Yadav, S. S., Maxted, N., Dulloo, M. E., Guarino, L. and Smith, P. (2015). *Crop Wild Relatives and Climate Change*, Hoboken, NJ: John Wiley & Sons Inc.

Rockström, J., Steffen, W., Noone, K., Persson, A., Chapin, F. S., Lambin, E. F., Lenton, T. M., Scheffer, M., Folke, C., Schellnhuber, H. J., Nykvist, B., de Wit, C. A., Hughes, T., van der Leeuw, S., Rodhe, H., Sörlin, S., Snyder, P. K., Costanza, R., Svedin, U., Falkenmark, M., Karlberg, L., Corell, R. W., Fabry, V. J., Hansen, J., Walker, B., Liverman, D., Richardson, K., Crutzen, P. and Foley, J. A. (2009). A safe operating space for humanity, *Nature* 461(7263), 472–475.

Sadiki, M., Jarvis, D. I., Rijal, D., Bajracharya, J., Hue, N. N., Camacho-Villa, T. C., Burgos-May, L. A., Sawadogo, M., Balma, D., Lope, D., Arias, L., Mar, I., Karamura, D., Williams, D., Chavez-Servia, J. L., Sthapit, B. and Rao, V. R. (2007). Variety names–an entry point to crop genetic diversity and distribution in agroecosystem. In: Jarvis, D. I., Padoch, C. and Cooper, H. D. (Eds), *Managing Biodiversity in Agricultural Ecosystem*, New York: Bioversity International and Columbia University Press, pp. 34–76.

Skarbø, K. and VanderMolen, K. (2016). Maize migration: key crop expands to higher altitudes under climate change in the Andes, *Clim. Dev.* 8(3), 245–255.

Spooner, D., van Treuren, R. and de Vicente, M. C. (2005). Molecular markers for genebank management. IPGRI Technical Bulletin No 10, Rome, Italy: International Plant Genetic Resources Institute.

Sthapit, B., Rana, R. B., Subedi, A., Gyawali, S., Bajracharya, J.,Chaudhary, P., Joshi, B. R., Sthapit, S., Joshi, K. D. and Upadhaya, M. P. (2006). Participatory four cell analysis (FCA) for understanding local crop diversity. In: Sthapit, B. R., Shrestha, P. K. and Upadhyay, M. P. (Eds), *Good Practices: On-Farm Management of Agricultural Biodiversity in Nepal*, Rome: International Plant Genetic Resources Institute, p. p12-15.

Thormann, I. and Engels, J. M. M. (2015). Genetic diversity and erosion–a global perspective. In: Ahuja, M. R. and Jain, S. M. (Eds), *Genetic Diversity and Erosion in Plants: Indicators and Prevention*, Switzerland: Springer, pp. 263–294.

Van de Wouw, M., Kik, C., van Hintum, T., Van Treuren, R. and Visser, B. (2009). Genetic erosion in crops: concept, research results and challenges, *Plant Genet. Resour. Charact. Util.* 8(1), 1–15.

Vetelainen, M., Negri, V. and Maxted, N. (2009). European landraces on farm conservation, management and use. Bioversity Technical Bulletin No. 15, Rome. Italy: Bioversity International.

Vincent, H., Wiersema, J., Kell, S. P., Fielder, H., Dobbie, S., Castañeda-Álvarez, N. P., Guarino, L., Eastwood, R., León, B. and Maxted, N. (2013). A prioritized crop wild relative inventory to help underpin global food security, *Biol. Conserv.* 167, 265–275.

von Wettberg, E. J. B., Chang, P. L., Başdemir, F., Carrasquila-Garcia, N., Korbu, L. B., Moenga, S. M., Bedada, G., Greenlon, A., Moriuchi, K. S., Singh, V., Cordeiro, M. A., Noujdina, N. V., Dinegde, K. N., Shah Sani, S. G. A., Getahun, T., Vance, L., Bergmann, E., Lindsay, D., Mamo, B. E., Warschefsky, E. J., Dacosta-Calheiros, E., Marques, E., Yilmaz, M. A., Cakmak, A., Rose, J., Migneault, A., Krieg, C. P., Saylak, S., Temel, H., Friesen, M. L., Siler, E., Akhmetov, Z., Ozcelik, H., Kholova, J., Can, C., Gaur, P., Yildirim, M., Sharma, H., Vadez, V., Tesfaye, K., Woldemedhin, A. F., Tar'an, B., Aydogan, A., Bukun, B., Penmetsa, R. V., Berger, J., Kahraman, A., Nuzhdin, S. V. and Cook, D. R. (2018). Ecology and genomics of an important crop wild relative as a prelude to agricultural innovation, *Nat. Commun.* 9(1), 649.

Walter, K. S. and Gillett, H. J. (Eds.) (1998). *1997 IUCN Red List of Threatened Plants*, Gland: IUCN.

Wetterich, F. (2003). Biological diversity of livestock and crops: useful classification and appropriate agri-environmental indicators. In: OECD (Ed.), *Agriculture and Biodiversity: Developing Indicators for Policy Analysis*, Proceedings from an OECD Expert meeting, Switzerland, November 2001, pp. 40–55.

Widawsky, D. and Rozelle, S. (1998). Varietal diversity and yield variability in Chinese rice production. In: Smale, M. (Ed.), *Farmers, Gene Banks and Crop Breeding: Economic Analyses of Diversity in Wheat Maize and Rice*, Dordrecht: Springer, pp. 159–172.

Wilson, E. O. (1992). *The Diversity of Life*, London, UK: The Penguin Press.

Chapter 4

Improving the global exchange of germplasm for crop breeding

Selim Louafi, UMR AGAP Institut, CIRAD, University of Montpellier, INRAE, Institut Agro, Montpellier, France; and Eric Welch, Arizona State University, USA

1 Introduction

Different norms, rules and practices (referred as institutions) organize the exchange of germplasm to address broader global challenges such as advancement of science and innovation, food security, sustainable agriculture and global equity. Some of these institutions are now embedded in various treaties and national regulations. This chapter demonstrates that these regulations are not as successful as they could be because they only partially integrate the complexity of the germplasm exchange environment. In order to better understand how germplasm exchange could be improved, it is important to go beyond the often-employed legalistic approach to examine the social contexts in which exchange takes place.

Increasing the mobilization of germplasm needed to address new global challenges requires collaborative initiatives made up of multiple actors. Yet, these global initiatives are often highly complex. For example, in research teams that include individuals who are close friends, germplasm exchange might be based on long-term obligations and norms of reciprocity, while in other newer teams, exchange must rely on explicit negotiated agreements. Additionally, exchange can occur among a set of actors organized as a club that concentrates resources, sets rules and provides access only to members. But it can also operate in open communities with a fluid membership that can set

http://dx.doi.org/10.19103/AS.2020.0085.20

the conditions of exchange depending on the context. Exchange of germplasm can occur within homogenous groups of actors who share similar objectives, or within heterogeneous groups where substantive differences in capacity, knowledge and objectives lead to imbalances in learning, reputational and financial outcomes among actors. Importantly, the norms, rules and practices pertaining to property rights, benefit sharing and openness can also dictate collaboration structures that are more or less homogeneous, closed or equitable.

These examples demonstrate just few ways in which the policies are embedded in the social context of germplasm exchange can vary. Nevertheless, how policies are embedded in collaborative initiatives will likely significantly affect the exchange of germplasm by, for example, increasing trust or reducing transaction costs. This chapter aims to increase understanding about collaboration dynamics such that new approaches can reduce the 'institutional misalignment' between policies and practices to improve mobilization of genetic diversity across heterogeneous actors and contexts. In particular, we ask: How do collaborative teams manage germplasm access, exchange and use? What factors explain why exchange is facilitated or blocked in various collaborative settings? What biases frame germplasm exchange and what are the consequences?

This chapter is based on a literature review and empirical work conducted by the authors in various collaborative contexts and for various crops and resources. It is hoped that a better understanding of these factors could help the policy community to design more effective rules and regulations and the germplasm user community to develop institutions that improve mobilization of genetic diversity. The first section describes the importance of germplasm exchange and the central role of collaboration. The second section describes a framework that describes how different factors may impact exchange of germplasm. The third section examines the consequences of not including these factors and discusses possible ways forward.

2 Exchange of germplasm for crop breeding: what are we talking about?

The use of genetic diversity in crop breeding is an essential component of improved agricultural productivity through the development of new varieties. These new varieties are very often developed by reshuffling existing alleles within a subset of well-performing genetic material, as it is in general a costly and time-consuming process to develop less-advanced material to attain higher performance levels. However, in the context of rapidly changing environments, recourse to the genetic variability of landraces and/or crop wild relatives is often required. This germplasm can be used to introduce specific traits into breeding populations or for base-broadening activities, which usually require long time and efforts.

Published by Burleigh Dodds Science Publishing Limited, 2021.

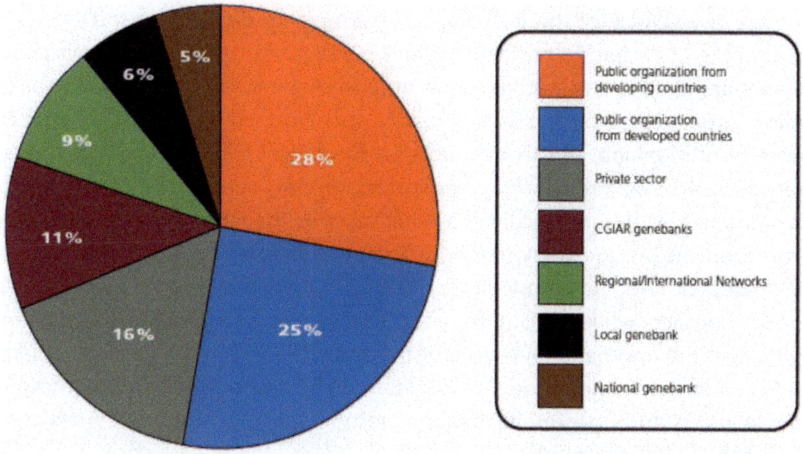

Figure 1 Sources of germplasm used by breeders in national breeding programs. Source: Food and Agriculture Organization of the United Nations. Reproduced with permission.

Breeders access germplasm from multiple sources: their own working collections, other germplasm pools managed by breeders, publicly available varieties, genebank collections, farmers' fields and the market (commercial seed and vegetative propagating material). The relative proportion of these different sources is difficult to determine as it may greatly vary depending on the crops, the type of breeder and regulatory considerations. One survey of breeders working in national programs in developing countries conducted in 2008 through the national information sharing mechanism on Global Plan of Action implementation of the FAO (2010, see Fig. 1) found that the main sources are public organizations[1] from developing or developed countries followed by private sector and genebanks from the Consultative Group of International Agronomic Research Centers (CGIAR). This diversity of sources is supported by other studies that have shown similar levels of distribution (Welch et al., 2017).

Flow of germplasm accessions from genebanks has been estimated to be several tens of thousands of transfers annually (FAO, 2010). However, most of that exchange is between genebanks for conservation and characterization purposes (Dulloo et al., 2010). Even though genebanks aim to facilitate exchange of material to advance research and breeding, there is little evidence that the material held in these collections is being used productively by breeding programs or farmers (Louafi and Manzella, 2018). From the perspective of breeders, only about one-fifth to one quarter of germplasm used in research comes from genebanks.

1 Within the category 'public organization', the survey does not differentiate whether the material comes from genebank or breeder's working collections.

The modalities for the exchange of germplasm depend on the crop and on the type of exchange partners. Over the years, many divergent views about ownership, control and access to germplasm have been expressed in varied policy forums (Tsoumani, 2020). Broadly speaking, starting from a common heritage of mankind, the Convention on Biological Diversity has introduced the concept of sovereign rights while recognizing some collective rights to indigenous people and local communities over their genetic resources and associated knowledge and know-how. The International Treaty on Plant Genetic Resources for Food and Agriculture (ITPGFRA) has developed a global public good approach while recognizing farmers' rights and intellectual property rights, and the International Union for the Protection of New Varieties of Plants (UPOV) and the World Trade Organization (WTO) recognized the granting of private rights through intellectual property rights[2]. Consequently, different approaches about germplasm ownership, control and access coexist, both in policy fora and are undertaken in practice by different groups.

International and national rules, regulations and laws have established increasing levels of control over the access, exchange and use of biological materials (Welch et al., 2013; Bretting, 2007; Sebastian and Payumo, 2006, ten Kate, 2002; Laird et al., 2020). These new institutions introduce new access and exchange procedures and obligations for managing and monitoring transfers to third parties such as memoranda of agreement, prior informed consent and material transfer agreements. The consequence of this proliferation of formal institutional norms is that breeders can no longer act according to their own set of preferred rules and norms when accessing genetic material. Since the signature of the Convention of Biological Diversity in Rio in 1992, they are increasingly expected to enter in complex, long and uncertain negotiation processes that go well beyond their own usual norms and practices. Although the regulations are intended to establish norms that facilitate international exchange and stimulate effective use of resources to solve complex problems while protecting the rights of multiple different stakeholders, they also increase transaction costs.

In the food and agriculture sector, the Multilateral System of Access and Benefit Sharing (MLS) of the ITPGRFA is a global expression of a collective management system for common pooling of genetic resources internationally (Girard & Frison, 2018; Louafi, 2012). Historically, crops and plant genetic resources (PGR) have been widely exchanged throughout the world, and many people in many different places have contributed in one way or another

2 The agricultural sector is characterized by the coexistence of at least two intellectual property systems: the patent system and the plant variety protection system. These two systems are harmonized at the international level by the World Trade Organization's (WTO) Trade-Related Intellectual Property Agreements (TRIPS) and the World Intellectual Property Organization's (WIPO's) affiliated UPOV Agreement. This later aims to find a balance between protection of innovations and free access to germplasm through the breeder's exemption which allows the subsequent use of an innovation protected by a plant variety certificate (PVP) for research purposes (Dutfield, 2011).

to the development of today's crop genetic diversity (Schloen et al., 2011). Consequently, current crop production relies significantly on the use of once-exotic species and all countries depend in part on genetic diversity that originated elsewhere (Khoury et al., 2016). The MLS is a reflection and recognition of this strong interdependency of countries and stakeholders regarding the material they hold. However, the extent to which the MLS has successfully enabled global exchange and use of agricultural genetic resources is unclear.

In part this is because the MLS and other global institutions such as the CBD often ignore the complex social context in which these legal and regulatory solutions are applied. The legal (laws, regulations, rules and contracts) and market-based (royalties from IPRs and benefit sharing agreements) policy approaches fit poorly with the broader set of motivations underlying movement of germplasm such as to increase knowledge, conserve biodiversity, combat climate change or reduce hunger. Germplasm exchange is also embedded in reputational, verification and reciprocity norms. Indeed, it has been shown that replicability of scientific findings (verifiability) and collegial expectations of two-way flow of resources over time (reciprocity) are important motivations for the conservation and exchange of genetic resources (Dedeurwaerdere et al., 2013). Additionally, germplasm exchange is seldom isolated from other shared resources such as technology, information, assistance and advice. Individual transactions are often part of longer term and broader collaboration relationship in which this broad set of resources is shared over time. These motivations and realities shape the exchange behavior of breeders and other researchers as much if not more than market and other rationales underlying many existing global policy solutions.

To provide evidence for the complexities of the exchange process, we present findings from a survey of FAO/CGRFA National focal points and GRFA stakeholders conducted by Louafi and Welch in 2017 for the Commission on Genetic Resources for Food and Agriculture (Louafi and Welch, 2018). Survey respondents were asked to indicate if their exchanges were generally embedded within a larger collaborative relationship or agreement. The results show that it is most often the case that respondents exchange as part of an established collaborative relationship. For example, in the plant sector, only 16% of exchanges are not part of an existing collaborative relationship (see Fig. 2). This is true for domestic and foreign exchanges alike and also across country groupings (OECD/non-OECD) and subsectors (Figs. 3 and 4).

To a large extent, these collaborative initiatives could be viewed as problem-oriented organizations designed to pool resources and knowledge, reduce transaction costs and enhance cooperation among a wide range of actors across different locations to create new knowledge and innovations while maximizing the benefits to all. Shaping these collaborative structures in a way that effectively organizes the complex institutional environment for germplasm access exchange and use could complement (or even sometimes substitute to) legal instruments or monetary incentives. The next section describes an analytical framework that provides a way to approach the substantial variations observed in these collaboration goals, structures and processes.

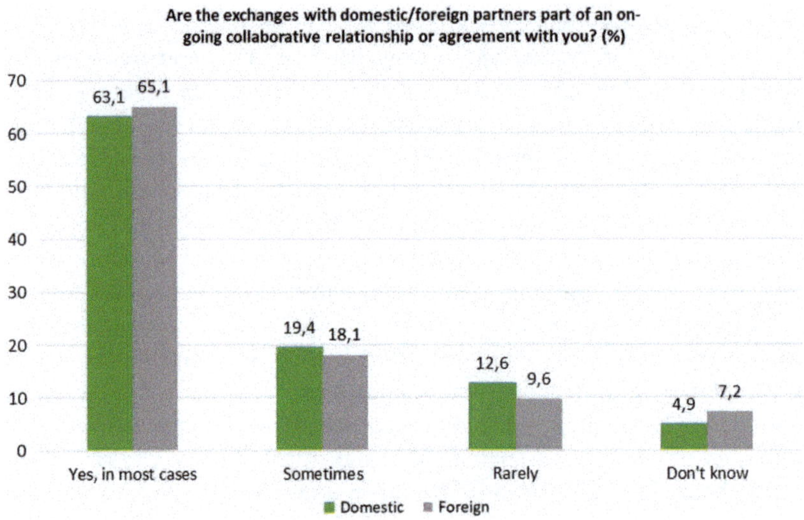

Figure 2 Collaboration pattern with domestic and foreign partners. Source: Food and Agriculture Organization of the United Nations. Reproduced with permission.

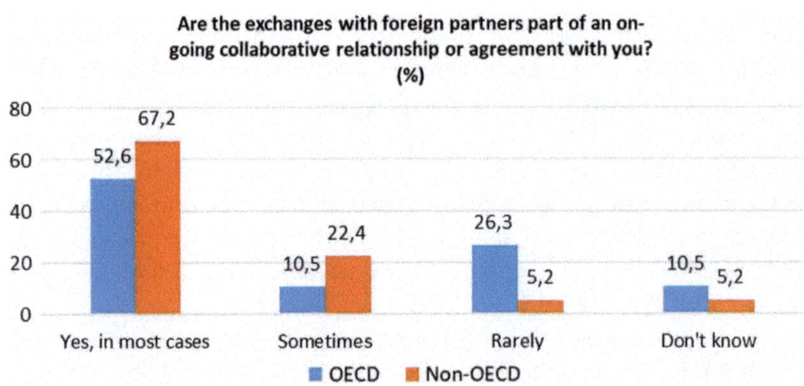

Figure 3 Collaboration pattern with foreign partners (OECD/non-OECD). Source: Food and Agriculture Organization of the United Nations. Reproduced with permission.

Published by Burleigh Dodds Science Publishing Limited, 2021.

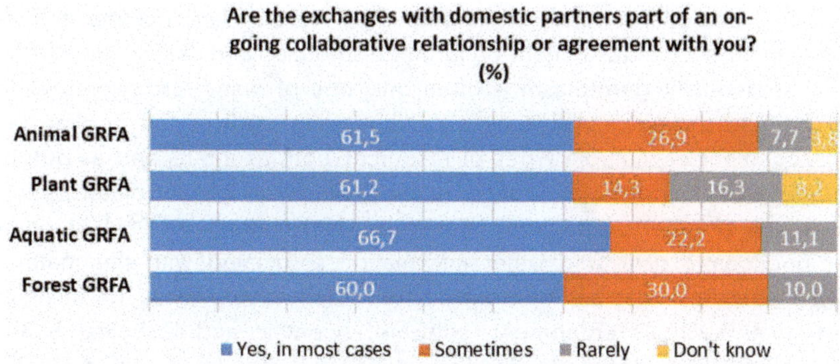

Figure 4 Collaboration pattern with domestic partners by subsectors. Source: Food and Agriculture Organization of the United Nations. Reproduced with permission.

3 Opening the black box of collaborations in plant genetic resources for food and agriculture

Collaboration implies a sustained relationship that involves a flow of different types of resources (technical, organizational, knowledge) among individuals and organizations with different capacities, objectives and perspectives about, for example, openness of science, equity and reciprocity. Because germplasm exchange between two or more entities often takes place within a collaborative context and includes multiple transactions over time, we propose an analytical framework that considers characteristics of resources, institutions, organizations, relationships and individual actors as determinants of germplasm exchange. Aspects of each of these factors can help explain why sometimes exchange is facilitated or blocked. Ultimately, a better understanding of these factors could help the policy community to design more effective rules and regulations. The following sections address each factor in turn.

3.1 Resources

Characteristics of resources affect exchange. First, not all species are equally valuable or equally available. Some species have higher value due to high market demand for new varieties (staple or major crops), low availability of genetic diversity or difficult conservation or complex reproduction. Besides, even different materials of same species are not equally valuable or equally available. For example, landraces with unique alleles or combinations of alleles may be more valuable and therefore more difficult to transfer or require substantial compensation to access. The ITPGRFA recognizes that all material is not equal. For example, the so-called 'Plant Genetic Resources for Food and Agriculture under Development' category gives to the provider the discretion to add additional terms and conditions to the standard material transfer agreement, such as restrictions on further distribution and use[3].

Additionally, germplasm is often only one of many resources used or exchanged through collaboration. Collaboration involves the pooling and management of multiple types of resources both as inputs and as outputs produced by research. A first set of resources includes genetic materials such as seeds or other propagation materials, plant material or DNA, genomic or phenotypic data, and other information associated with the material. Other resources play an important role in collaboration. These are technical resources such as equipment, software, scientific and technical human capital resources, organizational or administrative resources, such as conflict resolution services, institutional and governance resources such as material sharing standards, and social capital available through relations between individual collaborators.

Multiple dependencies exist among resources within a collaboration. The value of one particular resource may depend on its relationship to other shared resources. Additionally, new resources may be produced from combinations of different shared resource (Frischmann et al., 2014). This is for example the case of: the sharing of research results and information derived from the use of germplasm; the provision of access to technologies necessary for enhanced germplasm use; the creation of training and capacity building opportunities for increasing breeding capacities.

3.2 Institutions

Facilitating germplasm exchange is justified by the need to address higher societal goals such as food security and sustainable agriculture. Despite broad support of these goals by the global agricultural community, regulatory regimes and societal norms that govern germplasm exchange are fragmented and built upon different and often conflicting logics. The international treaty is one example of a regulatory regime that places high importance on the collective and equity logics that support germplasm exchange. These two logics value effective use of shared resources and reciprocity in order to strengthen the long-term cooperative capacities of stakeholders.

The collective logic stems from the high interdependence[4] on plant germplasm among countries that necessitates pooling of material and global collaboration to achieve global social objectives. The equity logic stems from the desire to accommodate and address differences in values, power and capacity across countries and stakeholders. Practically speaking, germplasm is

3 ITPGRFA, Article 12.3(e); SMTA, Article 6.5.
4 Interdependence is defined as when outcomes are based on a combination of parties' efforts. Interdependence involves mutual and complementary arrangements, reduces risks and encourages cooperation (Molm, 2003).

held, valued and used by a broad range of stakeholders around the world. Each plays an indispensable role in conserving germplasm and making it available for use. Yet, no single actor has the capacity to conserve and provide all potentially useful germplasm. At the same time, not all countries or actors have the same ability to benefit from germplasm use. Tremendous inequalities exist in terms of human capacity, access to technology and access to information necessary to effectively manage and use germplasm.

Collaboration structures and the resources within them are governed by other logics besides the collective and equity ones. Multiple logics such as the conservation, safety, market, academic or the ethics logics underlie regulations, norms and standards that focus on some aspects of the material and some specific activities. Because diverse institutional logics are supported and emphasized by different stakeholders, conflicts can arise that may create barriers to access or increase transaction costs of germplasm exchange and use (Welch et al., 2019; López Noriega et al., 2013). In this context of germplasm exchange - conflicting logics, multiple stakeholders and fragmented governance - the context of has changed over time. The post-WWII era in which access to germplasm was centralized through genebanks and provided mainly on an informal basis has gone. It has been replaced by a more fragmented and opportunistic environment in which diverse groups across different countries exchange multiple resources under different conditions.

3.3 Organizations

Data and material come from a range of different mid-level organizations (e.g. teams, funded projects, and consortia), agreements and arrangements that provide varying levels of access to and discretion over exchange and use of germplasm for research (Reichman et al., 2016; Ulhir and Schröder, 2007).

In the genetic diversity and breeding sector, structure and composition of collaborative initiatives can take several forms: closely tied and homogenous teams, communities of practice, research organizations (NARS, universities, institutes, companies), inter-organizational initiatives (public-private partnerships, network, consortia), global collaborative initiatives of widely geographically spread organizations (programs, funded projects). Organizations may also differ in their level of specialization and concentration of resources, in whether the decision and control structure is hierarchical or distributed, and in the amount of discretion and control entities have for distribution, monitoring and tracking of exchange and use.

Collaboration and exchange complexity arises because individuals and organizations belong simultaneously to several collaborative initiatives each with its own goals, incentives and organizations. Management decisions taken at the level of these organizations could be collectively discussed and adapted to respond to the needs and goals of the collective. As such, they offer some flexibility in prioritizing institutional logics, interpreting policy rules and designing governance systems.

3.4 Relations

Flows of resource inputs to research occur within exchange networks that connect recipients and providers through different types of relationships, some of which are stronger and more trusted than others. Researchers who conduct research on a particular species are typically colleagues or friends: they usually know each other well. With the exception of few major crops, crop networks are often relatively small and the number of people engaged in germplasm exchange within these communities is even smaller. Collaborators within these communities may have the same disciplinary background, been trained at the same university, worked in the same organization or attended the same conferences.

Germplasm exchange is affected by the structure, composition and relations in collaboration and resource exchange networks, which may be composed of individuals, organizations or combinations of the two. Network structure includes network size and connectedness. For example, findings from a collaboration network analysis in cacao genetic resources reveals a cohesive community connected through central actors that have different and complementary profiles as resource and service providers (Louafi et al., 2017). In many cases, genebanks play a central role in the global collaborative structure because they are key providers, sometimes primary providers, of the germplasm. Network connectivity (e.g. density) can increase joint action, avoid resource conflicts and facilitate the development of common resource regulations (Bodin and Crona, 2009). In a global context, community building activities that generate greater connectedness can increase the exchange of information, knowledge and resources across diverse individuals and organizations (Fusi et al., 2018). For instance, the Association on the Study and Information on Coffee (ASIC) organizes a conference every two years for the global coffee research and innovation community. This conference offers the opportunity for information and expertise exchange but also for the development of greater global connectivity among coffee researchers. Similar association or initiatives exist for all kind of crops.

Collaboration and resource networks are composed of different types of actors that have different of capacities, backgrounds, types of expertise, perceptions and interests. Compositional diversity in the network will affect network cohesion such that divergent network composition would work against global collaboration creating instead (Louafi et al., 2017): (i) regional and biological divides, due to numerous factors such as regional specificity of pest and diseases, quality and flavor driving different markets, the existence of regional/national regulatory frameworks, or simply geographical distances; (ii) disciplinary divides and more particularly, the persistent division between conservation/diversity characterization, genomics, breeding and bio-informatics, each of which values different types of genetic material and associated information ; and (iii) market divides, specialty/high premium versus the bulk markets and, more generally, the private company strategies with regard to genetic resources-related research and partnerships.

Similarly, tie strength, measured as the frequency of interaction, duration of acquaintance, or emotional closeness (Wellmann, 1999; Krackhardt, 1992), is often associated with greater trust, longer term reciprocity and mutual confiding. These are important qualities of a resource exchange network. Individuals are more willing to exchange germplasm with close colleagues and friends due to higher trust and norms of reciprocity (Fusi et al., 2018; Cook and Whitmeyer, 1992). In fact, close colleagues may be more willing to exchange resources, even when the formal access is restricted by national regulations or industry competition (Shibayama and Baba, 2011). Similarly, researchers may be more likely to provide returns from the use of resources from close colleagues in order to maintain and build stronger collaborative relationships.

In sum, collaboration patterns can predict social processes, such as germplasm exchange and governance outcomes, including adaptive capacity. This section focused on network structure and composition. However, it is also equally relevant to assess attitudes and behavior at the level of the nodes, including the individual actors.

3.5 Actors

Attributes of actors may affect the way they engage in collaboration with each other and more specifically, their willingness to share resources. Attributes such as rank, discipline, sector and size of an individual's data or material collection, are likely to affect willingness to share or ability to easily access germplasm. In the context of genomic epidemiology network, Chokshi et al. (2006) have shown that lack of trust about what others will do with the data, discomfort with transparency, high costs of collection, first rights of primary researchers and students on the project, age, discipline and attitudes toward open science are all factors that explain data sharing.

Stakeholder groups holding and using germplasm include farmers, communities, associations and producers at the local level, public institutions at national and international levels, and private enterprises and farmers' and breeders' cooperatives at national and international levels as well as networks at local, national, regional and international level. They all fulfill different tasks in a complex network of actors, and are to various degrees involved in the

Published by Burleigh Dodds Science Publishing Limited, 2021.

maintenance and conservation of genetic diversity, the exchange and direct use of genetic material, and research and product development based upon genetic resources.

Individual attitudes toward germplasm sharing are formed through a socialization process and the incentive structure that conditions activity. Divergent of attitudes toward openness are usually observed between individuals from infrastructure science (e.g. genebanks), discovery science (e.g. breeders, geneticists) and bio-informatics. By contrast, scientists who are trained in research labs where there is a strong culture of 'open science', teamwork and sharing might be more likely to engage in resource sharing practices that promote this approach.

Similarly, the usual social and cultural norms of mutual assistance, reciprocity and solidarity associated to agrarian societies tend to encourage the free flow of seeds and germplasm between farmers or with farmers and other actors. In reality, many nuances have been introduced by several empirical studies to this perceived frictionless circulation of planting material (see Coomes et al., 2015 for a review of this literature). Findings from Kiptot et al. (2006) and Kawa et al. (2013) (cited by Coomes et al., 2015) showed, for example, that farmers with leadership positions and greater ethnobotanical knowledge (but not necessarily higher cultivar diversity) are found to be more likely to give out seeds than others.

4 The germplasm exchange fallacies

The different dimensions described in the previous section recognize a complex context of germplasm exchange in which tensions may arise between conflicting institutional logics, norms and values that are being expressed differently depending on the social and organizational context of exchange and the broader resource system in which germplasm is embedded.

This complex picture reveals at least three biases that limit germplasm exchange for research and breeding:

1 *Single-logic bias*. The existing regulatory framework focuses almost exclusively on access to germplasm based on legal status and monetary incentives to protect innovation or share benefits derived from germplasm. Such an approach fails to include other equally important determinants of exchange.
2 *Single-resource bias*. The exchange of germplasm tends to be addressed in isolation from other resources. Such an approach fails to maintain the integrity of the resource system in which germplasm is embedded and fails to account for the functional interdependence between the separate elements of the resource system.

3 *Single-transactional bias*. Germplasm exchange tends to be approached as a single transaction that occurs between individual actors or organizations. Such approach fails to account for the broader social environment within which mutually beneficial relationships build norms of trust and reciprocity in collaboration and resource exchange networks.

4.1 The single-logic bias

The 'problem' of germplasm management has been traditionally framed at the global level as a conservation-based concern for crop diversity erosion linked to the genetic homogenization induced by the development of industrial forms of agriculture. In response, effort has sought to coordinate conservation activities to collect and conserve germplasm in ex situ-centralized genebanks and simultaneously stimulate use in research and breeding programs. Because regulatory frameworks have been designed based on conservation and innovation logics, they rely on technical solutions (ex situ conservation) and incentive structures (monetary through the granting of IPR to recoup the investment made in research and breeding) relevant for a small range of resources, actors and motivations. This way of framing the problem and designing solutions does not consider how and why people engage in producing more diverse new germplasm. It also ignores inequality in the capacity to use and produce diversity for their different needs and interests.

To be fair, some space has been carved out in existing policy frameworks that attempt to reconcile different institutional logics. As mentionned earlier, the Treaty recognizes specifically the collective dimension of innovation in breeding and the need for coordination between different communities. The Treaty recognizes 'that plant genetic resources for food and agriculture are a common concern of all countries' such that all countries largely depend on plant genetic resources for food and agriculture that originated elsewhere. The MLS of ABS is the expression of this need for collective action at the global level to manage a pool of common but distributed resources across multiples countries and entities (Louafi and Bhatti, 2012). It comprises a large number of plant genetic resources held by varied entities throughout the world, and aims 'both to facilitate access to plant genetic resources for food and agriculture, and to share, in a fair and equitable way, the benefits arising from the utilization of these resources, on a complementary and mutually reinforcing basis'.[5]

The facilitation of access and sharing of benefits is managed by agreeing collectively on standard conditions of access and benefit-sharing for all material included in the MLS and on establishing a Global Benefit Sharing Fund, which

5 Article 11.1 of Treaty.

disbursement is made according to procedures and priorities also agreed collectively. Besides, the pooling of benefits enables the realization of *collective* benefits independently of whether economic return is generated through the development of a particular variety or technology. Increasing knowledge, information and data produced through phenotyping, genotyping or evaluation are benefits that can be widely shared as many countries make use of the same species, establish similar production systems and struggle with the same biotic and abiotic stressors (Schloen et al., 2011). The Treaty also addresses equity beyond simply sharing of monetary benefits. It recognizes non-monetary benefits as important elements in the Treaty context to enable less-endowed actors to use the genetic diversity (i.e. have access to the benefit of 'facilitated access').

Even though the Treaty is to date the global framework that has managed to integrate the multiple logics, the process of Treaty implementation and its political dynamics still tend to often prioritize the legal and monetary logics over equity and collective logics. As a result, actors - individuals, groups, teams and countries - have engaged the Treaty opportunistically vis-à-vis other existing frameworks such as the UPOV and the Nagoya Protocol. This lack of shared commitment has negatively impacted the legal solution set up by the Treaty to facilitate exchange for research and innovation. Ultimately, no global framework has succeeded in integrating multiple logics in a functional way perceived as legitimate by all countries.

4.2 The single-resource bias

The more comprehensive picture provided by our framework also focuses attention on maintaining the integrity and functional interdependence of the resource system in which germplasm is embedded and avoiding the single resource bias that tends to structure the regulatory and legal approaches to facilitating germplasm exchange.

Resources may be subject to different property regimes across or within countries. It is indeed important to note that some resources may not be necessarily under the full control of the actual holder and that the regulatory and institutional controls are not necessarily uniform across countries. In this context, pooling various resources and agreeing on common rules within the collaboration provides one way to manage the complexity inherent to this system of interdependent resources (Kamau and Winter, 2013). Thus, facilitating germplasm exchange requires managing the resource system in which germplasm, as one resource unit, is embedded.

The Treaty does recognize information exchange, capacity building and access to technology for the good functioning of the multilateral system (as part of the so-called non-monetary benefit sharing components). However, no real operational mechanisms are in place to manage these components and germplasm in an integrated way. This gives the impression that the monetary and benefit sharing components can be managed separately or that they are secondary to germplasm, the main resource that the Treaty aims to regulate. Yet joint management is essential for both function and equity: facilitated access is only made possible as long as the necessary capacities, information or technology are also available. Making information available and strengthening the technical, knowledge and financial capacities of the various members is essential to increase the use of the material and information pooled (i.e. sharing the benefits of the facilitated access to germplasm offered by the Treaty).

4.3 The transaction bias

Germplasm exchange takes place within a complex organizational and relational context that determines how policies will be ultimately implemented. The production, management and exchange of germplasm is an inherently social phenomenon, taking place over a wide range of scales and within a complex, overlapping variety of supply channels. Organizations operate as hubs for organizing the complex and sometimes conflicting institutional environment related to germplasm. The level of internalization and integration of the different logics and policy constraints may vary greatly depending on the organization or network. The process of forging mutually beneficial relationships over a long period of time through repeated interactions and ongoing collaboration could be instrumental in facilitating exchange. Those relationships provide the social structure through which actors exchange materials, set expectations for compensation and manage regulatory constraints. López Noriega et al. (2013) noted that 'larger research projects, in which transfers of genetic materials are supportive of broader research objectives, can be instrumental. Scientists involved in international germplasm improvement networks noted that long-term co-operative links (for example, through the INGER network led by the IRRI) are less affected by the reduced willingness of countries to share germplasm'.

A typology of the various organizational models in response to the multiplicity of institutional logics and the constraints on access is difficult to draw given the complexity, interlinked and context-specific nature of collaborative initiatives in plant genetic resources (Fusi et al., 2018). From a broad perspective and from looking only at the way the trade-off between efficiency and inclusiveness is managed, initiatives tend to differ along a spectrum between the following two extremes:

- *Club approach*. A club is composed of a relatively homogeneous set of members who agree to centralized governance structures that set formal rules on access to and use of germplasm. High connectivity can facilitate exchange among members but lead to network closure that constrains exchange and collaboration between members and non-members, and with other clubs. This can lead to an overall decrease in the circulation of genetic diversity.

- *Inclusive approach*. An inclusive model encourages distributed organizational forms and capacities through a web of managed ties across which resources flow according to *ad hoc* procedures and rules, and greater informality based on trust and reciprocity. Actors invest in the technical capacity needed to ensure effective use of shared resources by connected stakeholder groups and countries that are less endowed. This model favors a stronger approach to ensuring an open system while managing interdependencies and creating greater recognition across all actors that exchange, research and innovation take place on equitable terms. The social diversity gathered will likely lead to more genetic diversity exchanged. Such model requires energetic and professional management to establish and sustain trust-building mechanisms and minimize opportunistic behavior.

Of course, these two ideal-type models are seldom found in reality. Eventually, the way the trade-offs and tensions are managed depends on joint commitment and management, which ultimately reconciles the fragmented institutional landscape.

5 Conclusion

This chapter offers a renewed perspective on the issue of germplasm exchange. Based on the increasing number of empirical studies, it describes the reasons why facilitation of exchange cannot be confined to a simple set of legal access rules. By developing an analytical framework that includes several dimensions beyond regulation, it sheds some light on the fallacies and biases surrounding the debate on the enhancement of germplasm exchange. It creates the foundation for a more comprehensive approach to governance of germplasm that recognizes the collaborative context and acts upon the relationship between germplasm and the diversity of institutional logics, germplasm and its resource system, and germplasm and its social environment.

Facilitating germplasm exchange starts by first recognizing that difficulties arise from the reconciliation of various institutional logics in relation to germplasm management, use and exchange. It also requires recognition the nature of the germplasm itself: human-made and shaped by the actors and their interactions. The multiplex interdependences across various resources must be carefully managed. Finally, it also requires action at the organizational and relational levels to manage regulatory constraints, reduce transaction costs and set expectations across members.

Governance systems that recognize and address these complexities is particularly important in a context in which facilitating germplasm exchange is no longer devoted solely to breeding for yield improvement, but must also address the diversity of objectives depending on specific contexts and actors' needs and capacities. Achieving more sustainable agriculture requires not only more improved varieties, but also recognizing different production models and greater investment in human capacity across a more widely distributed innovation system that can address specific needs.

6 Where to look for further information

A good general overview to the International Treaty on Plant Genetic Resources for Food and Agriculture is Halewood, M., Lopez Noriega, I. and Louafi, S. (Eds), (2012), *Crop Genetic Resources as a Global Commons*. Earthscan Publications, London.

Two more recent reference books include:

- Tsioumani, E. (2020). *Fair and Equitable Benefit-Sharing in Agriculture: Reinventing Agrarian Justice* (p. 186). Taylor & Francis.
- Girard, F., & Frison, C. (Eds.). (2018). *The Commons, Plant Breeding and Agricultural Research: Challenges for Food Security and Agrobiodiversity*. Routledge.

More institutional resources about access and benefit sharing could be find at the following organizations:

- Convention on Biological Diversity, The access and benefit-sharing clearing-house: https://absch.cbd.int.
- International Treaty on Plant Genetic Resources for Food and agriculture: http://www.fao.org/plant-treaty/en/.
- UNDP/GEF, GLOBAL ABS Community: https://community.abs-sustainabledevelopment.net.
- IUCN, An explanatory guide to the Nagoya Protocol on access and benefit-sharing: https://www.iucn.org/content/explanatory-guide-nagoya-protocol-access-and-benefit-sharing.
- South Center: Genetic resources and TK: https://www.southcentre.int/category/issues/innovation-development/genetic-resources-tk/.
- The ABS Capacity Development Initiative, http://www.abs-initiative.info.
- Alliance Bioversity International and CIAT: https://www.bioversityinternational.org/research-portfolio/policies-for-plant-diversity-management/.

7 References

Bodin, Ö. and Crona, B. I. (2009). The role of social networks in natural resource governance: what relational patterns make a difference? *Global Environmental Change* 19(3), 366–374.

Bretting, P. K. (2007). The US National Plant Germplasm System in an era of shifting international norms for germplasm exchange. Acta Horticulturae 760, 55–60. doi:10.17660/ActaHortic.2007.760.5.

Chokshi, D. A., Parker, M. and Kwiatkowski, D. P. (2006). Data sharing and intellectual property in a genomic epidemiology network: policies for large-scale research collaboration. *Bulletin of the World Health Organization* 84(5), 382–387. https://doi.org/10.2471/blt.06.029843.

Cook, K. S. and Whitmeyer, J. M. (1992). Two approaches to social structure: exchange theory and network analysis. *Annual Review of Sociology* 18(1), 109–127.

Coomes, O. T., McGuire, S. J., Garine, E., Caillon, S., McKey, D. B., Demeulenaere, E., Jarvis, D., Aistara, G., Barnaud, A., Clouvel, P., Emperaire, L., Louafi, S., Martin, P., Massol, F., Pautasso, M., Violon, C. and Wencelius, J. (2015). Farmer seed networks make a limited contribution to agriculture? Four common misconceptions. *Food Policy* 56, 41–50.

Dedeurwaerdere, T., Broggiato, A., Louafi, S., Welch, E. W. and Batur, F. (2013). Governing global scientific research commons under the Nagoya protocol. In: Morgera, E., Buck, M. and Tsioumani, E. (Eds), *The 2010 Nagoya Protocol on Access and Benefit-Sharing in Perspective: Implications for International Law and Implementation Challenges*. Martinus Nijhoff Publishers, Leiden, pp. 389–421.

Dulloo, M. E., Hunter, D. and Borelli, T. (2010). Ex situ and in situ conservation of agricultural biodiversity: major advances and research needs. *Notulae Botanicae Horti Agrobotanici Cluj-Napoca* 38(2), 123–135.

Dutfield, G. (2011). The role of the International Union for the Protection of New Varieties of Plants (UPOV). *Intellectual Property Issue Paper* (9), 24 p.

FAO. (2010). *The Second State of the World's Plant Genetic Resources for Food and Agriculture*, Rome, Italy.

Frischmann, B. M., Madison, M. J. and Strandburg, K. J. (2014). *Governing Knowledge Commons*. Oxford University Press, New-York.

Fusi, F., Manzella, D., Louafi, S. and Welch, E. (2018). Building global genomics initiatives and enabling data sharing: insights from multiple case studies. *Omics: A Journal of Integrative Biology* 22(4), 237–247.

Kamau, E. C. and Winter, G. (2013). *Common Pools of Genetic Resources: Equity and Innovation in International Biodiversity Law*. Routledge, Abingdon.

Kawa, N. C., McCarty, C. and Clement, C. R. (2013). Manioc varietal diversity, social networks, and distribution constraints in rural Amazonia. *Current Anthropology* 54(6), 764–770.

Khoury, C. K., Achicanoy, H. A., Bjorkman, A. D., Navarro-Racines, C., Guarino, L., Flores-Palacios, X. and Ramírez-Villegas, J. (2016). Origins of food crops connect countries worldwide. *Proceedings of the Royal Society. Series B: Biological Sciences* 283(1832), 20160792.

Kiptot, E., Franzel, S., Hebinck, P. and Richards, P. (2006). Sharing seed and knowledge: farmer to farmer dissemination of agroforestry technologies in western Kenya. *Agroforestry Systems* 68(3), 167–179.

Krackhardt, D. (1992). The strength of strong ties: the importance of Philos in organizations. In: Nohria, N. and Eccles, R. (Eds), *Networks and Organizations: Structure, Form, and Action*. Harvard Business School Press, Boston, MA, pp. 216–239.

Laird, S., Wynberg, R., Rourke, M., Humphries, F., Muller, M. R. and Lawson, C. (2020). Rethink the expansion of access and benefit sharing. *Science* 367(6483), 1200–1202.

López Noriega, I., Halewood, M., Galluzzi, G., Vernooy, R., Bertacchini, E., Gauchan, D. and Welch, E. (2013). How policies affect the use of plant genetic resources: the experience of the CGIAR. *Resources* 2(3), 231–269.

Louafi, S. (2012),. Collective action challenges in the implementation of the Multilateral System of the International Treaty, what roles for the CG Centres? In: Halewood, M., Lopez Noriega, I. and Louafi, S. (Eds), *Crop Genetic Resources as a Global Commons*. Earthscan Publications, London, pp. 321–339.

Louafi, S. and Bhatti, S. (2012). Efforts to get the multilateral system up and running, A review of activities coordinated by the Treaty's Secretariat. In: Halewood, M., Lopez Noriega, I. and Louafi, S. (Eds). *Crop Genetic Resources as a Global Commons*. Earthscan Publications, London, pp. 188–197.

Louafi, S. and Manzella, D. (2018). International treaty: heterogeneity and equity in global resources management. In: Girard, F. and Frison, C. (Eds), *The Commons, Plant Breeding and Agricultural Research: Challenges for Food Security and Agrobiodiversity, EarthScan Food and Agriculture*. Routledge, London, p. 302.

Louafi, S., Meter, A., Laliberté, B. and Medina, V. (2017). Design and implementation of a global collaborative framework on cacao genetic resources research: incentives, constraints and institutional structures. In: Proceedings of the first International Symposium on Cocoa Research ISCR 2017. Lima: ICCO, 22 p. International Symposium on Cocoa Research–ISCR 2017: Promoting Advances in Research to Enhance the Profitability of Cocoa Farming. 1, Lima, Pérou, 13–17 November 2017.

Louafi, S. and Welch, E. (2018). Access and benefit-sharing for genetic resources for food and agriculture: survey findings. CGRFA-17/19/3.2/Inf.4. CGRFA, FAO: Rome.

Molm, L. D. (2003). Theoretical comparisons of forms of exchange. *Sociological Theory* 21(1), 1–17.

Reichman, J. H., Dedeurwaerdere, T. and Uhlir, P. F. (2016). *Governing Digitally Integrated Genetic Resources, Data, and Literature: Global Intellectual Property Strategies for a Redesigned Microbial Research Commons*. Cambridge University Press, Cambridge.

Schloen, M., Louafi, S. and Dedeurwaerdere, T. (2011). Access and benefit-sharing for genetic resources for food and agriculture. Current use and exchange practices, commonalities, differences and user community needs. Report from a multi-stakeholder expert dialogue. Background Study Paper no. 59. FAO, Rome.

Sebastian, L. and Payumo, J. G. (2006). NARES capacity in relation to international treaties and conventions on intellectual property rights, agricultural biotechnology, and plant genetic resources management. *Asian Journal of Agriculture and Development* 4(1), 1–24.

Shibayama, S. and Baba, Y. (2011). Sharing research tools in academia: the case of Japan. *Science and Public Policy* 38(8), 649–659.

ten Kate, Kt (2002). Global genetic resources: science and the convention on biological diversity. *Science* 295(5564), 2371–2372.

Uhlir, P. F. and Schröder, P. (2007). Open data for global science. *Data Science Journal* 6, OD36–OD53.

Welch, E. W., Fusi, F., Louafi, S. and Siciliano, M. (2017). Genetic resource policies in international collaborative research for food and agriculture: a study of USAID-funded innovation labs. *Global Food Security* 15, 33–42.

Welch, E. W., Shin, E. J. and Long, J. V. (2013). Potential effects of the Nagoya protocol on the exchange of non-plant genetic resources for scientific research: actors, paths, & consequences. *Ecological Economics* 86, 136–147.

Welch, E. W., Taggart, G., Feeney, M. K. and Siciliano, M. (2019). Navigating the labyrinth: academic scientists' responses to new regulatory controls on biological material inputs to research. *Environmental Science and Policy* 101, 136–146.

Wellmann, B. (1999). *Networks in the Global Village: Life in Contemporary Communities*. Westview Press: Boulder, CO.

Part 2

Protecting plant genetic diversity: in-situ and on-farm strategies

Chapter 5

Key steps in conservation and use of plant genetic resources: an overview

Nigel Maxted and Joana Magos Brehm, University of Birmingham, UK

1 Introduction

Plant genetic resource (PGR) conservation and use can only hope to address climate change and existential emergency of human population if wasteful duplication of effort is reduced and its efficiency is increased (Maxted et al., 2016). Less resources are wasted on discrete, stand alone, even academic PGR conservation and use activities that fail to result in ultimate conservation implementation and use of the conserved resources. More effective joined-up integration of conservation planning with conservation implementation and conservation implementation with the utilization of conserved resource is required. As such, an overview of the key steps in conservation and the use of PGR is essential to help secure the future of humankind. Such a holistic approach to PGR conservation/use continuum is needed to address the post-2020 conservation priorities currently being drafted and ensure the UN Sustainable Development Goals for people and planet are achieved by 2030 (Fig. 1) (UN, 2015).

Biodiversity conservation aims to: *maintain the diversity of living organisms, their habitats and the interrelationships between organisms and their environment* (Spellerberg and Hardes (1992), while genetic resource conservation focusses explicitly on the conservation of the taxon's full breadth of genetic diversity and links the conserved resources to utilisation in breeding, so aims to *maintain the full breadth of genetic diversity of living organisms, their habitats and the*

http://dx.doi.org/10.19103/AS.2020.0085.04

Figure 1 UN Sustainable Development Goals for people and planet by 2030 (UN, 2015).

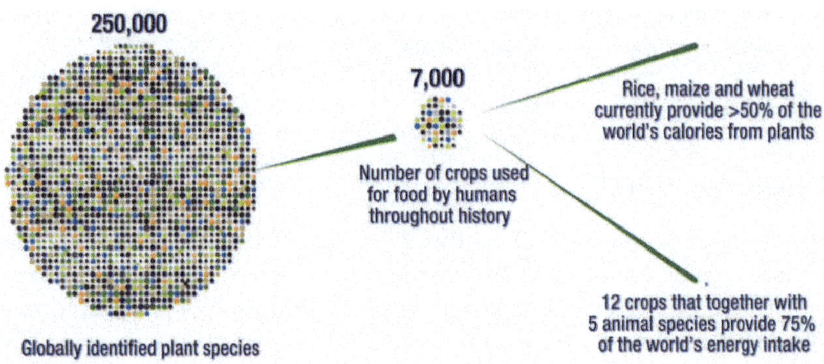

Figure 2 The world's reliance on a narrow diversity of food crops (FAO, 1995).

interrelationships between organisms and their environment, as well linking the conserved resource to its utilisation for the benefit of humankind. Traditionally, PGR conservation has focussed almost exclusively on crops and, lately, their wild related species, a relatively small subset of all plant species (Fig. 2).

This subset of plant species might be more precisely described as PGR for Food and Agriculture (PGRFA) because the focus is on conserving genetic diversity for use in food and agricultural production. A distinction re-enforced by the IPGRI (currently The Alliance of Bioversity International and CIAT Alliance) (1993) definition of PGR is: *Plant genetic resources are the taxonomic and genetic diversity of plants that is of value as a resource for the present and future generations of people*. Therefore, PGR is broader than PGRFA and also includes taxa and genetic diversity that are used for other purposes than Food and Agriculture (Fig. 3):

- **Modern cultivars** - Genetically uniform or clonal crop varieties bred by plant breeders and currently sown by farmers that become a genetic resource once commercially obsolete.
- **Obsolete cultivars** - Former cultivars that are no longer commercially grown and do not appear on national variety lists, but which may possess genes useful to plant breeders.
- **Breeding lines, clones, populations and genetic stocks** - Material used by plant breeders to develop modern cultivars by means of crossbreeding or use of biotechnology tools.
- **Crop landraces** (LR) - Genetically diverse crop varieties that are the product of traditional seed saving systems, not modern plant breeding, commonly associated with local adaptation and traditional agricultural practices in more marginal agricultural environments.
- **Weedy races** - Wild species that occur as part of crop-weed complexes either as a result of hybridisation between the crop and wild species, the crop and wild species being evolved from the same ancestor or as the crop's progenitor, often found in gene centres but which hybridize freely with the crop and may introgress useful genes from wild species.
- **Crop wild relatives (CWR)** - Wild species that are relatively closely related to a crop and may be crossed with it either using conventional or genetic engineering techniques to introduce desirable traits from the wild species to the crop.

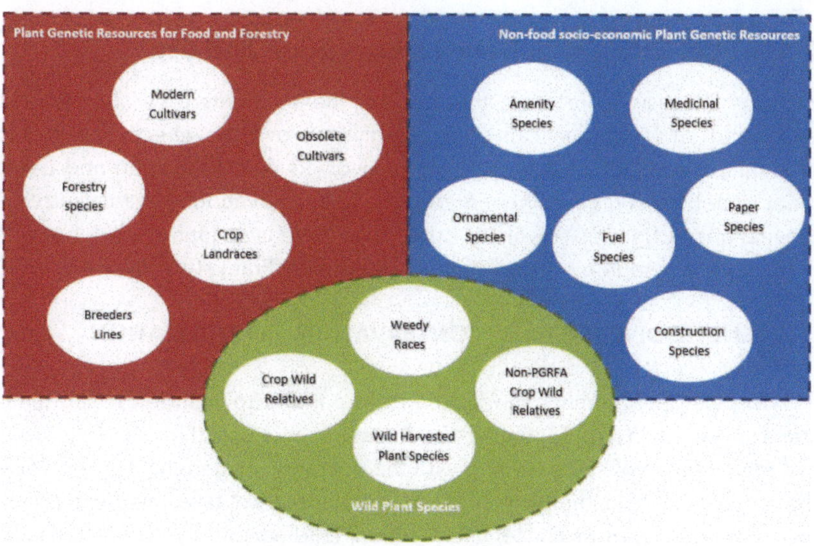

Figure 3 The diversity of plant genetic resources (Maxted et al., 2020).

- **Wild harvested plant species** – Wild species that are gathered directly from the wild and used as food, medicines, religious purposes and so on.
- **Non-food socio-economic species** – Species whose value is associated with non-agricultural exploitation, such as species with medical, forestry, recreational or ornamental value.
- **Other wild species** – Species of less-immediate utilisation potential in terms of trait acquisition but which form the basis of natural communities.

There is a fundamental distinction between biodiversity and PGR conservation, the former is primarily focussed on (a) conservation of whole communities, with a focus on 'keystone species' that define a community and without which coherence is lost, (b) conservation success is viewed in terms of maintenance of viable taxon populations, and (c) conservation is non-utilitarian, 'each species deserves a future in its own right', whereas PGR conservation is primarily focussed on (a) conservation on the full range of genetic (allelic) variation within taxa, (b) conservation success is viewed in terms of maximizing the range of genetic (allelic) variation within taxa, (c) conservation is often explicitly utilitarian, linked directly to some form of human utilization. Having drawn attention to this distinction, the CBD explicitly links conservation and use in Article 1, '*The objectives of this convention . . . are the conservation of biological diversity, the sustainable use of its components and the fair and equitable sharing of the benefits arising out of the utilisation of genetic resources.*' The distinction is growing less categorical and a continuum between the two approaches might better characterize today's reality but clear examples of each are still evident.

2 A model of plant genetic conservation and use

Humans like using protocols that set out the elements of a complex task and using such schema can undoubtedly improve the effectiveness of its implementation; so, what are the steps involved in conservation and use of plant genetic resources? The aim of genetic conservation is to maximize the maintenance of genetic diversity, but further there is an intimate link between plant genetic resource, conservation and utilisation (Frankel and Bennett, 1970):

GENETIC DIVERSITY >>> **CONSERVATION** >>> **UTILIZATION**

This simple schematic model can be expanded to include the sequential steps involved in plant genetic conservation and use (Fig. 4). The model includes a series of steps starting with the full range of genetic diversity for all the target plant taxa, through the prioritisation of target taxa, the planning and the implementation of conservation action, leading through characterisation and evaluation, and utilisation in the development of novel crop varieties by

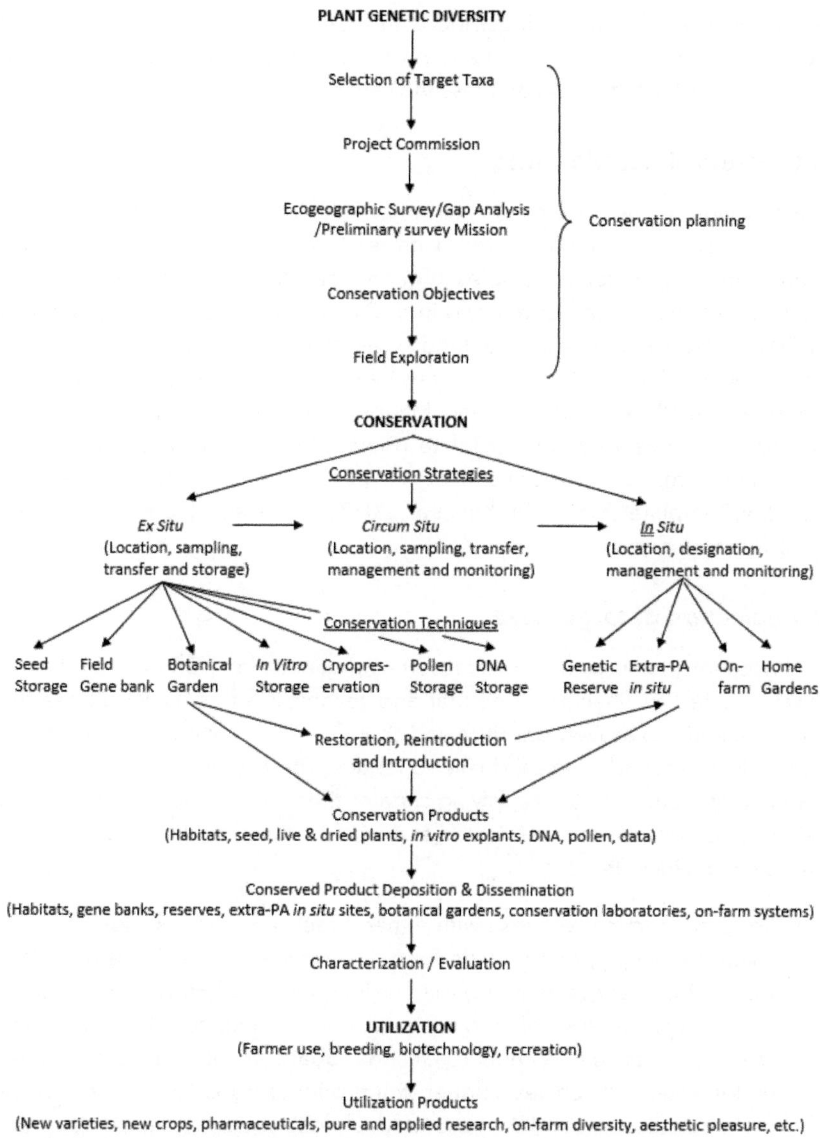

Figure 4 Model of plant genetic conservation and use (Maxted et al., 2020).

farmers and/or breeders. The application of this model is at the core of food security, poverty alleviation and the well-being of humankind. It is easy to apply and applicable for all PGR taxa. Although the model implies a linear process (plant genetic resources conserved to conserved resource utilisation), the process itself is rarely truly linear. Not only are there multiple ways of planning the conservation actions, but also practical implementation of these actions will often involve a series of feedback loops. Indicators of conservation success are generally defined and reviewed and proposed conservation actions are then refined and adjusted based on these indicators.

3 Conservation planning

Conservation planning aims at systematically identify areas for active *in situ* conservation complemented with *ex situ* collections of genetically representative population samples in genebanks (Magos Brehm et al., 2019). Various tools have been developed in order to guide and facilitate countries in CWR conservation planning but these can be used or adapted to any plant group (Magos Brehm et al., 2019). These are: the 'Interactive Toolkit for CWR Conservation Planning' (Box 3 and Magos Brehm et al., 2017a), the 'CWR Checklist and Inventory Data Template' (Bioversity International and University of Birmingham, 2017; Thormann et al., 2017) and the 'Occurrence Data Collation Template' (Magos Brehm et al., 2017b) (also see Magos Brehm et al., 2019 for more detail).

3.1 Selection of target taxa

In any country or province, the resources available for PGR conservation is always limited in financial, temporal and technical terms, which forces the conservationist to actively prioritize and choose which conservation target taxa will be first-, second- and third-level priorities. The prioritization is based on taxon-comparative value on applying certain criteria and those with the highest value are actively conserved (Maxted et al., 1997; Hunter and Heywood, 2011). These criteria include:

- *Socio-economic use* – Taxa with either direct or indirect socio-economic use, for example, that provide food, fuel, medicines, building materials, tools, adornment, recreation and so on for humankind will be given priority. For instance, in terms of crop socio-economic value, rice LR conservation will be prioritized over rhubarb LR and rhubarb LR will be prioritized over undomesticated species. Further, when prioritizing CWR, both the value of the related crop and the ease to which traits might be transferred from the CWR to the crop, using the gene pool concept (Harlan and De Wet, 1971) or the taxon group concept (Maxted et al., 2006), will be used for prioritization.
- *Current conservation status* – When reviewing a taxon for conservation action, current conservation activities must be considered. If sufficient genetic diversity is already conserved from a range of agricultural systems, ecological habitats and geographical locations using both *in situ* and/or *ex situ* techniques, then additional conservation efforts may not be necessary.

Details of what material is already conserved can be obtained from catalogues, databases and other records of the holdings of gene banks, botanical gardens and *in situ* conservation areas, for example, EURISCO (www.ecpgr.cgiar.org/resources/germplasm-databases/eurisco-catalogue/) for European PGR *ex situ* collections; GENESYS (www.genesys-pgr.org) for global PGR *ex situ* collections; GRIN (www.npgsweb.ars-grin.gov/gringlobal) for US PGR *ex situ* collections; Botanic Gardens Conservation International (www.bgci.org/plant_search.php) for global botanic garden *ex situ* collections; Protected Planet an online interface for the World Database on *in situ* Protected Areas (https://www.protectedplanet.net/); and the Global Biodiversity Information Facility, which provides various types of biodiversity data including *ex situ* and *in situ* conservation data (https://www.gbif.org/).

- *Perceived threat* - Plant taxa or LR that are known to have decreased in population numbers or genetic diversity should be prioritized over taxa or LR with stable or increasing populations. Ideally, time series data on population trends or genetic diversity for most species or LR would be available as a basis for the assessment, but such data are rarely available. The most widely used means of assessing threat for wild plant taxa is the IUCN Red List Categories and Criteria (IUCN, 2012), and it provides a standardized system for the assessment of wild taxa according to their extinction risk (see http://www.iucnredlist.org/).
- *Taxonomic and genetic distinction* - It can be argued that with limited conservation funding available, one approach to maximize the breadth of diversity conserved is the use of taxonomic hierarchy to conserve a few examples from each higher-level taxon, assuming conservation of closely related taxa will involve significant allelic duplication. Remotely related taxa will share a lesser proportion of alleles and genes than closely related taxa so by maximizing the conservation of the most remotely related taxa we can maximize the overall conservation of genetic diversity.
- *Ecogeographic distinction* - It would appear reasonable to assume that plant species that are rare, in terms of geographical and ecological range, are under greater threat from genetic erosion and extinction than those that are widespread in many distinct habitats (Ford-Lloyd et al., 2008). Conservation priority should therefore be given to species that are restricted in their distribution or habitat requirements, such as endemics, whose distribution is restricted to a region.
- *Native status* - Native defined as being naturally found in a country or region, or at least present for the last 500 years, may be thought of as similar to ecogeographic distinction, though some view political defined country borders as a poor basis for biodiversity characterization. However, it can be argued that native species should be given priority for conservation,

given each country's national sovereignty and responsibility for conserving these species within its national territory. But, on the other hand, introduced species that are non-invasive may have gained well socio-economic or even cultural importance over the possibly hundreds of years since their original arrival and may be well-prioritized for conservation. For example, CWR that were introduced to a country/region over 500 years ago may have evolved over those years' relevant traits that may be useful in crop improvement.

- *Biological importance* – Not all species within an ecosystem play an equally important role in maintaining overall ecosystem functions or processes, and these species are usually termed keystone species. They are essential to the overall integrity of the ecosystem and to the survival of other species and their removal fundamentally and disproportionately change the ecosystem. Therefore, keystone species are often prioritized for conservation action.
- *Cultural importance* – Species may be prioritized because of their symbolic or religious significance in local or national culture, as, for example, the Cedar of Lebanon (*Cedrus libani* A. Rich.) in Lebanon, the Banyan tree (*Ficus benghalensis* L.) in India, the Swamp cypress (*Taxodium distichum* (L.) Rich.) in Mexico and (*Bauhinia guianensis* Aubl.) among the Waimiri Indians of Brazil (Maxted et al., 1997).
- *Relative cost of conservation* – The relative cost of either collection and *ex situ* storage will affect the selection of target taxa, with taxa requiring less-expensive conservation implementation being prioritized. Using a similar argument for *in situ* conservation, the number of target taxa is possible to conserve in a site and will also affect the choice of taxa, with taxa that can be actively conserved alongside each other being a higher priority than those requiring distinct reserves of their own.
- *Conservation sustainability* – Conservation, whether *in situ* or *ex situ*, should be, by definition, long-term, and requires a relatively large initial investment of resources, whether sampling and collecting seed samples in a remote location or establishing a new reserve. There would therefore be little value in collecting and storing seed, establishing a genetic reserve or encouraging farmers to practise on-farm conservation unless the conservation activity was likely to succeed and be sustainable in at least from the medium to long term.
- *Legislation* – Species protected under international, regional or national legislation (which are usually those under greatest threat) will also be given high priority for active conservation, as the signatories of conventions or treaties (e.g. Conventional on International Trade in Endangered Species of Wild Fauna and Flora, 1973; International Treaty on Plant Genetic Resources for Food and Agriculture, 2001 – Annex 1) are required to implement their obligation.

- *Ethical and aesthetic considerations* - The ethical justification for conservation reflects the sympathy, responsibility and concern that the general public feel towards fellow species and ecosystems. This may be considered to favour so-called 'flagship' species such as giant pandas, tigers, whales or orchids, though in recent years education of the public in the value of food diversity is having an impact as noted by the increase of LR in mainstream supermarkets (Veteläinen et al., 2009b).
- *Priorities of conservation agencies* - The mandate and priorities of the conservation agency commissioning the work will impact the choice of taxa to conserve. The priorities of a country's Ministries of Agriculture, Forestry and the Environment are likely to be quite different with first focussing on conserving crops and their wild relatives, the second on timber trees and the third having a more general ecosystem-based, rare or threatened taxon remit.
- *Climate change* - The relative impact of climate change on particular taxa is increasingly being used as a means of identifying priorities, the relative impact being assessed most commonly using some form of species distribution modelling (SDM).

The various prioritization criteria discussed above provide relative 'value' to plant taxa and thus aid the efficient selection of priority taxa for conservation. It is by weighing each in accordance with the mandate of the commissioning agency that the conservationist will be able to determine relative conservation priorities more objectively. However, for PGR taxa three criteria are commonly used: socio-economic value of the crop, crop relatedness for CWR taxa (or the potential for crop improvement as measured by the relatedness of the wild relative to the crop) and relative threat assessment (Kell et al., 2017); although native status is also often used in national PGR conservation planning (Iriondo et al., 2016). Whatever prioritization methodology and criteria are used, the total number of target taxa is likely to be adjusted to a number that can be actively conserved using the available financial and human resources.

It seems today every news broadcast brings more examples of climate extreme events and their devastating impact on biodiversity, which of course includes PGR. It is predicted as an increase of the average global temperature, melting glaciers and polar ice, rise of sea level and change of oceanic currents and acidity; increased storms, hurricanes, droughts, rain, changes in snow cover and earthquakes; hastening desertification or irrigation and extreme salination; and causing agriculture failures, species extinction and ecosystem disruption (IPCC, 2014, 2018). As such, climate change is directly impacting crop production but also indirectly by threatening the extinction or erosion of PGR, the gene and allelic diversity, required to mitigate the impact of climate change on crop production (Maxted et al., 2020). Therefore, climate smart conservation of PGR is essential, meaning that climate change assessments of competing conservation options are required as a precursor to any conservation implementation. Given a choice of *ex situ* PGR collection sites or *in situ* PGR genetics reserve/on-farm locations the competing options should be modelled and the most climate smart option prioritized, worst potentially impacted sites are first collected and least impacted sites are selected for genetics reserve/on-farm action. IPCC (2007) concluded that taxa vary in their susceptibility to climate change due to their inherent biological traits related to their life history, ecology, behaviour, physiology and genetics. Foden et al. (2009, 2013) proposed that a species' susceptibility to climate change can be estimated based on specific biological traits (life history, ecology, behaviour, physiology and genetic diversity) and this assessment might guide species prioritization. However, a more direct means of assessing the impact of climate change on plant diversity is through SDM. Climate envelope models use environmental data to identify taxon specific ecological niches (Phillips et al., 2006) to postulate the locations where a species has been found or are absent, to infer its climatic requirements. These inferred requirements can then be used to compare a taxon's current niche with future climate scenario models to predict the longevity of potential protected area (PA) sites, the likely shift and extent of suitable niche environments available as the climate changes (Hijmans and Graham, 2006) and the probability that suitable conditions will remain within the boundaries of sites managed. To illustrate SDM application in conservation planning an example from Oman (Al Lawati et al., 2016) is provided in Box 1.

Box 1 Climate change species distribution modelling in Oman (Al Lawati et al., 2016).

Lucerne (*Medicago sativa* L.) was chosen to illustrate the procedure for climate change modelling. Current climatic data were obtained from WorldClim (Hijmans et al., 2005) and future climate projections from CCAFS (www.ccafs-climate.org). The SRES-A2 emission scenario was selected as the most realistic and used in similar studies (Ramirez-Villegas et al., 2014). The General Circulation Model used was the UKMO-HadCM3. The 13 variables used to create the potential species distribution maps were selected using objective methods which included a test for collinearity and principal component analysis (PCA) which resulted in the removal of correlated variables. The resulting variables were checked by an Omani national expert for appropriateness in predicting the distribution of the species. MaxEnt (Phillips et al., 2006) was used to create the predicted distribution at a resolution of 2.5 arc-minutes (~5 km^2 cell size at the equator). Validation of the model was achieved using a random test percentage of 30%. Modelled distributions were restricted to the 'equal training sensitivity and specificity' threshold. The remaining parameters were kept at default.

The three time projections for *M. sativa* are shown in Fig. 5a-c . Figure 5a shows the distribution of *M. sativa* under current climatic conditions with a distribution of 12% throughout the country and predicted presence in three protected areas. Figure 5b the predicted distribution under climatic conditions in 2020 with 17.4% distribution area and presence in one protected area. Figure 5c shows a significantly reduced predicted distribution in 2050 with 2.3% distribution area and presence within one protected area.

To ensure *in situ* conservation of *M. sativa* it is recommended to set up active on-farm conservation management with local communities within the Al Jabal Al Akhdar Scenic Reserve in the Ad Dakhliyah

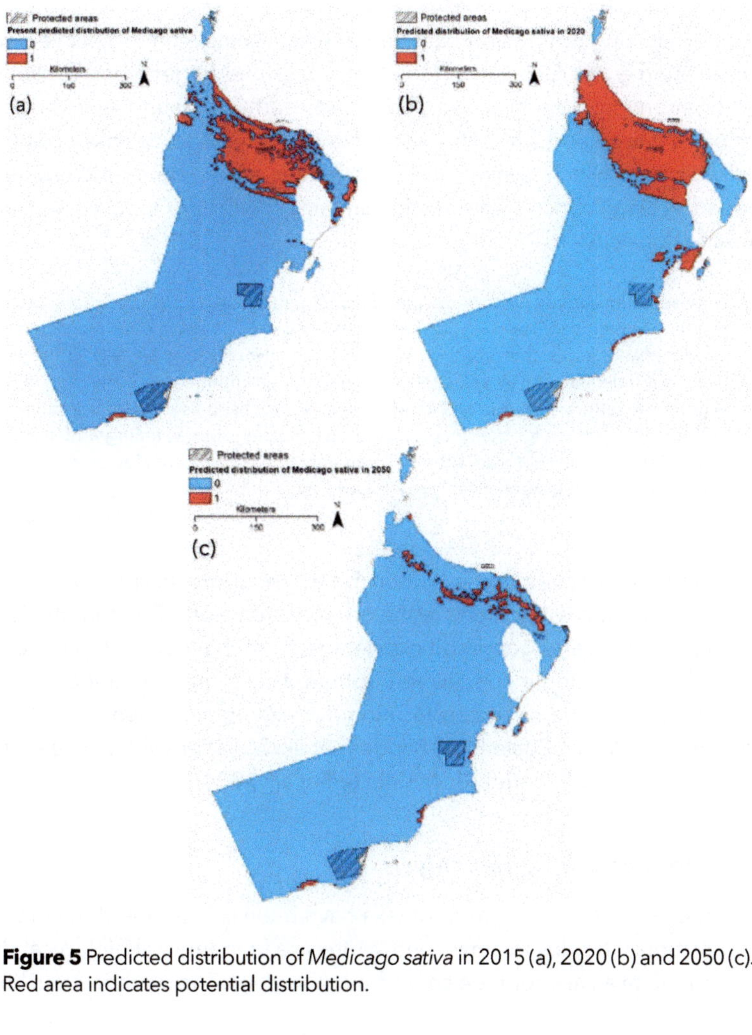

Figure 5 Predicted distribution of *Medicago sativa* in 2015 (a), 2020 (b) and 2050 (c). Red area indicates potential distribution.

region. This area appears to be the 'core' of the *M. sativa* distribution as populations are predicted to be present here under all three time projections up to 2050. It would also be a priority to increase active population management in the south near Salalah and *A. javanica* adjacent to the Yemen border because they both appear to persist over the prediction period and may contain important and different genetic diversity to the populations in the north. Further the areas where population are projected to go extinct due to climate change should be sampled and store *ex situ* soon before the populations go extinct.

3.2 Project commission

Some form of commission statement establishing the objectives of the conservation actions, specifying the target taxa and target areas, outlining the conservation problem(s) to be addressed, perhaps stating how target taxa are to be utilized and where the conserved material is to be safely duplicated, and indicating which conservation techniques are to be employed, is often needed. A clear, concise commission statement will help to focus subsequent conservation activities:

> A conservation strategy is commissioned for the genus *Vigna* L. in Sub-Saharan Africa by the International Plant Genetic Resources Institute. The survey has the objective of identifying areas that contain novel genetic diversity not already conserved that could be utilised in selection or breeding programmes for the benefit of African agriculture. The report should contain a prioritised list of *ex situ* and *in situ* conservation actions for the genus, including collecting routes, timing and suitable local contacts. It should also attempt to identify those *Vigna* species of immediate and medium-term potential value to African agriculture.
>
> *Maxted et al. (2004)*

The commission statement is written by those contracting the conservation work or those who will undertake the conservation work. The commission may vary in taxonomic and geographic coverage, from the systematic conservation action plan for a single species throughout its geographic range to a range of target taxa from a restricted location; for example, *Phaseolus lunatus* L. in the Central Valley of Costa Rica (Baudoin et al., 2004) or wild cowpea relatives (*Vigna* spp.) in sub-Saharan Africa (Maxted et al., 2004).

3.3 Ecogeography/gap analysis

Once the target taxon or group of taxa have been selected and delimited, the conservationist begins to amass and synthesize fundamental biological data to help formulate an appropriate conservation strategy. The synthesis and analysis of these data enable the conservationist to make vital decisions concerning: which combination of *in situ* and *ex situ* conservation techniques to apply, what sampling strategy to adopt, where to site the reserve or store the germplasm, what collecting route should be followed, and when the collecting window (period when seed is ripe and suitable for collection) occurs. If the basic biological data for this species data indicates that it has previously been found on stony slopes at the edge of the Fergana valley in Uzbekistan, then further material of this species is likely to be found under similar conditions, and is less likely to be found in different habitat types or far distant regions.

The process of collating and analysing geographical, ecological, taxonomic and genetic data for use in designing conservation strategies is referred to as 'ecogeography' and was defined in their review by Castañeda-Álvarez et al. (2011) as:

> a process of gathering and synthesizing information on ecological, geographical, taxonomic and genetic diversity. The results are predictive and can be used to assist in the formulation of complementary *in situ* and *ex situ* conservation priorities.

Ecogeographic studies involve the use of large and complex data sets obtained from the literature and from the compilation of passport data associated with herbarium specimens and germplasm accessions (see Box 2). The data compiled are essentially of four kinds: ecological, geographic, taxonomic and genetic. These data are synthesized and analysed (see Box 3) to generate three basic products: the database, which contains the raw data for each taxon, the conspectus, which summarizes the data for each taxon, and the report, which discusses the contents of the database and conspectus, as well as proposing appropriate conservation strategies or action plan. Additionally, the results obtained with the ecogeographical and gap analyses, that is, the conservation recommendations, are central to the development of national strategies or national strategic action plans for the conservation and use of the resources (see Box 3 and Magos Brehm et al., 2019).

The process of establishing what conservation strategies to propose often involves a clear understanding of the target taxon's ecogeographic envelope, but also what current conservation actions are in place for the taxon. The latter is facilitated by gap analysis (Burley, 1988; Margules, 1989; Margules and Pressey, 2000), and today ecogeography and gap analysis are often used in parallel. Gap analysis was originally applied to conservation planning of indigenous forests (particularly on small islands rich in endemic species), and has subsequently been extended to agrobiodiversity conservation (Maxted et al., 2008; Ramírez-Villegas et al., 2010; Parra-Quijano et al., 2012) encompassing conservation assessment of both *in situ* and *ex situ* genetic diversity and has been applied for diverse taxa, for example, African *Vigna* (Maxted et al., 2004), *Hordeum* (Vincent et al., 2012), *Phaseolus* (Ramírez-Villegas et al., 2010), *Glycine* in Australia (González-Orozco et al., 2012), and temperate forage and pulse legume species (Maxted et al., 2012b). Ecogeographic and gap analysis techniques enhance the efficiency of genetic conservation because they enable the conservationist to clearly identify the geographical regions and ecological niches that the taxon inhabits; therefore, not only can areas with high numbers of target taxa be identified, but also, areas that contain high taxonomic or genotypic diversity, uniqueness of habitat, and economic or breeding importance.

Box 2 Some sources of taxonomic and ecogeographic information and literature (Maxted et al., 2020).

- *CWR Diversity* – a global quarriable portal providing access to (i) the Harlan and de Wet CWR Inventory; (ii) the CWR Global Occurrence Database of germplasm accessions, specimens and population occurrence data; and (iii) the CWR Atlas which provides CWR species distributions and *ex situ* collecting priorities maps (http://www.cwrdiversity.org/).
- CWR Global Portal – a global query able portal providing access to (i) experts working with CWR, (ii) CWR-related projects, (iii) CWR images, (iv) national or regional CWR checklists / inventories, (v) CWR conservation strategies, (vi) CWR presentations divided by themes (genetic diversity, conservation, climate change), (vii) CWR publications (papers, books, theses, newsletters), and (viii) the Interactive Toolkit CWR Conservation Planning which has further CWR resources (http://www.cropwildrelatives.org/).
- CWR Conservation Planning Toolkit - provides a tool to aid the planning of CWR conservation (Magos Brehm, J., Kell, S., Thormann, I., Gaisberger, H., Dulloo, M. E. and Maxted, N. (2017a). Interactive Toolkit for Crop Wild Relative Conservation Planning version 1.0. University of Birmingham, Birmingham, UK and Bioversity International, Rome, Italy. Available at: www.cropwildrelatives.org/conservation-toolkit/).
- *Global Biodiversity Information Facility (GBIF)* – provides extensive access to global taxon nomenclature and population occurrence data (http://www.gbif.org/).
- *Tropicos* – a global quarriable database of taxonomic, occurrence and literature data (http://www.tropicos.org/).
- *International Plant Names Index (IPNI)*: a database of the names and associated basic bibliographical details of all seed plants. Its goal is to eliminate the need for repeated reference to primary sources for basic bibliographic information about plant names. The data are freely available and are gradually being standardized and checked. IPNI is the product of collaboration between The Royal Botanic Gardens, Kew, The Harvard

University Herbaria, and the Australian National Herbarium (http://www.ipni.org/).

- *National Geospatial-Intelligence Agency (NGA)* - NGA enables georeferencing of locality data based on the locality name (http://geonames.nga.mil/gns/html/).
- *Protected Planet* - a world database of protected areas which can be queried interactively or downloaded (http://protectedplanet.net/).
- *The IUCN Red List* - a database of taxonomic, conservation status and distribution and relative threat information on plants, fungi and animals that have been globally evaluated using the IUCN Red List Threat Categories and Criteria (http://www.iucnredlist.org/).
- *The Plant List* - a working list of vascular plants (flowering plants, conifers, ferns and their allies and mosses and liverworts) produced by the Royal Botanic Gardens, Kew and Missouri Botanical Garden. It contains 350 699 accepted names, alongside synonyms and unresolved names (http://www.theplantlist.org/).
- *Plant Book* - a comprehensive listing of higher plant names with basic botanical information compiled by D. J. Mabberley (2017) and published by Cambridge University Press.
- *Standard World Floras* - Frodin, D. G., (2011). Guide to the Standard Floras of the World, 2nd Edition, Cambridge University Press, Cambridge.

Box 3. Interactive Toolkit for CWR Conservation Planning (Magos Brehm et al., 2017a).

The Interactive Toolkit is designed to provide guidance to plan and implement systematic and active *in situ* and *ex situ* conservation of CWR at national level. The conservation recommendations that result from this national CWR conservation planning process can and should be used to develop national strategic action plans (NSAP) (or national strategies) for the conservation and sustainable use of CWR. It can be used by local institutions (e.g. gene banks, universities, research institutes), individual scientists, NGOs, agencies responsible for planning and implementing NSAP or national strategies, such as national agricultural or environmental agencies, and FAO national focal points. There is no single method for planning CWR conservation or for developing an NSAP for the conservation and use of CWR. This is mainly due to factors concerning financial resources, availability of baseline biodiversity data, the local community where the NSAP is to be implemented, as well as the focal area and remit of the agencies that are responsible for formulating and implementing the NSAP. Nevertheless, CWR conservation planning can be viewed as a series of steps and decisions that follow the same basic pattern in all countries.

The Toolkit thus includes a series of 12 modules (plus an introductory module to 'National systematic CWR conservation planning'), each corresponding to a step in conservation planning. Each module comprises an introduction to that particular step, methodology that explains how to undertake that step, examples and applied use which include case studies showcasing how that particular CWR conservation planning step has been undertaken in different scenarios, the list of references used to compile the texts in that module and a list of additional materials and resources (bibliographic references, PowerPoint presentations, relevant projects, recordings and video files, software tools, web links and relevant social networks) that help the user visualize and understand how to undertake that particular step. The central element of the Toolkit are the interactive flowcharts present in each methodology section which are generally composed of a series of yes/no questions, helping the user move through the various phases and choose the options that are most appropriate given the user's national context. Detailed explanations of each phase are linked to the interactive flowchart. In addition, there is a static version of the flowchart which can be consulted at any time and a full written text that exactly matches the information presented in the flowchart. The Toolkit can be used either online or offline (by exporting its content) and in order to access the modules of the Toolkit the user needs to register using an email address and providing some additional information used for statistical purposes only.

This Toolkit and its protocols, examples and resources should thus be viewed as a framework and an aid for planning CWR conservation. It can be used for the entire conservation planning process or for individual steps. It should be noted that the Toolkit was designed for CWR conservation planning, but it can be adapted to any plant group.

The collation and analysis of ecogeographic/gap analysis data is complex, but good introductions to the topic are provided by Maxted et al. (1995), Guarino et al. (2006), Scheldeman and van Zonneveld (2010), Castañeda Álvarez et al. (2011), van Zonneveld et al. (2012), Maxted et al. (2012a), Magos Brehm et al. (2017a) and Maxted et al. (2020). The bulk of contemporary ecogeographic/gap analysis is based on passport data associated with *ex situ* gene bank accessions or more often herbarium specimens, which is unavoidable given the scarcity of other data for analysis. However, it does necessitate an assumption that intra-taxonomic genetic diversity is directly correlated with ecogeographic diversity; an assumption that may be often but not always be valid. The point may be illustrated with a classic study of the wild lentil gene pool by Ferguson et al. (1998), *Lens culinaris* subsp. *orientalis* is found from Western Turkey to Tajikistan, and if resources were available to establish three genetic reserves for the subspecies these could be logically located equidistant between Turkey, Iran and Tajikistan, assuming that geographic distance can be used as a proxy for genetic distance. However, genetic analysis found that *L. culinaris* subspecies *orientalis* genetic diversity was partitioned into 10 clusters (Fig. 6) and populations representing all 10 clusters are found in the Western element of the Fertile Crescent and outside this region little unique genetic diversity related to only one of the clusters was found. Therefore, in this case, ecogeographic data would be an inadequate proxy for genetic diversity as it would indicate *ex situ* sampling or *in situ* genetic reserve establishment across the geographic distribution rather than focussing both in the Western Fertile Crescent alone.

As well as obvious site/populations in hotspots, populations should be selected for conservation if they are found on the periphery of the target taxon's distribution or those containing morphological or ecological variants. Ecogeographic land characterization (ELC) maps may be used for this purpose (Parra-Quijano et al., 2008, 2012), that is, to identify the various ecogeographic scenarios in which a species occurs, which - assuming they are a good proxy for genetic diversity - reflect the adaptations of the studied species that enable it to thrive in that particular set of ecological conditions. As a result, ELC maps help to identify important diversity within a taxon, which links to the potential utilization of intra-CWR diversity for crop improvement. Additionally, atypical populations or those existing under atypical conditions may possess genes or alleles that are unknown or extremely rare in the target taxon's centre of diversity, and this material possibly contains genetic variation of special use to breeders, for example, pest or disease resistance, or adaptation to climate. For example, the garden pea wild relative *Vavilovia formosa*, which is a rare CWR found primary over 2000 m in and around the Caucasus Mountains growing on limestone scree, is not actively conserved but several attempts to collect seed for *ex situ* conservation have proven unsuccessful, so the conservation

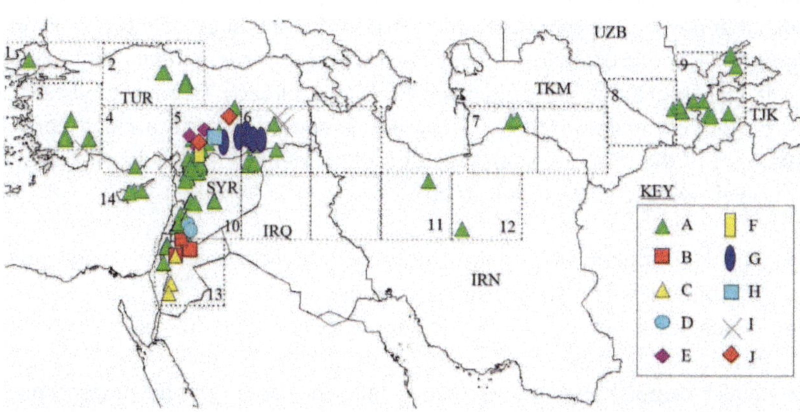

Figure 6 Distribution of *Lens culinaris* subsp. *orientalis* divided into 10 genetic diversity clusters (Ferguson et al., 1998a).

recommendation is to establish an *in situ* genetic reserve for the species in the Caucasus, namely in north of Yerevan in Armenia, which is the site with the easiest access.

3.4 Field surveys

If the available ecogeographic data for the target taxon are limited, there will not be sufficient background biological evidence to formulate an effective conservation strategy. In this case, it is necessary to undertake an initial survey mission to gather sufficient, novel ecogeographic data to formulate the conservation strategy. The survey mission may be in the form of 'coarse grid surveying', which involves travelling throughout a likely target region and sampling sites at relatively wide intervals over the whole region. The precise size of the interval between sites depends on the level of environmental diversity across the region, but it may, for example, involve sampling every 10–50 km. This might be followed by 'fine grid surveying' in the initial area highlighted as containing the target taxon at say a sampling interval of 1–20 km. Further, SDM can be produced to predict where further populations might be found (e.g. Engler et al., 2004). In this way the actual surveying can be based on potential distribution maps which then help the surveying to be more targeted.

3.5 Conservation objectives

The various conservation planning activities build the evidence-base for the target taxon, summarizing its level of genetic variation within and between populations, local population structure, breeding system, taxonomy and ecogeographic envelop of the target taxon, as well as many other biological details, and concludes by stating the conservation objectives in the conservation strategy or action plan. These necessarily highlight the populations of the target taxon that need to be more actively conserved and that contain the maximum amount of genetic diversity in order to effectively conserve the target taxon.

If clear conservation objectives have not already been established in the project commission, the conservationist should include a clear, concise statement of the proposed actions to be implemented. The statement should include answers to such questions as:

- Which populations require more active conservation using which combination of conservation techniques?
- Which local farmers/communities can play a part in the conservation activities?
- Which populations require close monitoring, demographic or genomic?
- Which stakeholders will be actively involved in *in situ* population maintenance?

- Is a national or international collecting team required to collect the priority target taxa?
- How can the conserved resource be made accessible to the local and remote use community?
- Is a more detailed study required before any of these questions can be answered?

4 Conservation: strategies and techniques

There are two basic strategies to conservation, each comprising a range of techniques that can be adopted to conserve plant genetic resources. *In situ* strategies involve the (a) location of populations of the target taxon, (b) the designation of populations for active and long-term conservation where it is found, (c) the management of that population to maintain demographic numbers and genetic diversity over time, (d) the monitoring of the target taxon population to ensure demographic numbers and genetic diversity are maintained over time, and (e) facilitating access to the conserved resource for utilization, while *ex situ* strategies involve the (a) location of populations of the target taxon, (b) the sampling of the populations, (c) the transfer of the sample away from the site where it was originally found to a remote location for active, long-term conservation, (d) the maintenance of the population sample to sustain demographic numbers and genetic diversity over time, and (e) facilitating access to the conserved resource for utilization. The CBD (1992) clearly establish definitions of the two strategies:

> *Ex situ* conservation means the conservation of components of biological diversity outside their natural habitats.
> *In situ* conservation means the conservation of ecosystems and natural habitats and the maintenance and recovery of viable populations of species in their natural surroundings/and, in the case of domesticates or cultivated species, in the surroundings where they have developed their distinctive properties.

The CBD states that each contracting party shall adopt *ex situ* conservation measures 'predominantly for the purpose of complementing *in situ* measures'. However, in relation to PGR conservation there is a good argument for the two strategies being of equal importance (Maxted et al., 2020). The distinction between PGR conservation, or for that matter any socio-economic species conservation, and wider biodiversity conservation, is not the conservation of the resource for the sake of it, but the link between the conserved resource and sustainable use, as clearly indicated in Fig. 4. So, if *in situ* conservation allows the diversity to evolve naturally in the wild, *ex situ* conservation does not, but it is a good source for when this diversity is needed for crop improvement or any other conservation technique (reintroduction, population reinforcement etc.).

4.1 Conservation techniques

The two conservation strategies may each be further subdivided into several specific techniques (Table 1), each one resulting in the active conservation of the genetic diversity of plants. Each technique is defined in Table 2 and their advantages and disadvantages are summarized in Table 3.

There is an extensive literature associated with each of the conservation techniques and insufficient room here to elaborate each technique in detail but further information can be gained on *in situ* conservation of CWR diversity from Tuxhill and Nabhan (2001), Heywood and Dulloo (2005), Iriondo et al. (2008), Hunter and Heywood (2011), and Maxted et al. (2020); on-farm conservation of LR diversity from Brush (2000), Veteläinen et al. (2009a), Jarvis et al. (2016); home gardens and community-based conservation of LR diversity from Eyzaguirre and Linares (2004), De Boef et al. (2013); seed storage conservation of CWR and LR diversity from Guarino et al. (1995, 2011), Hawkes et al. (2000), Smith et al. (2003), ENSCONET (2009), the FAO (2014), Vernooy et al. (2017); *in vitro* storage of CWR and LR diversity from Reed et al. (2004), Savolainen et al. (2006); DNA and pollen storage of CWR and LR diversity from Adams (1997), Carmen de Vicente and Andersson (2006); and botanic gardens of CWR diversity from Heywood and Wyse-Jackson (1991), Blackmore and Oldfield (2017).

4.2 Complementary conservation

Although *in situ* and *ex situ* conservation techniques have their individual advantages and disadvantages, they should not be seen as being alternatives or in opposition to one another, but rather as being complementary. As is stressed in Article 9 of the CBD (1992) the two strategies or various techniques should backup one another. The actual combination of strategies and techniques applied for each target taxon will depend on the conservation aims, the type

Table 1 PGR conservation strategies and techniques (Maxted et al., 2020)

Strategies	Techniques
In situ involves:	
a. location	• genetic reserve
b. designation	• extra-PA *in situ*
c. management	• on-farm
d. monitoring	• home gardens
e. access	• seed storage
Ex situ involves:	
a. location	• *in vitro* storage
b. sampling	• DNA storage
c. transfer	• pollen storage
d. maintenance	• field gene bank
e. access	• botanic garden

Table 2 Definition of PGR conservation techniques (Maxted et al., 2020)

Strategies	Techniques	Definition
In situ conservation	Protected area (genetic reserve) conservation	The location, management and monitoring of (genetic) diversity in natural wild populations within defined areas designated for active, long-term conservation.
	Extra PA *in situ* conservation	The location, management and monitoring of genetic diversity of natural wild populations in informally managed *in situ* conservation sites.
	On-farm conservation	The sustainable management of genetic diversity of locally developed traditional landraces with associated wild and weedy species or forms by farmers within traditional agricultural, horticultural or agri-silvicultural cultivation systems.
	Home garden conservation	The sustainable management of genetic diversity of locally developed traditional crop varieties by individuals in their back-yard gardens.
Ex situ conservation	Seed storage	The collection of seed samples in one location and their transfer to a formal or less formal community gene bank for storage; orthodox samples are usually dried to a suitably low moisture content, then kept at sub-zero −18°C temperatures, increasingly cryogenic at −196°C. Community gene banks may not freeze seed samples preferring to keep them for shorter periods at +4-5°C.
	DNA storage	The collection of DNA and storage in appropriate, usually refrigerated conditions.
	Pollen storage	The collection of pollen and storage in appropriate, usually refrigerated conditions.
	In vitro storage	The collection and maintenance of explants in a sterile, pathogen-free environment (suitable for recalcitrant taxa that cannot be conserved in gene banks).
	Field gene bank storage	The collection of large number of individuals, seeds or living material from one location and its transfer and housing or planting in a second site for conservation (suitable for recalcitrant taxa that cannot be conserved in gene banks).
	Botanic garden (arboretum)	The collection of a relatively small number of individuals, seeds or living material (tree species for arboreta) from one location and its transfer and housing or planting in a second site for public education purposes.

Table 3 Relative advantages and disadvantages of *in situ* and *ex situ* conservation (Maxted et al., 2020)

Strategy	Advantages	Disadvantages
In situ	• Dynamic conservation in relation to biotic and abiotic changes • Permits species/pathogen interactions and co-evolution • Provides easy access for evolutionary and genetic studies • Appropriate for 'recalcitrant' or intermediate seeded species where seed cannot be frozen and retain viability • Target taxa conserved alongside associated taxa within a single reserve	• Materials less easily available for utilisation • Vulnerable to natural and human mediated catastrophes, for example, fire, vandalism, urban and industrial development, climate change • Appropriate management regimes often poorly understood • Require routine active management and monitoring • Limited genetic diversity can be conserved in any single location
Ex situ	• Medium- and long-term storage is feasible • Limitless genetic diversity can be conserved in any single location • Easy access for evaluation for resistance to pests and diseases • Easy access to plant breeding, farmers and other forms of utilisation • Little routine maintenance once material is conserved (except for field gene banks and botanic garden accessions)	• Impossible to store seeds of recalcitrant or intermediate species, unless via in vitro or field gene banks • Practical limitation on storage space for diversity of recalcitrant or intermediate species in field gene banks • Freezes evolutionary development in relation to environmental changes • Genetic diversity is potentially lost with each regeneration cycle • In vitro storage may result in loss of diversity and mutation

of species being conserved, the resources available and whether the species has utilisation potential (Hawkes et al., 2000). The efficacy of adopting such a 'holistic' or integrated approach using multiple techniques to conserve the target taxon is now well established. When formulating a genetic conservation strategy, the conservationist will need to address not only genetic, but also practical and political questions (Maxted et al., 2020), such as:

- What are the species storage characteristics?
- What do we know about the species breeding system?
- Do we want to store the germplasm in the short, medium or long term?
- Where is the germplasm located and how accessible is it/does it need to be for user access?
- Are there specific legal issues relating to access to the conserved resource?
- How effective is the conservation infrastructure (e.g. gene bank, genetic reserve or on-farm site)?

- What conservation backup is necessary/desirable?
- How might the resource be best exploited?
- What skills and financial resources are available to facilitate conservation and use implementation?

Given answers to these questions, the appropriate combination of techniques to conserve the gene pool can be applied in a pragmatic and balanced complementary manner (Hawkes et al., 2000). A different balance would be required for different crops, wild species or regions (see Fig. 7). Figure 7a shows the hypothetical situation where the conservation target is a cereal species and the bulk of the gene pool is conserved using seed storage and genetic reserve conservation, lesser proportion of other techniques and there is no field genebank conservation; Figure 7b refers to sweet potato, a typical root crop, for which field gene banks, in vitro, genetic reserve and on-farm conservation predominate; Figure 7c is a temperate fruit tree and field gene banks, in vitro and genetic reserve prevail; and Figure 7d is a typical CWR species where seed storage, extra PA *in situ* and genetic reserve conservation are usually carried out and there is no field gene bank or in vitro storage. The precise combination

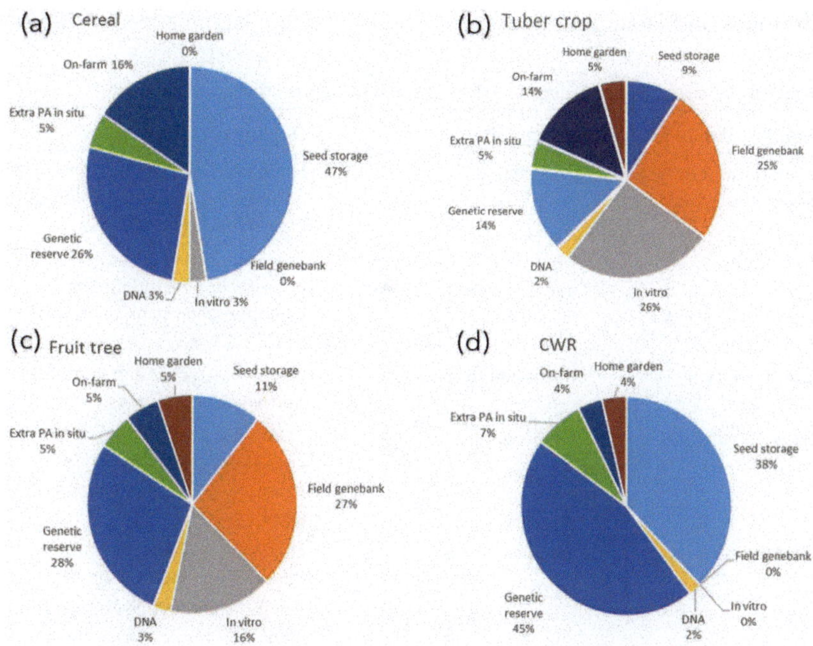

Figure 7 Hypothetical representation of the proportions of the gene pool conserved using eight *in situ* and *ex situ* conservation techniques for different plant groups; (a) cereal crop, (b) tuber crop, (c) fruit tree and (d) CWR (from Maxted et al., 2020).

of techniques is formulated afresh for each species or group of species, demonstrating the flexibility of this approach.

4.3 Conservation products

The products of conservation activities are conserved germplasm (seeds, embryos), live plants, dried plants (herbarium specimens), in vitro cultures and conservation data. As discussed above, *ex situ* conserved orthodox seed is commonly held in gene banks at sub-zero temperatures and low moisture content to prolong their life. Live plants are conserved in genetic reserves, field gene banks, botanical gardens and research facilities. Germplasm that is stored in a suspended form such as tissue, pollen or DNA is stored as cultures in specialist laboratory facilities. Dried plant voucher specimens are held in herbaria; they are commonly linked to specific samples of germplasm to provide identification validity and should as much as possible be representative of the conserved populations. Conserved material is ideally associated with a range of passport data, which details the taxonomic, geographical and ecological provenance of the material. All passport data associated with conserved material should be entered into a database and made available not only for its management but also for the formulation of future conservation priorities and strategies, and any form of utilization (summarized in Table 4).

Table 4 The storage and duplication sites of conservation products

Conservation product	Storage site	Duplication site
Plant germplasm (seed, vegetative organs etc.)	Gene bank	National, regional and international gene banks, duplication with other conservation techniques
Live plants	Field gene bank, botanical garden, genetic reserve, on-farm	Duplication with other conservation techniques, for example, gene bank storage of seed
Dried plants	Herbarium or museum	National, regional and international herbaria or museums
Explants or plantlets	Tissue culture	Duplication with other conservation techniques, for example, gene bank storage of seed
DNA and pollen	Various cultures	Duplication with other conservation techniques, for example, gene bank storage of seed
Conservation data	Conservation database	Duplication with other national, regional and international conservation agencies

Whether the conservation products are conserved *in situ* or *ex situ* they should be duplicated in more than one location to ensure their long-term security (FAO, 2014). The distribution of duplicate sets of material avoids accidental loss of the material due to fire, economic or political difficulties, warfare or other unforeseen circumstances. Duplication of the passport data is extremely important and relatively easy, and the commissioning agency, relevant host country institutes and other interested parties should hold copies. Safety duplication of conservation products should be routine – 'don't put all your conservation eggs in one basket'.

5 Conservation into utilization

As already stressed, for PGR conservation there is an intimate link between conservation and utilization. The products of conservation, whether they are 'living' or 'suspended', should be made available for utilization by humankind, either now or in the future. Plants provide an extremely wide range of useful products throughout the world. A mixture of direct harvesting from the wild or cultivation, ranging from basic subsistence farming to sophisticated agricultural systems, supplies food, medicines and a wealth of raw materials. As well as the more obvious plant products such as food, medicines and construction materials provide a wide variety of resources used in industry and commerce. For instance, plant extracts are used in the manufacture of glues, soaps, cosmetics, dyes, plastics, lubricants and polishes. In addition, wild plants are exploited for their aesthetic and medicinal qualities (e.g. wild orchids, sweet peas, evening primrose and ginseng); the worldwide annual commercial value of products derived from PGR is estimated at US$ 500–800 billion per year (Table 5).

Table 5 Ballpark estimates of the annual markets for PGR (ten Kate and Laird, 1999)

PGR use sector	Market (US$ bn) low	Market (US$ bn) high
Agricultural production (commercial seed sales)	300+ (30)	450+ (30)
Pharmaceuticals	75	150
Biotechnology in fields other than healthcare and agriculture	60	120
Botanical medicines	20	40
Ornamental horticultural products	16	19
Personal care and cosmetic products	2.8	2.8
Crop protection products	0.6	3
Rounded totals	**500**	**800**

The main uses of plant genetic resources by humankind are as follows (Cook, 1995):

- Food: crop species including beverages.
- Food additives including processing agents and other additives.
- Animal foods: fodder or forage species.
- Materials: such as wood, fibres, cork, cane, tannins, latex, resins, gums, waxes, oils, etc.
- Poisons.
- Medicines: human and veterinary.
- Environmental: these include species that are ornamental, recreational, hedges, shade plants, windbreaks, soil improvers, plants for regeneration, erosion control, indicator species (e.g. pollution, underground water).
- Gene donors: plants that contain desirable traits that can be transferred to other species to improve their use.

In the farming and agriculture context, genetic resources can be used directly, for instance, by farmers in their continual role in selecting seed from 'better' plants as planting material for the next cultivation cycle or in the selection of breeder lines for novel varieties, where the material is grown in field trails and little, if any, breeding intervention occurs. The conserved material may also be directly used in reintroduction programmes where the LR has been lost locally due to civil unrest or the application of perverse government incentives that encourage the alteration of traditional practices. However, PGRFA is more commonly used indirectly by breeders in crop improvement programmes. Commonly, the first stage of utilisation involves recording genetically controlled characteristics (characterisation). Genetic material may also be tested in diverse environmental conditions to evaluate and screen for tolerance to harsh conditions or be deliberately infected with diseases or pests to screen for biotic resistance (evaluation). The biotechnologist may screen accessions for single genes, which once located may be transferred into a host organism, while the biochemist (bioprospector) screens for chemical products that may be of use to the pharmaceutical industry. The products of utilisation are therefore numerous, including new varieties, new crops, improved breeds and pharmaceuticals, as well as a 'good' environment for recreational activities.

As already emphasized, PGR conservation and use are inextricably linked. In fact it can be argued that the survival of human life on earth depends on the continued use of wild fauna and flora, therefore non-use of plant diversity is not an option. However, we can no longer afford to carelessly eradicate species or genetic diversity through over-exploitation. The concept of sustainable utilisation refers to non-exhaustive use of natural resources, particularly addressed in Article 10 of the CBD:

Each Contracting Party shall, as far as possible and as appropriate:

(a) Integrate consideration of the conservation and sustainable use of biological resources into national decision-making;
(b) Adopt measures relating to the use of biological resources to avoid or minimize adverse impacts on biological diversity;
(c) Protect and encourage customary use of biological resources in accordance with traditional cultural practices that are compatible with conservation or sustainable use requirements;
(d) Support local populations to develop and implement remedial action in degraded areas where biological diversity has been reduced; and
(e) Encourage cooperation between its governmental authorities and its private sector in developing methods for sustainable use of biological resources.

6 Conclusion

Just as the CBD drew attention to the explicit link between conservation, utilization and human well-being, so each is now necessarily linked to the sustainability of the finite genetic resources at its base. We currently can, by definition, have no conception of which plant species will be useful in the future. For instance, which CWR contain those vital alleles required to fight the crop diseases yet to evolve? If we wish to maintain the current wealth of natural plant diversity and the options for the future they offer, we need to classify that diversity, much of which still remains to be described, ensure that the diversity is conserved and characterized, reduce human mis-management of the environment that threatens diversity and promote its sustainable use. Continuing to ignore these fundamental issues threatens the future of humankind itself and will undoubtedly result in our own extinction.

7 Where to look for further information

7.1 Main recommended additional texts:

A comprehensive introduction to the subject of plant genetic conservation and use is provided by Nigel Maxted, Danny Hunter and Rodomiro Ortiz Rios' *Plant genetic conservation* (see Maxted et al., 2020 in the References for full details). Other single source text include: Frankel et al., 1995, Guarino et al., 2011 and Hunter et al., 2017.

7.2 Useful references for specific subjects:

Policy context: CBD, 1992; Frison et al., 2011; Moore and Tymowski, 2005 and Halewood et al., 2013.

Plant conservation planning: Castañeda Álvarez et al., 2011; Parra-Quijano et al., 2012; Maxted et al., 2012; Maxted et al., 2013; Jackson et al., 2013; Jarvis et al., 2016; Magos Brehm et al., 2017a; and Ramírez-Villegas et al., 2010.

Home garden conservation: Eyzaguirre and Linares, 2004.

In Situ CWR conservation: Maxted et al., 1997; Heywood and Dulloo, 2005; Iriondo et al., 2008; and Hunter and Heywood 2011.

On-farm LR conservation: Brush, 2000; Veteläinen et al., 2009a and b; and Jarvis et al., 2016;

Ex situ CWR conservation: Hawkes et al., 2000; Smith et al., 2003; Reed et al., 2004; ENSCONET, 2009; Carmen de Vicente and Andersson, 2006; and Blackmore and Oldfield, 2017.

Community conservation: Friis-Hansen and Sthapit, 2000; Cunningham, 2001; De Boef et al., 2013; and Vernooy et al., 2017.

7.3 Major journals for the subject are:

Biological Conservation
Conservation Biology
Crop Science
Crop Wild Relative
Diversity and Distributions
Euphytica
Genetic Resources
Genetic Resources and Crop Evolution
Journal of Applied Ecology
Landrace
Plant Genetic Resources: Characterization and Utilization
Plant Systematics and Evolution
Seed Science and Technology
Theoretical and Applied Genetics
The Plant Journal

7.4 Some useful websites are:

Botanical Garden Conservation International (BGCI) - HYPERLINK "http://www.rbgkew.org.uk/BGCI/" http://www.rbgkew.org.uk/BGCI/ http://www.biodiv.org/.

Consultative Group In Agricultural Research (CGIAR) - www.cgiar.org/.
Convention on Biological Diversity (CBD) - http://www.biodiv.org/.
European Cooperative Programme on Plant Genetic Resources (ECP/GR) - http://www.cgiar.org/ecpgr/platform/index.htm).
Food and Agricultural Organisation of the United Nations - http://www.fao.org/.
Genetic Resources Information Network (GRIN) - http://www.ars-grin.gov/npgs/acc/acc_queries.html.
Interactive Toolkit for Crop Wild Relative Conservation Planning - http://www.cropwildrelatives.org/conservation-toolkit/.
Species 2000 - http://www.sp2000.org/.
United Nations Environnent Programme (UNEP) - http://www.unep.unep.org/.
Union for the Protection of New Varieties of Plant (UPOV) - http://www.upov.int/.
World Conservation Monitoring Centre (WCMC) - http://www.wcmw.org.uk/.
World Conservation Union (IUCN) - http://www.iucn.org/.
World Information and Early Warning System on Plant Genetic Resources (WIEWS) - www.fao.org/ag/agp/pgr/wiews/.
World Wide Fund For Nature (WWF) - http://www.wwf.org/.

8 References

Adams, R. P. (1997). Conservation of DNA: DNA banking. In: Callow, J. A., Ford-Lloyd, B. V. and Newbury, H. J. (Eds), *Biotechnology and Plant Genetic Resources*. Pp. 163–174. CAB International, Oxford.

Al Lawati, A. H., Al Saady, N., Ghaloub, H. A., Patzelt, A., Phillips, J., Maxted, N., Al Balushi, A. H., Al-Maqbali, D., Al Naabi, A. S., Al Busaidi, K. A., Al Kharusi, M., Al Shukaili, M. S., Al Jabri, A. A. and Al Nabhani, H. M. (2016). *Plant Agrobiodiversity Conservation Strategy for the Sultanate of Oman*. Oman Animal and Plant Genetic Resources Center, The Research Council of the Sultanate of Oman, Muscat, Sultanate of Oman.

Baudoin, J.-P., Rocha, O., Degreef, J., Maquet, A.. and Guarino, L. (2004). Ecogeography, demography, diversity and conservation of *Phaseolus lunatus* L. in the Central Valley of Costa Rica. In: *Systematic and Ecogeographic Studies on Crop Genepools 12*. International Plant Genetic Resources Institute, Rome, Italy.

Bioversity International and University of Birmingham (2017). CWR checklist and inventory data template v.1, doi:10.7910/DVN/B8YOQL. Harvard Dataverse. Available at: https://dataverse.harvard.edu/dataset.xhtml?persistentId=doi:10.7910/DVN/B8YOQL.

Blackmore, S. and Oldfield, S. (2017). *Plant Conservation Science and Practice: The Role of Botanic Gardens*. Cambridge University Press, Cambridge.

Brush, S. B. (2000). *Genes in the Field: On-Farm Conservation of Crop Diversity*. Lewis Publishers, Boca Raton.

Burley, F. W. (1988). Monitoring biological diversity for setting priorities in conservation. In: Wilson, E. O. and Peter, F. M. (Eds), *Biodiversity*. Pp. 227–230. National Academy Press, Washington, DC.

Carmen de Vicente, M. and Andersson, M. S. (Eds) (2006). *DNA Banks - Providing Novel Options for Gene Banks?* International Plant Genetic Resources Institute, Rome, Italy.

Castañeda-Álvarez, N. P., Vincent, H. A., Kell, S. P., Eastwood, R. J. and Maxted, N. (2011). Ecogeographic surveys. In: Guarino, L., Ramanatha Rao, V. and Goldberg, E. (Eds), *Collecting Plant Genetic Diversity: Technical Guidelines. 2011 Update*. Bioversity International, Rome, Italy. Available at: http://cropgenebank.sgrp.cgiar.org/index.php?option=com_content&view=article&id=679 (Accessed 09.01.2020).

CBD (1992). *Convention on Biological Diversity: Text and Annexes*. Pp. 1–34. Secretariat of the Convention on Biological Diversity, Montreal, Canada.

Cook, F. E. M. (1995). *Economic Botany Data Collection Standards*. Royal Botanic Gardens, Kew.

Cunningham, A. (2001). *Applied Ethnobotany People, Wild Plant Use and Conservation*. Earthscan Publications, UK.

De Boef, W. S., Peroni, N., Subedi, A. and Thijssen, M. H. (Eds) (2013). *Community Biodiversity Management: Promoting Resilience and the Conservation of Plant Genetic Resources*. Pp. 353–364. EarthScan, UK.

Engler, R., Guisan, A. and Rechsteiner, L. (2004). An improved approach for predicting the distribution of rare and endangered species from occurrence and pseudo-absence data. *Journal of Applied Ecology* 41(2): 263–274.

ENSCONET (2009). *ENSCONET seed collecting manual for wild species*. Royal Botanic Gardens, Kew, UK and Universidad Politéchnica de Madrid, Madrid, Spain. Available at: www.kew.org/sites/default/files/ENSCONET_Collecting_protocol_English.pdf (Accessed 09.01.2020).

Eyzaguirre, P. B. and Linares, O. F. (2004). *Home Gardens and Agrobiodiversity*. Smithsonian Books, Washington, DC.

FAO (1995). *Dimensions of Need: An Atlas of Food and Agriculture*. Food and Agriculture Organization of the United Nations, Rome, Italy.

FAO (2014). *Gene Bank Standards for Plant Genetic Resources for Food and Agriculture* (rev. edn.). Food and Agriculture Organization of the United Nations, Rome, Italy. Available at: http://www.fao.org/3/a-i3704e.pdf (Accessed 06.01.2020).

Ferguson, M. E., Ford-Lloyd, B. V., Robertson, L. D., Maxted, N. and Newbury, H. J. (1998). Mapping the geographical distribution of genetic variation in the genus *Lens* for the enhanced conservation of plant genetic diversity. *Molecular Ecology* 7(12): 1743–1755.

Foden, W. B., Butchart, S. H. M., Stuart, S. N., Vié, J. C., Akçakaya, H. R., Angulo, A., DeVantier, L. M., Gutsche, A., Turak, E., Cao, L., Donner, S. D., Katariya, V., Bernard, R., Holland, R. A., Hughes, A. F., O'Hanlon, S. E., Garnett, S. T., Şekercioğlu, C. H. and Mace, G. M. (2013). Identifying the world's most climate change vulnerable species: a systematic trait-based assessment of all birds, amphibians and corals. *PLoS ONE* 8(6): e65427. doi:10.1371/journal.pone.0065427.

Foden, W. B., Mace, G., Vié, J.-C., Angulo, A., Butchart, S., Devantier, L., Dublin, H. T., Gutsche, A., Stuart, S. and Turak, E. (2009). Species susceptibility to climate change impacts. In: Vié, J.-C., Hilton-Taylor, C. and Stuart, S. N. (Eds). *Wildlife in a Changing World – An Analysis of the 2008 IUCN Red List of Threatened Species*. Pp. 77–88. IUCN, Gland, Switzerland.

Ford-Lloyd, B. V., Kell, S. P. and Maxted, N. (2008). Establishing conservation priorities for crop wild relatives. In: Maxted, N., Ford-Lloyd, B. V., Kell, S. P., Iriondo, J., Dulloo,

E. and Turok, J (Eds). *Crop Wild Relative Conservation and Use*. Pp. 110–119. CAB International, Wallingford.

Frankel, O. H. and Bennett, E. (1970). *Genetic Resources in Plants – Their Exploration and Conservation*. Blackwell Publishing, Oxford.

Frankel, O. H., Brown, A. H. D. and Burdon, J. J. (1995). *The Conservation of Plant Biodiversity*. Pp. 1–299. Cambridge University Press, Cambridge.

Friis-Hansen, E. and Sthapit, B. (2000). *Participatory Approaches to the Conservation and Use of Plant Genetic Resources*. International Plant Genetic Resources Institute (IPGRI), Rome, Italy.

Frison, E. A., Cherfas, J. and Hodgkin, T. (2011). Agricultural biodiversity is essential for a sustainable improvement in food and nutrition security. *Sustainability* 3(1): 238–253.

Frodin, D. G. (2011). *Guide to Standard Floras of the World: An Annotated, Geographically Arranged Systematic Bibliography of the Principal Floras, Enumerations, Checklists and Chorological Atlases of Different Areas* (2nd edn.). Cambridge University Press, Cambridge, UK.

González-Orozco, C. E., Brown, A. H. D., Knerr, N., Miller, J. T. and Doyle, J. J. (2012). Hotspots of diversity of wild Australian soybean relatives and their conservation *in situ*. *Conservation Genetics* 13(5): 1269–1281. doi:10.1007/s10592-012-0370-x.

Guarino, L., Maxted, N. and Chiwona, E. A. (2006). A methodological model for ecogeographic surveys of crops. IPGRI Technical Bulletin No. 9. Pp. 1–58. IPGRI, Rome.

Guarino, L., Ramanatha Rao, V. and Goldberg, E. (Eds) (2011). Collecting plant genetic diversity: technical guidelines. Bioversity International, Rome. Available at: http://cropgene bank.sgrp.cgiar.org/images/file/procedures/collecting2011/ (Accessed 30.10.2020).

Guarino, L., Ramanatha Rao, V. and Reid, R. (Eds) (1995). *Collecting Plant Genetic Diversity: Technical Guidelines*. CAB International, Wallingford, UK.

Halewood, M., Lopez Noriega, I. and Louafi, S. (Eds) (2013). *Crop Genetic Resources as a Global Commons: Challenges in International Law and Governance*. Earthscan Publications, London.

Harlan, J. R. and de Wet, J. M. J. (1971). Towards a rational classification of cultivated plants. *Taxon* 20(4): 509–517.

Hawkes, J. G., Maxted, N. and Ford-Lloyd, B. V. (2000) *The ex situ Conservation of Plant Genetic Resources*. Kluwer Academic Publishers, Dordrecht, Netherlands.

Heywood, V. H. and Dulloo, M. E. (2005). *In Situ Conservation of Wild Plant Species – A Critical Global Review of Good Practices*. International Plant Genetic Resources Institute, Rome.

Heywood, V. H. and Wyse-Jackson, P. (Eds) (1991). *Tropical Botanical Gardens: Their Role in Conservation and Development*. Academic Press, London.

Hijmans, R. J., Cameron, S. E., Parra, J. L., Jones P. G. and Jarvis, A. (2005). Very high resolution interpolated climate surfaces for global land areas. *International Journal of Climatology* 25: 1965–1978.

Hijmans, R. J. and Graham, C. H. (2006). The ability of climate envelope models to predict the effect of climate change on species distributions. *Global Change Biology* 12(12): 2272–2281.

Hunter, D. and Heywood, V. H. (Eds) (2011). *Crop Wild Relatives: A Manual of in situ Conservation (Issues in Agricultural Biodiversity)*. Pp. 414. Earthscan Publications, UK.

Hunter, D., Guarino, L., Spillane, C. and McKeown, P. (2017). *Handbook of Agricultural Biodiversity*. Routledge, UK.

IPCC (Intergovernmental Panel on Climate Change) (2007). Summary for Policymakers. In: Parry, M. L., Canziani, O. F., Palutikof, J. P., van der Linden, P. J. and Hanson, C. E. (Eds), *Climate Change 2007: Impacts, Adaptation and Vulnerability; Contribution of Working Group II to the Fourth Assessment Report of the Intergovernmental Panel on Climate Change*. Pp. 7–22. Cambridge University Press, Cambridge, UK.

IPCC (2014). Climate change 2014: synthesis report: longer report. Available at: http://www.ipcc.ch/pdf/assessment-report/ar5/syr/SYR_AR5_LONGERREPORT.pdf (Accessed 03.01.2020).

IPCC (2018). Summary for policymakers. In: Masson-Delmotte, V., Zhai, P., Pörtner, H.-O., Roberts, D., Skea, J., Shukla, P. R., Pirani, A., Moufouma-Okia, W., Péan, C., Pidcock, R., Connors, S., Matthews, J. B. R., Chen, Y., Zhou, X., Gomis, E., Lonnoy, E., Maycock, T., Tignor, M. and Waterfield, T. (Eds), *Global warming of 1.5°C. An IPCC Special Report on the impacts of global warming of 1.5°C above pre-industrial levels and related global greenhouse gas emission pathways, in the context of strengthening the global response to the threat of climate change, sustainable development, and efforts to eradicate poverty*. Available at: https://www.ipcc.ch/site/assets/uploads/sites/2/2019/05/SR15_SPM_version_report_LR.pdf (Accessed 03.01.2020).

IPGRI (1993). *Diversity for Development*. International Plant Genetic Resources Institute, Rome, Italy.

Iriondo, J. I., Fielder, H., Fitzgerald, H., Kell, S. P., Labokas, J., Magos Brehm, J., Negri, V., Phillips, J., Rubio Teso, M. L., Sensen, S., Taylor, N. and Maxted, N. (2016). National strategies for the conservation of crop wild relatives. In: Maxted, N., Dulloo, E. M. and Ford-Lloyd, B. V. (Eds), *Enhancing Crop Gene Pool Use: Capturing Wild Relative and Landrace Diversity for Crop Improvement*. Pp. 161–171. CAB International, Wallingford, UK.

Iriondo, J. M., Maxted, N. and Dulloo, E. (Eds) (2008). *Conserving Plant Genetic Diversity in Protected Areas: Population Management of Crop Wild Relatives*. CAB International, Wallingford.

IUCN (2012). *IUCN Red List Categories and Criteria. Version 3.1* (2nd edn.). International Union for Conservation of Nature and Natural Resources, Gland, Switzerland and Cambridge, UK. iv + 32 pp. Available at: https://portals.iucn.org/library/sites/library/files/documents/RL-2001-001-2nd.pdf (Accessed 09.01.2020).

Jackson, M., Ford-Lloyd, B. and Parry, M. (Eds). *Plant Genetic Resources and Climate Change*. pp. 114–136. CAB International, Wallingford, UK.

Jarvis, D. I., Hodgkin, T., Brown, A. H. D., Tuxill, J., Lopez Noriega, I., Smale, M. and Sthapit, B. (2016). *Crop Genetic Diversity in the Field and on the Farm: Principles and Applications in Research Practices*. Yale University Press, New Haven and London, UK.

Kell, S. P., Ford-Lloyd, B. V., Magos Brehm, J. M., Iriondo, J. M. and Maxted, N. (2017). Broadening the base, narrowing the task: prioritizing crop wild relative taxa for conservation action. *Crop Science* 57(3): 1042–1058.

Mabberley, D. J. (2017). *Plant Book: A Portable Dictionary of Plants, Their Classification and Uses* (4th edn.). Cambridge University Press, Cambridge.

Magos Brehm, J., Kell, S., Thormann, I., Gaisberger, H., Dulloo, E. and Maxted, N. (2017a). Interactive Toolkit for Crop Wild Relative Conservation Planning Version 1.0.

University of Birmingham, Birmingham. Bioversity International, Rome. Available at: http://www.cropwildrelatives.org/conservation-toolkit/ (Accessed 20.10.2020).

Magos Brehm, J., Kell, S., Thormann, I., Gaisberger, H., Dulloo, E. and Maxted, N. (2017b). *Occurrence Data Collation Template v.1*, doi:10.7910/DVN/5B9IV5. Harvard Dataverse. Available at: https://dataverse.harvard.edu/dataset.xhtml?persistentId=doi:10.7910/DVN/5B9IV5 (Accessed 20.10.2020).

Magos Brehm, J., Kell, S. P., Thormann, I., Gaisberger, H., Dulloo, M. E. and Maxted, N. (2019). New tools for crop wild relative conservation planning. *Plant Genetic Resources: Characterization and Utilization* 17(2): 208–212. doi:10.1017/S1479262118000527.

Margules, C. R. (1989). Introduction to some Australian developments in conservation evaluation. *Biological Conservation* 50(1–4): 1–11.

Margules, C. R. and Pressey, R. L. (2000). Systematic conservation planning. *Nature* 405(6783): 243–253.

Maxted, N., Kell, S. P., Ford-Lloyd, B. V., Dulloo, E. and Toledo, Á (2012a). Toward the systematic conservation of global crop wild relative diversity. *Crop Science* 52(2): 774–785.

Maxted, N., Castañeda Álvarez, N. P., Vincent, H. A. and Magos Brehm, J. (2012b). Gap analysis: a tool for genetic conservation. In: Guarino, L., Ramanatha Rao, V. and Goldberg, E. (Eds), *Collecting Plant Genetic Diversity: Technical Guidelines. 2011 Update*. Bioversity International, Rome. Available at: http://cropgenebank.sgrp.cgiar.org/index.php?option=com_content&view=article&id=678 (Accessed 09.01.2020).

Maxted, N., Kell, S. P. and Magos Brehm, J. (2013). Crop wild relatives and climate change. In: Jackson, M., Ford-Lloyd, B. V. and Parry, M. (Eds), *Plant Genetic Resources and Climate Change – A 21st Century Perspective*. Pp. 114–136. CAB International, Wallingford, UK.

Maxted, N., Dulloo, M. E., Ford-Lloyd, B. V., Iriondo, J. and Jarvis, A. (2008). Genetic gap analysis: a tool for more effective genetic conservation assessment. *Diversity and Distributions* 14: 1018–1030.

Maxted, N., Ehsan Dulloo, M. and Ford-Lloyd, B. V. (Eds) (2016). *Enhancing Crop Gene Pool Use: Capturing Wild Relative and Landrace Diversity for Crop Improvement*. CAB International, Wallingford, UK.

Maxted, N., Ford-Lloyd, B. V., Jury, S., Kell, S. P. and Scholten, M. A. (2006). Towards a definition of a crop wild relative. *Biodiversity and Conservation* 15(8): 2673–2685.

Maxted, N., Hargreaves, S., Kell, S. P., Amri, A., Street, K., Shehadeh, A., Piggin, J. and Konopka, J. (2012b). Temperate forage and pulse legume genetic gap analysis. *Bocconea* 24: 5–36.

Maxted, N., Hawkes, J. G., Guarino, L. and Sawkins, M. (1997). The selection of taxa for plant genetic conservation. *Genetic Resources and Crop Evolution* 44(4): 337–348.

Maxted, N., Hunter, D. and Ortiz Rios, R. O. (2020). *Plant Genetic Conservation*. 560 pp. Cambridge University Press, Cambridge.

Maxted, N., Mabuza-Dlamin, I. P., Moss, H., Padulosi, S., Jarvis, A. and Guarino, L. (2004). *An Ecogeographic Survey: African Vigna. Systematic and Ecogeographic Studies of Crop Gene pools 10*. International Plant Genetic Resources Institute, Rome.

Maxted, N., van Slageren, M. W. and Rihan, J. (1995). Ecogeographic surveys. In: Guarino, L., Ramanatha Rao, V. and Reid, R. (Eds), *Collecting Plant Genetic Diversity: Technical Guidelines*. Pp. 255–286. CAB International, Wallingford.

Moore, G. and Tymowski, W. (2005). *Explanatory Guide to the International Treaty on Plant Genetic Resources for Food and Agriculture*. IUCN Environmental Policy and Law Paper No. 57.

Parra-Quijano, M., Draper, D. and Torres, E. (2008). Ecogeographical representativeness in crop wild relative *ex situ* collections. In: Maxted, N. Ford-Lloyd, B. V., Kell, S. P., Iriondo, J., Dulloo, E. and Turok, J. (Eds) *Crop Wild Relative Conservation and Use*. Pp. 249–273. CAB International, Wallingford, UK.

Parra-Quijano, M., Iriondo, J. M. and Torres, E. (2012). Ecogeographical land characterization maps as a tool for assessing plant adaptation and their implications in agrobiodiversity studies. *Genetic Resources and Crop Evolution* 59(2): 205–217.

Phillips, S. J., Anderson, R. P. and Schapire, R. E. (2006). Maximum entropy modelling of species geographic distributions. *Ecological Modelling* 190(3/4): 231–259.

Ramírez-Villegas, J., Khoury, C., Jarvis, A., Debouck, D. G. and Guarino, L. (2010). A gap analysis methodology for collecting crop gene pools: A case study with *Phaseolus* beans. *PLoS ONE* 5(10): 1–18.

Ramírez-Villegas, J., Cuesta, F., Devenish, C., Peralvo, M., Jarvis, A. and Arnillas, C. A. (2014). Using species distribution models for designing conservation strategies of Tropical Andean biodiversity under climate change. *Journal for Nature Conservation* 22(5): 391–404.

Reed, B. M., Engelmann, F., Dulloo, M. E. and Engels, J. M. M. (2004). Technical guidelines for the management of field and *in vitro* germplasm collections. In: *IPGRI Handbooks for Gene banks No. 7*. International Plant Genetic Resources Institute, Rome, Italy.

Savolainen, V., Powell, M. P., Davis, K., Reeves, G. and Corthals, A. (2006). *DNA and Tissue Banking for Biodiversity and Conservation: Theory, Practice and Uses*. Royal Botanic Gardens, Kew, UK.

Scheldeman, X. and van Zonneveld, M. (2010). Training manual on spatial analysis of plant diversity and distribution. Bioversity International, Rome, Italy. Available at: http://www.bioversityinternational.org/e-library/publications/detail/training-manual-on-spatial-analysis-of-plant-diversity-and-distribution/ (Accessed January 2020).

Smith, R. D., Dickie, J. B., Linington, S. H., Pritchard, H. W. and Probert, R. J. (2003). *Seed Conservation: Turning Science into Practice*. Royal Botanic Gardens, Kew.

Spellerberg, I. F. and Hardes, S. (1992). *Biological Conservation*. Cambridge University Press, Cambridge.

ten Kate, K. and Laird, S. A. (1999). *The Commercial Use of Biodiversity: Access to Genetic Resources and Benefit Sharing*. Earthscan Publications, London.

Thormann, I., Kell, S., Magos Brehm, J., Dulloo, M. E. and Maxted, N. (2017). *CWR Checklist and Inventory Data Template v.1*. doi:10.7910/DVN/B8YOQL. Harvard Dataverse. Available at: https://dataverse.harvard.edu/dataset.xhtml?persistentId=doi:10.7910/DVN/B8YOQL (Accessed 09.01.2020).

Tuxhill, J. and Nabhan, G. P. (2001). *People, Plants and Protected Areas: A Guide to In Situ Management*. Earthscan Publications, UK.

United Nations (2015). Transforming our world: the 2030 Agenda for Sustainable Development. Paragraph 54 of United Nations Resolution A/RES/70/1 of 25 September 2015. UN, New York.

van Zonneveld, Mv, Scheldeman, X., Escribano, P., Viruel, M. A., Van Damme, P., Garcia, W., Tapia, C., Romero, J., Sigueñas, M. and Hormaza, J. I. (2012). Mapping genetic diversity of cherimoya (*Annona cherimola* Mill.): application of spatial analysis for conservation and use of plant genetic resources. *PLOS ONE* 7(1): e29845. doi:10.1371/journal.pone.0029845.

Vernooy, R., Sthapit, B. and Bessette, G. (2017). *Community Seed Banks: Concepts and Practices - Facilitator Handbook*. Bioversity International, Rome, Italy.

Veteläinen, M., Negri, V. and Maxted, N. (Eds) (2009a). *European Landraces: On-Farm Conservation, Management and Use. Bioversity Technical Bulletin 15*. Pp. 1-359. Bioversity International, Rome, Italy.

Veteläinen, M., Negri, V. and Maxted, N. (2009b). A European strategic approach to conserving crop landraces. In: Veteläinen, M., Negri, V. and Maxted, N. (Eds). *European Landraces: On-Farm Conservation, Management and Use. Bioversity Technical Bulletin 15*. Pp. 305–325. Bioversity International, Rome, Italy.

Vincent, H., von Bothmer, R., Knüpffer, H., Amri, A., Konopka, J. and Maxted, N. (2012). Genetic gap analysis of wild *Hordeum* taxa. *Plant Genetic Resources* 10(3): 242–253.

Chapter 6

Key issues facing genebanks in preserving crop genetic diversity ex situ: overview of the range of challenges

Paula Bramel, formally Crop Trust, Germany

1 Introduction

There has been a long history of collection and conservation of plant genetic resources ex situ in botanical gardens, private gardens or fields, non-governmental organizations, public genebanks and research collection (Fu, 2017; Hay, 2019). This approach was initiated in the 1960s to counter the alarm raised internationally for the genetic erosion taking place in farmers' fields and in nature. Pistorius (1997) and Fu (2017) describe the history of the coordinated efforts taken internationally to counter this loss of genetic diversity through assembling ex situ collections for use by crop breeders. Ex situ conservation has a focus on a more centralized approach to maintaining locally acquired accessions securely with increased ease of access by research for the long term. The accessions are considered more available and accessible to users, especially when the associated accession level information is also available and shared. This approach to conservation is seen as a secure option for the future as the local genetic diversity is being challenged with local changes in land use, increased urbanization, land degradation, changes in agricultural practises, availability of improved varieties, changes in market preference and the impact of climate change that are increasing the risk for loss of crop diversity.

One of the largest initiatives in the past to collect threatened crop diversity and secure its conservation through an international effort was through the

http://dx.doi.org/10.19103/AS.2020.0085.12

establishment of the International Board for Plant Genetic Resources (IBPGR) in 1974. Thormann et al. (2019) summarizes the history of these collection efforts that resulted in the collection of more than 200 000 samples of locally threatened landraces, crop wild relatives and other material in 136 countries. Many of these collections were made in collaboration of national and international experts and the accession were conserved in an internationally linked system of genebanks known as the Registry of Base Collections (RBC) that was set up to conserve and make available a subset of this material to users. Hansen et al. (1984) reported on a survey of the status of the genebanks in the IBPGR network. In 1974 there were only a handful of seed stores and genebanks focussed on ex situ conservation, but by 1979 there were 54 genebanks and by 1984, when they did the survey, there were 100 genebanks conserving seed with 55 of these meeting the standards that had been internationally agreed. There was also a recognition that less than optimal storage conditions and management was leading to non-availability or accessibility of the accession to the users for exchange. The RBC were established with key guiding principles and architectural pillars that were that the genebanks accepts responsibility for the management of the regional or global base collection for a given crop with well-defined storage conditions and infrastructure that ensure conservation for the long or medium term. They must also make the accessions available to bona fide users.

IBPGR designated security base collections for major cereals, food legumes, vegetables and root/tuber crops. They supported more than 500 collection expeditions. By the end of the 1990s the RBC included 52 genebanks located in all continents holding 80 genera in 250 species for 144 000 accessions but not all unique since some were duplicates. The collections held by the RBC were distributed to more than 500 other genebanks in more than 30 years. Thormann et al. (2019) included a list of RBC by crop, country and institution. There are now 1750 genebanks or collections worldwide conserving about 7.4 million accessions

The policy environment for exchange of germplasm has changed significantly since IBPGR first established the network and made the collections in the 1980s and 1990s (Halewood et al., 2013; Moore 2013; Van Hinton and Brink, 2019). Thormann et al. (2019) assessed the probability that these accessions are available under the International Treaty for Plant Genetic Resources for Food and Agricuture (ITPGRFA) multilateral system. They concluded that about 82% of the accessions have a high probability that they are included and thus are accessible and available to users. Thormann et al. (2015) attempted to link the RBC accessions to current accession in the genebanks. They were able to link only 25% of the accessions. Their survey found that in general passport information is inadequate and not comprehensive enough to make the link but also the historical record on the source of an accession is inadequate in many of the genebanks. Thus the status of these accessions is not secure, documented or accessible.

The status of conservation and availability of the RBC accessions is just one example of a larger issue for ex situ conservation. There have also been other initiatives in the past to collection and secure the conservation of unique regional, national or local crop diversity or to secure a specific crop diversity globally. Moore (2013) describes the history of the international and regional cooperation in ex situ conservation. There is a recognition in the establishment of the multilateral system in the ITPGRFA that Plant Genetic Resources for Food and Agricuture (PGRFA) is a 'common concern for all countries, in that all countries depend upon very largely PGRFA that have originated elsewhere'. The global nature of crop production and consumption means that there is a global interdependency of food supply and genetic resources. This interdependency was also recognized in the first and second report on the state of the world's Plant Genetic Resources for Food and Agriculture (FAO, 1998, 2010). Khoury et al. (2015) demonstrated that national food supplies and production systems are highly interdependent in relation to plant genetic resources and this interdependence has increased in the past 50 years due to globalization of food systems. Thus, a global system for the conservation of this diversity as a global public good is critical to ensure availability and access to the resource for all future generations. Hodgkin et al. (2013) established a framework for this global system that is needed to secure the future conservation and use of ex situ collections.

Moore (2013) describes the history of a number of international (such as the RBC or the CGIAR genebanks), regional genebanks or networks (such as Southern Africa Development Community (SADC), Plant Genetic Resources Center (SPGRC) (established in 1988) or Center for Pacific Crop and Trees (CePaCT)), and crop-specific networks such as International Coconut Genetic Resources Network (COGENT) for coconut genetic resources or Global Musa Genetic Resources Network (MUSANET) for Musa genetic resources. These regional or crop-specific networks were established as opportunities to improve ex situ conservation and achieve economy of scale through cooperation in such areas as safety duplication; to increase efficiency and effectiveness of conservation; or to establish regional facilitation for ex situ conservation as well as distribution. Many of these are still ongoing but suffer from inadequate resources to fully meet the perceived need.

For example, SPGRC was established as the Southern Africa Development Community (SADC) regional genebank and finally setup in 1989. National plant genetic resources centres were also established and investments were made into the SADC region into germplasm collections and conservation facilities. These accessions were to be duplicated within the regional genebank to secure conservation for the long term. The regional genebank was also to facilitate national availability of the accessions to users. In all the past regional or crop-specific initiatives, the main focus initially was on collection of unique but threatened local diversity with a focus to secure it ex situ at the regional or global level. All these initiatives currently struggle to maintain and make available this diversity due to inadequate resources, technical constraints or policy issues. The past investment into the collection of threatened diversity into ex situ collections has resulted in a large number of collections held by a diversity of institutes internationally or nationally that have some unknown level of redundancy that require sufficient resources to maintain and make available to users. The sustainability of the current system for ex situ conservation is unclear as is the future accessibility (Fu, 2017).

The long-term nature of ex situ conservation and use are management, technological and policy challenges (Fu, 2017). There are lessons to be learned from the past efforts at international, regional and national level that illustrate the challenges of the long-term nature of conservation and ensuring present and future use, especially given the uncertainty in assess and benefits sharing term and conditions; inadequate annual resources for conservation as the perceived priority and urgency for conservation wanes; the technical challenges for maintaining accession level genetic integrity with genebank management systems; and the need to secure conservation increases as the local diversity is eroded or lost.

The Crop Trust leads the development of global crop strategies that have attempted to identify key global actions to be taken by the crop communities to secure the global resources for 26 crops (https://www.croptrust.org/our-work/supporting-crop-conservation/conservation-strategies/). Most of the global conservation strategies have been based upon a survey of the key collection holders to determine the composition of the collections as well as its status in terms of key indicators for secure conservation and availability for use. Bramel and Upadhyaya (2018) and Bramel (2017) discuss the global conservation system and urgent further needs for food legumes and wheat. For both these crops and many of the others, there were significant issues for the long-term maintenance of the collections and the limited accession level documentation being shared.

Heywood (2017) reviewed the major achievement made in conserving plant diversity since the 1980s. The major achievements have included the substantial growth in the number of genebanks and other ex situ facilities as well as the inclusion of wild species as seed samples stored in them. The development of new approaches to bridge the gap between ex situ and in situ conservation was also recognized as a key achievement.

The Crop Trust has also been conducting reviews of international, regional and national genebanks (https://www.genebanks.org/resources/genebank-reviews/). There are many common issues that are identified in the crop strategies and in the reviews. So while there is a recognition that there is a critical need to facilitate access and benefit sharing to enhance the use of the collections, there is less appreciation of the urgency to also secure the conservation of the existing accessions given the many constraints that the genebanks are facing. Some of the key ones will be discussed further in the next sections.

2 Issues for the composition of ex situ collections

McCouch et al. (2012) and Bramel and Krishnan (2019) discuss how the history of an ex situ collection can influence the current accession level composition in relation to duplication, relatedness, level of genetic diversity, representativeness and meeting user needs. This historical bias could be due to changes in the focus for conservation over time (such as breeder material versus landraces), opportunities for acquisition that have been driven by short-term external needs (e.g. a set of breeders lines or genetic stocks that were acquired for a specific project but still being maintained as part of the collection) or active safety duplication for other genebanks (such as the RBC in the 1990s). The result of this past acquisition could result in a collection with a high degree of redundancy within or across other collections that no longer meet the needs for local conservation or for the research users of the future. There are finite resources available for long-term ex situ conservation and this needs to be considered in the future curation of a collection. Heywood (2017) recognized that ex situ collections often have limited value for conservation because of a lack of a focussed accession policy; failure to follow established sampling and storage protocols; and activity plans that are poorly prepared and/or implemented.

Bramel and Krishnan (2019) suggest that there is a need to adopt a more nuance approach to curation that accounts for the finite resources available, the dynamic nature of the users' demands for germplasm and the extent of uniqueness or redundancy being conserved in an individual collection or in a global system of collections conserving a crops' diversity. Hodgkin et al. (2013) also identified a number of issues for the global system in relation to individual collection composition, such as the representativeness or comprehensiveness of conservation of global diversity due to the unique contributions of the individual collections. Getting an understanding of this completeness has never been easy given the poor availability of accession level passport data and the many gaps in this data. This assessment also needs to account for in situ-level diversity estimates that are not available readily for any crop.

Bramel and Krishnan (2019) discussed the opportunities of using genetic profiles of accession for better assessing the genetic diversity within and between collections as well as in the wild. This systematic assessment can be used to ensure adequate representativeness of key diversity in ex situ collections in terms of the degree and distribution of diversity. This improved knowledge of the nature of genetic diversity conserved will lead to a global understanding of the completeness of the conservation for individual crops. It will lead to the identification of redundancy within or between collections at the accession level to improve the composition and management of ex situ collections. It can also be used to identify local, national and global gaps in collections that need to be urgently filled.

This approach can also be used to determine the parentage or source of unknown accessions. This offers an opportunity to increase the utilization of all accessions. An example of the application of molecular approaches to diversity assessment to elucidate similarities and parental relationships is found in Ordidge et al. (2018) for apples in an international collection. De Ron et al. (2018) was also able to ascertain the adequacy of the genetic variability for food legumes conserved in both ex situ collections and in situ in isolated gardens and small fields in Portugal, Spain, Italy and Greece.

So while there is an increasing number of tools available to use to address these key research questions, the application in an individual collection will be enhanced by initiatives that allow for across collection sharing of the genetic profiles (Bramel and Volk, 2019). Ferguson et al. (2019) reported on a comprehensive assessment of global crop diversity for cassava utilizing diversity sets from two international collections and a landrace collection from east, southern and central Africa. The 'Seeds of Discovery' initiative (https://seedsofdiscovery.org/) and the rice-focussed '3000 genomes' project (http://iric.irri.org/resources/3 000-genomes-project) are two examples aiming to study the vast accession level diversity stored in international maize, wheat and rice genebanks.

3 Issues for routine operations for conservation

Many collections were initially established with external donor funds as part of projects or global initiatives as discussed earlier. Hay (2019) gives an excellent description of the development of ex situ seed conservation that started with the early Plant Introduction Station in some countries, such as Russia, the UK, Europe, the United States, Australia and New Zealand. These were referred to as the first generation genebanks. Eventually as food security and global hunger became key international concerns, the international collections were established along with regional and more national collections with a focus on conserving crop diversity. These collections could have been research collections without any systematic focus for conservation or it may have been a genetic resources unit with a specific conservation objective. The mission and mandate for many of these institutions have changed over time with expanded mandates in terms of crops, research focus, the addition of animal, microbial, forest genetic resources or others.

Many of these changes in the mandate, mission and objectives for the collection or the institutes resulted in changes in priority and resources for the routine conservation of the ex situ collection that limited the availability of resources for conservation or reduce the availability to users. These changes are resulting in key constraints to the long-term conservation and sustained availability of the accession and the accession level information due to a lack of consistent, sustained annual funds for routine operations. Hufford et al. (2019) concluded that one of the key future issues for crop diversity was the need for better support of genetic conservation efforts including genebanks, as an insurance policy for the adaptability and resilience of agro-and managed ecosystems.

In FAO (2014), updated genebank standards were published that set out the conditions that genebanks should follow for routine operations such as collection, drying, seed cleaning, storage, viability monitoring, regeneration and distribution for orthodox seed as well as field genebanks, in vitro cultures and cryopreservation. The FAO (2010) reported on the status of conservation and availability of crop diversity from the world's genebanks. The report was a compilation of individual country reports that described a global situation where most genebanks were not able to meet the international standards for secure conservation due to chronic underfunding. It also reported on a global system with a high degree of redundancy so this global situation was also not rational in terms of the need to conserve a comprehensive set of accession securely for the future.

Over time, this chronic underfunding of genebanks results in backlogs for key operations such as viability monitoring or regeneration. This can result in increased risk for loss of the genetic integrity of the accessions. Hay (2019) discussed the key factors that influence seed longevity. Seed can be stored for very long time for certain species with high quality, healthy seed that are conserved in the optimal conditions. This has not been the case for many genebanks and thus there is a need to routinely monitor viability based upon an understanding of the species-specific requirement. There are significant backlogs in routine operations for testing and retesting for monitoring viability in base samples in most genebanks today, especially the many national collections held in resource poor countries. Many collections were established in the 1990s and the original samples have been stored in less than optimal base collection conditions. These samples are now 30 years old or more. In many cases, the age of accessions that were acquired or collected by a genebank are skewed towards the early years of the collection or to specific initiative undertaken during its history. This skew in the age of accession is or will soon be leading to significant backlogs for regeneration, especially if genebanks are not routinely regenerating base samples.

Many genebanks do not routinely assess the plant or seed health of the accession in storage or in the field. Thus they are conserving accessions where the impact of seed-borne diseases on viability or in the distribution of the seed is not accounted for in the monitoring for viability testing or anticipating regeneration. Establishing a routine process to comprehensively monitor seed or plant health is not practical when one considers collections that consist of a large number of diversity of species where regeneration or multiplication is being done in a number of different environments. Thus there is a need to develop a more pragmatic approach that will focus more on ensuring that seed or plants are meeting the general standards for conservation and phytosanitary requirements for distribution.

Maintaining the availability of an accession is about maintaining a high level of viability, acceptable seed health and an adequate number of seeds for distribution as well as regeneration or multiplication. The monitoring of seed quantity involves estimates of the weight of a given quantity of seed, such as 1000 seeds, and the weights for the active and base collection sample. This data is not being taken routinely for all accessions in many genebanks. It is also not linked to the distribution data so that any seed sent out is deducted from the samples. This needs to be an established procedure for genebanks to reduce the risk to genetic integrity for the accession from regeneration of small seed lots.

Although safety duplication has always been recognized a very important aspect of secure conservation in the international genebank standards (FAO, 2014), many genebanks have not undertaken the safety backup of their accession and the associated information to the degree expected. There are many constraints in undertaking safety duplication for a collection, many relate to the lack of resources but also the low priority given this for a genebank. There are national and institutional policy barriers in conserving the accession outside the country, even if conserved as a 'black box'. There are issues related to the quality and availability of seed for conserving in another site. There are needs to establish formal agreements that include an elaboration on the responsibility for the receiving genebank. There are also constraints at the receiving genebank such as adequate space in the cold rooms, inadequate resources for maintenance or phytosanitary restriction for the transfer. The establishment of the Svalbard Global Seed Vault has made it possible for genebanks to secure their collections at an additional level, but there is still a need for genebanks to establish primary site for safety duplication.

Safety duplication for field collections or in vitro collections are an urgent need for many genebanks where many accession are only held at one site and face many risks. COGENT (2018) discusses the urgent need for safety duplication of coconut genetic resources due to a lack of funding, policy and phytosanitary issues related to safe transfer protocols. For instance, about 80% of the cultivars being conserved in 15 countries are only found in one country. The current network system for conservation is not being secured and when genebanks were asked how to better safety duplicate the accession, they concluded that a backup should be applied to a selected set of priority accessions that represented global diversity rather than secure every cultivar. This is clearly one approach but it does require a well-connected crop community that has sustained funding for conservation, monitoring and distribution.

There has been has a history of collaboration globally in the conservation of banana genetic resources (http://www.musanet.org/). The International Transfer Centre (ITC) was set-up in 1994 to serve as part of a global system to improve the ease and safe movement of germplasm but now it also serves as a site for safety duplication of the many field genebanks. The Musa Global Conservation Strategy (MusaNet, 2016) reported that despite the availability of the ITC and over 50 ex situ collection globally, there were only 26 collections with accessions duplicated in the ITC and only four of these had more than 80% of their accession duplicated. The surveyed collections did indicate they had safety duplication in other in vitro or field genebanks beside the ITC. This is a surprising result when you consider the free availability of the ITC as a global service to the individual collections. Some of the reasons given for the lack of safety duplication were the lack of funds for virus indexing, equipment and laboratories for establishing the clean and in vitro cultures that would allow the secure movement. There were also constraints in terms of getting permits to move germplasm outside the country, phytosanitary risk and the uncertainty the collection holders had in getting material back when needed. The lack of safety duplication is also reported for other key tree crops, such as coffee (Bramel and Krishnan, 2019), apples (Bramel and Volk, 2019) and tea (Bramel and Chen, 2019). Thus for perennial crop genetic diversity, there is a need to consider a more global approach to safety duplication but this is clearly complicated by the need for additional resources to manage the phytosanitary risk with the germplasm movement and the policy constraints.

The availability and use of accession level information is critical for the monitoring and use of genetic resources. Van Hinton and Brink (2019) described the types of data being used by genebanks and its users as passport data, management data, phenotypic data and -omics data. They describe the availability of data in genebanks and the need for better information quality management standards. They also describe the policy challenges for access to genebank information. Fu (2017) concluded that genebank documentation systems are inadequate and need to be updated. In many genebanks, the accession level data is not digitized and management data is not used for monitoring by information systems. Some genebanks still consider the sharing of accession level information as not a priority. The development of GRIN-Global for individual genebank information systems and Genesys as a platform for the global publishing of accession level information offers many more options for the future (https://www.genebanks.org/the-platform/information-systems/).

For long-term conservation, there are specific needs to annually maintain the quality and rate of flow for the routine operations. As has been seen from the issues described above, there are risk to viability, availability and the genetic integrity of the conserved accessions when operations do not meet international standards (FAO, 2014). Hay (2019) concluded that while the basic methodology of seed banking was not complex, the insurance value of ex situ conservation depended upon both the availability of viable seed and the associated accession level information. This would require future research on how to make conservation and distribution more secure, efficient and cost-effective with greater optimization and curation of collections to higher standards.

Rao et al. (2006) reported clear and detailed procedural guidelines for routine genebank protocols for seed handling. These protocol covered in the manual included a standard flow diagram for the operation and general details on each step. This is a very useful guide for curators and genebank staff but there will still be a need to optimize the flow and protocols for the specific species conserved and the environments where there are operating. Quality management for genebanks depends upon a clear flow of routine operation with adequate equipment, consumables, infrastructure and documentation for secure and cost-effective conservation. Having clear secure routine operations with standard operating procedures (SOP) that are tested, written and updated as needed is a key aspect of the quality management as described by Crop Trust, https://www.croptrust.org/2015/topics/qms/.

The quality management system (QMS) approach to management at the Centre for Genetic Resources the Netherlands (CGN) is described at: https://www.wur.nl/en/Research-Results/Statutory-research-tasks/Centre-for-Genetic-Resources-the-Netherlands-1/CGN-Quality-Management-System.htm. Meeting the goal of QMS to improve efficiency and reliability of the genebank operations meant to identify the various processes and validate them. The CGN concluded that there were many processes that had not been recognized before and so the establishment of the QMS leads to these being validated. For the future they see the QMS manual will facilitate staff capacity building and the modification of operational processes as needed.

One key aspect of QMS is staffing in relation to the expertise required, clear term of reference for each position, a plan for succession in case of a vacancy and a capacity building plan to keep staff updated on the latest methodologies and technology. The USDA-ARS has recognized that there is a need to anticipate the need for 30% turnover of staff in the next 5 years due to retirement (Volk et al., 2019). To manage the impact of this staffing change and the need to have a large number of newly trained experts in genetic resource curation, Volk et al.

(2019) outlines a plant genetic resources management training programme that includes a resource library, distant learning courses and specialized workshops. They also recognize the need to capture the knowledge and experience of the current curators and other staff in these training programmes before they retire. This is also the goal of succession planning in QMS systems.

Another key aspect of QMS for a genebank is to manage the risk to the genebank (https://www.genebanks.org/the-platform/quality-management/). Given the role of a genebank as insurance for future loss of diversity, it is important to assess the risk and develop a risk mitigation matrix as central framework for management and monitoring performance. This should include a business plan with contingency planning. There is also a need to increase the monitoring of genebanks, such as the key indicators and targets for performance as is being done for the GCIAR genebanks (https://www.genebanks.org/resources/genebanks-in-numbers/key-performance/). This monitoring system would allow for a better understanding of the status of genebanks globally and better planning for actions to mitigate risk.

In many genebanks, the conservation infrastructure is aging and deteriorating with very little investment into upgrades since the establishment of the genebank. This is a critical need for infrastructure such as seed dryers and cold stores as well as the need to increase efficiency with new technology and automation. The lack of investment into upgrades will lead to increased risk of loss of accession or entire collections. One key aspect of this is in the issues related to the cost of power, electricity and fuel for generators for many genebanks. There is a need to focus on upgrades in infrastructure that is more energy-efficient and utilizes more alternative sources of energy. This upgrade will not only reduce the risk to genebank operations from electrical failures but also increase cost-effectiveness for the future.

4 Issues for use of conserved genetic resources

Hufford et al. (2019) identified a key issue for the future for crop diversity related to the need for a significant international legal effort be made to counter the constraints that currently limit use and conservation research and education efforts. Oldham et al. (2013) concluded that patent activity was narrowly focussed on a small number of species but long-term needs will best be served by focussing biodiversity research and development on a broader range of species by enhancing the implementation of the Convention on Biological Diversity and its Nagoya Protocol. The obstacles to this are significant because of the global nature of the problem, the range of actors involved and the wide diversity of interests.

There are many constraints to the timely distribution of accession by individual genebanks, such as the lack of transparent policy, phytosanitary issues, shortage of high-quality seed or planting material, the cost of packaging and shipment. This is a particular challenge for national genebanks and for vegetatively propagated or recalcitrant seed crops. There is not feedback from and to users on genebank operations as well as information generated on the accessions. This does not build sustained links to users and partnership with users to secure information and knowledge gained on accession are also conserved and shared for the long term. Fu (2017) discussed the consequences of the lack of stakeholder engagement with genebanks and that this needs to be addressed to build greater investment into ex situ conservation.

As pointed out in the global conservation strategies for wheat, food legumes, apples, coffee and tea, (Bramel, 2017; Bramel and Upadhyaya, 2018; Bramel and Volk, 2019; Bramel and Krishnan, 2019; Bramel and Chen, 2019) there is limited active, sustained links between formal and informal conservers in the global system. Krishnan et al. (2019) showcase examples of collaboration of botanical gardens with academic institutions, non-profits, agricultural research organizations and genebanks to increase public engagement with food and agriculture. These collaborations will need to be enhanced in the future to meet the need to ensure support for crop biodiversity conservation among a public which is not familiar with agriculture.

5 Conclusion

Fu (2017) described an analysis of vulnerability as essential for the sustainability of genebank systems. He concluded that little is known about vulnerability of the genebanks worldwide since there are few reports on this implementation of this analysis. He developed a model for a sustainable genebank system with three key core elements (germplasm, management, and stakeholders), and six element that provide support. Fu (2017) utilized this very simple model to assess the impacts of the current stresses on sustainability. He concluded that currently genebanks are under stress from inadequate public investment, weakening political support and insufficient stakeholder engagement. There is also an impact of poor support for research and technology for conservation as well as a deterioration in staff capacity. Thus the risk of genetic erosion within and between genebanks is increasing and Fu (2017) identified a number of actions that can be taken by genebanks and others to mitigate this risk. The future sustainability of the genebank systems will ultimately depend upon increased investment and partnerships to reduce the loss of genetic diversity.

Fu (2017) reported on an analysis of the critical issues in conservation through the review and ranking for degrees of challenge through a review of literature. The top 10 critical issues were (in order of priority) catastrophic events and collapse of collections; lack of professional training in plant genetic resource conservation; lack of support for research on long-term conservation;

reduced support by stakeholders into long-term conservation in public genebanks; bias in current collection's diversity with the need to fill significant gaps that are at high risk of genetic erosion; the need to update genebank information systems, including the IT infrastructure as well as competencies, inadequate access to accession level information, especially evaluation and characterization; diluted political commitment to better genebank systems; loss of viability and consequent regeneration backlogs; and general worldwide underfunding of genebanks. These issues are very similar to what was reported in the earlier sections.

Fu (2017) concluded that the Global Crop Diversity Trust would need an endowment of between 3 and 15 billion dollars to ensure the adequate routine conservation of the anticipated 10 million accessions in the future in ex situ collections. Given the global interdependency for the use of PGRFA, there is an even greater urgency now in 2020s to facilitate the development of a global system for ex situ conservation that is secure, cost-effective, rational, sustainable and available for use into the far future.

6 Where to look for further information

There are a number of ongoing research efforts in ex situ conservation of genetic resources. Further information on accessions being conserved by genebanks can be found on Genesys at: https://www.genesys-pgr.org/. Further information on international standards on conservation and use of genetic resources can be found from: https://www.genebanks.org/the-platform/conservation-module/. There are a number of national sources of information on genetic resource conservation and use. These include the USDA-ARS and other national genebanks. Information related to policy can be found through the International Treaty for PGRFA: http://www.fao.org/plant-treaty/en/.

7 References

Bramel, P. 2017. Chapter 1. Wheat genetic resources: global conservation and use in the future. In: *Achieving Sustainable Cultivation of Wheat* (vol. 1) Langridge, P. (Ed.). Burleigh Dodds Science Publishing Limited, Cambridge, UK, 3–24.

Bramel, P. and Krishnan, S. 2019. Systematic assessment for conservation and utilization of crop genetic resources. In: *Sustaining Global Food Security: The Nexus of Science and Policy* Zeigler, R. S. (Ed.). CSIRO Publishing, Australia, 3–14.

Bramel, P. and Upadhyaya, H. 2018. Chapter 3. Conserving and characterizing the genetic diversity of grain legumes. In: *Achieving Sustainable Cultivation of Grain Legumes* (vol. 1) Sivasankar, S. (Ed.). Burleigh Dodds Science Publishing Limited, Cambridge, UK.

Bramel, P. J. and Chen, L. 2019. *A Global Strategy for the Conservation and Use of Tea Genetic Resources*. Global Crop Diversity Trust, Bonn, Germany, 56.

Bramel, P. J. and Volk, G. 2019. *A Global Strategy for the Conservation and Use of Apple Genetic Resources*. Global Crop Diversity Trust, Bonn, Germany.

COGENT 2018. A global strategy for the conservation and use of coconut. In: *Genetic Resources, 2018–2028*. Bourdeix, R. and Prades, A. (Comps). Bioversity International, Montpellier, France.

De Ron, A. M., Bebeli, P. J., Negri, V., Vaz Patto, M. C. and Revilla, P. 2018. Warm season grain legume landraces from the south of Europe for germplasm conservation and genetic improvement. *Frontiers in Plant Science* 9:1524.

Ferguson, M. E., Shah, T., Kulakow, P. and Ceballos, H. 2019. A global overview of cassava genetic diversity. *PLoS ONE* 14(11):e0224763.

Food and Agriculture Organization (FAO) 1998. *The State of the World's Plant Genetic Resources for Food and Agriculture*. Food and Agriculture Organization of the United Nations, Rome, Italy.

Food and Agriculture Organization (FAO) 2010. *The Second Report on the State of the World's Plant Genetic Resources for Food and Agriculture*. Food and Agriculture Organization of the United Nations, Rome, Italy.

Food and Agriculture Organization (FAO) 2014. *GeneBank Standards for Plant Genetic Resources for Food and Agriculture* (revised edn.). Food and Agriculture Organization of the United Nations, Rome, Italy.

Fu, Y.-B. 2017. The vulnerability of plant genetic resources conserved *ex situ*. *Crop Science* 57(5):2314–2328.

Halewood, M., Lopez Noriega, I. and Louafi, S. 2013. The global crop commons and access and benefit sharing laws: examining the limits of international policy support for the collective pooling and management of plant genetic resources. In: *Crop Genetic Resources as a Global Commons. Challenges in International Law and Governance* Halewood, M., Lopez Noriega, I. and Louafi, S. (Eds). Routledge, UK, 1–36.

Hansen, J., Williams, J. T. and Freund, R. 1984. *Institutes Conserving Crop Germplasm: The IBPGR Global Network of Genebanks*. IBPGR Secretariat, Rome, 25.

Hay, F. 2019. *Seed Banking as Future Insurance against Crop Collapse. Oxford Research Encyclopaedia of Environmental Science*. Oxford University Press, USA, 1–16.

Heywood, V. H. 2017. Plant conservation in the Anthropocene - Challenges and future prospects. *Plant Diversity* 39(6):314–330.

Hodgkin, T., Demers, N. and Frison, E. 2013. The evolving global system of conservation and use of plant genetic resources for food and agriculture: what is it and where does the Treaty fit in. In: *Crop Genetic Resources as a Global Commons. Challenges in International Law and Governance* Halewood, M., Lopez Noriega, I. and Louafi, S. (Eds). Routledge, UK, 344–367.

Hufford, M. B., Berny Mier Y Teran, J. C. and Gepts, P. 2019. Crop biodiversity: an unfinished magna opus of nature. *Annual Review of Plant Biology* 70:727–751.

Krishnan, S., Moreau, T., Kuehny, J., Novy, A., Greene, S. L. and Khoury, C. K. 2019. Resetting the table for people and plants: botanical gardens and research organizations collaborate to address food and agricultural plan blindness. *Plant, People, Planet* 1(3):157–163.

Khoury, C. K., Achicanoy, H. A., Bjorkman, A. D., Navarro-Racines, C., Guarino, L., Flores-Palacios, X., Engels, J. M. M., Wiersema, J. H., Dempewolf, H., Ramírez-Villegas, J., Castaneda-Alvarez, N., Fowler, C., Jarvis, A., Rieseberg, L. H. and Struik, P. C. 2015. Estimation of Countries' Interdependence in Plant Genetic Resources Provisioning

National Food Supplies and Production Systems. Research Study 8. Food and Agriculture Organization, Rome, 26.

McCouch, S. R., McNally, K. L., Wang, W. and Sackville Hamilton, R. 2012. Genomics of genebanks: a case study in rice. *American Journal of Botany* 99:401-423.

Moore, G. 2013. International and Regional cooperation in the implementation of the International Treaty on Plant Genetic Resources for Food and Agriculture. In: *Crop Genetic Resources as a Global Commons. Challenges in International Law and Governance* Halewood, M., Lopez Noriega, I. and Louafi, S. (Eds). Routledge, UK, 329-343.

MusaNet 2016. *Global Strategy for the Conservation and Use of Musa Genetic Resources* Laliberté, B. (Comp). Bioversity International, Montpellier, France.

Oldham, P., Hall, S. and Forero, O. 2013. Biological diversity in the patent system. *PLoS ONE* 8(11):e78737.

Ordidge, M., Kirdwichai, P., Baksh, M. F., Venison, E. P., Gibbings, J. G. and Dunwell, J. M. 2018. Genetic analysis of a major international collection of cultivated apple varieties reveals previously unknown historic heteroploid and inbred relationships. *PLoS ONE* 13(9):e0202405.

Pistorius, R. 1997. *Scientists, Plants and Politics: A History of the Plant Genetic Resources Movement*. Int. Plant Genet. Resour. Inst., Rome, Italy.

Rao, N. K., Hanson, J., Dulloo, M. E., Ghosh, K., Nowell, D. and Larinde, M. 2006. *Manual of Seed Handling in Genebanks*. Bioversity International, Rome, Italy.

Thormann, I., Engels, J. M. M. and Halewood, M. 2019. Are the old International Board for Plant Genetic Resources (IBPGR) base collections available through the Plant Treaty's multilateral system of access and benefit sharing? A review. *Genetic Resources and Crop Evolution* 66(2):291-310.

Thormann, I., Fiorino, E., Halewood, M. and Engels, J. M. M. 2015. Plant genetic resources collections and associated information as baseline resource for genetic diversity studies: an assessment of the IBPGR supported collections. *Genetic Resources and Crop Evolution* 62(8):1279-1293.

Van Hinton, T. and Brink, M. 2019. Technology and policy challenges to utilization of plant genetic resources. In: *Sustaining Global Food Security: The Nexus of Science and Policy* Zeigler, R. S. (Ed.). CSIRO Publishing, Australia, 36-48.

Volk, G. M., Namuth-Covert, D. and Byrne, P. F. 2019. Training in plant genetic resources management: a way forward. *Crop Science* 59:1-14.

Chapter 7

Techniques and key issues in collecting crop wild relatives

Michael Way, Royal Botanic Gardens, Kew, UK

1 Introduction

Much of the scientific basis for collecting germplasm from wild plants developed in the 1970s as the potential role for seed banking to confront global threats to plant diversity became clearer. This applied to both the preservation of plant genetic resources for agriculture and conservation of seed for recovery of threatened wild plants. Of the more than 7.4 million germplasm collections estimated by the FAO (2010) to be held in gene banks worldwide, Castañeda-Álvarez et al. (2016) reported a critical under-representation in gene banks of crop wild relatives (CWR), which are increasingly recognised as essential for future breeding and research efforts. This chapter reviews the main techniques for collecting seed of CWR, with an emphasis on current scientific challenges faced by CWR collectors, concluding with research issues that influence the quality and genetic diversity of a collection, and therefore its ultimate use by researchers.

Although much work remains to be done to define optimal collecting, handling, storage, and germination conditions, and also to improve the genetic

http://dx.doi.org/10.19103/AS.2020.0085.08

and resource efficiency of programmes, the essential conditions required to keep seeds with orthodox seed storage behaviour viable in conventional seed bank storage are now well established. The seed biology knowledge applied to collecting and conservation of seed, as reviewed by Hay and Probert (2013) and updated by Hay in this volume, has now been tested in practice across all taxonomic groups and environments, this is informed by several decades of large-scale practical seed banking of crop and wild species, for example, by the US Agricultural Research Service at the National Laboratory for Genetic Resources Preservation (NLGRP), and of wild species by the Royal Botanic Gardens, Kew, at the Millennium Seed Bank (MSB). Detailed guidance is now accessible for practitioners in manuals such as Collecting Plant Genetic Diversity (Guarino et al. 2011) and standards such as the FAO gene bank standards (FAO 2014), MSB Partnership collections standards (Breman and Way 2018) and the CPC Best Plant Conservation Practices (Center for Plant Conservation (undated)).

From the collectors' point of view, there are two major research constraints that have impacted our ability to make sufficient high-quality, genetically diverse CWR collections over the last two decades:

First, in contrast to the relatively uniform and well-documented flowering and fruiting of crops, populations of CWR typically retain important traits that have not been narrowed by crop breeding; indeed the collecting and use of CWR is required precisely to overcome that genetic bottleneck (Tanksley and McCouch 1997). Collectors therefore have needed to better understand the natural variation of CWR populations, in order to make good decisions about seed quality (Section 6), genetic sampling (Section 7) and seed handling (Section 8) resulting in viable germplasm with high genetic diversity that is representative of the source population.

Secondly, the detailed knowledge about relative importance, distribution, identification, collecting and handling techniques for CWR remained patchy and fragmented prior to the development of online tools including the prioritised Harlan and de Wet CWR Inventory (Vincent et al. 2013), digitised specimen data through the Global Biodiversity Information Facility (GBIF) (https://www.gbif.org/) and a range of catalogues and botanical databases reviewed by Garruccio (2011). One approach to integrating the essential information required by collectors in a consistent and useful form for fieldwork has been development of printed CWR target collecting guides covered in Section 3, which have proved useful in collecting regions where access to information is poor.

The main focus of this chapter is the collecting of seed, which has been shown to provide the most cost-effective (Li and Pritchard 2009) ex situ response to the erosion of genetic diversity at species level, for taxa-producing orthodox seed, that is, that tolerate conventional dry and cold storage conditions according to FAO & MSBP standards (FAO 2014; Breman and Way 2018). Wyse et al. (2018) estimated that 90% of 4888 species on the CWR inventory are capable of conventional seed bank storage; accordingly there are nearly 500 that require alternative techniques for germplasm collection and preservation, so the protocols for handling 'exceptional species' (Pence 2013) through cryopreservation and recovery need continual development to be able to extend taxonomic coverage in a cost-effective way.

In this chapter the collecting of other germplasm such as pollen, meristems, tubers, cuttings or live plants will not be covered in detail, but resources are suggested (see Section 10) to cover this material, which is generally more difficult and more costly to collect, transport, preserve and use than orthodox seeds, and also requires sampling from more individuals to capture equivalent genetic diversity. Future work needs to determine which exceptional species (Pence 2013) can be best preserved ex situ through use of these methods.

Collecting techniques and issues are explored through the example of the 'Adapting Agriculture to Climate Change: collecting, protecting and preparing crop wild relatives' (CWR Project) (https://www.cwrdiversity.org/) managed by the Global Crop Diversity Trust and the Millennium Seed Bank Partnership (MSBP) of the Royal Botanic Gardens, Kew. The CWR Project was launched in 2011 with US$50 million in funding from the Government of Norway, and has been implemented in partnership with national and international gene banks and plant breeding programmes around the world, to find and collect seed of wild relatives of 28 of the world's most important crop gene pools. In 2019, having completed nearly 3000 collecting-days in the field in 25 countries, the CWR Project reported the collection of seed from an unprecedented 4644 collections of 371 CWR species, with particular impact on CWR of *Solanum, Vicia, Medicago, Hordeum* and *Oryza* (Global Crop Diversity Trust 2019).

2 Targeting species and regions for collecting

The ongoing and predicted loss of wild CWR populations due to climate change (Jarvis et al. 2008) cannot be effectively countered by opportunistic specimen-collecting as an extension to existing field activities, as populations of potentially useful taxa will be overlooked and genetic diversity may be lost. Instead, the collection of high-quality, genetically diverse seed from populations of priority CWR requires thorough preparation, logistical support and capable personnel to be deployed into the field, often for remote expeditions which need to be planned in a cost-effective manner. Following global level commitments to CWR conservation recognised by the FAO International Treaty (FAO 2001) and the Global Strategy for Plant Conservation (CBD 2010), prioritisation of CWR conservation has successfully made advances at two critical levels. First, the taxa falling within the three accepted gene pools of major crops have been compiled and published as the prioritised Harlan and de Wet CWR Inventory (Vincent et al. 2013) available at: www.cwrdiversity

.org/checklist/. Secondly, the coverage of existing germplasm collections from across recorded native distribution of CWR taxa has been analysed following the Gap Analysis procedure demonstrated for CWR by Maxted et al. (2008), subsequently enhanced for *Phaseolus* beans (Ramírez-Villegas et al. 2010), then extended to model the global distribution of 1076 CWR taxa related to 81 crops by Castañeda-Álvarez et al. (2016). See Maxted in this volume for further developments of this important technique.

To enable conservation actions to proceed at the necessary scale without detailed genetic diversity data, the proxy of 'ecogeographical representativeness' is used (e.g. Parra-Quijano et al. 2011) to maximise the likely capture of population level genetic diversity. We know that there are exceptions where genetic diversity is not always partitioned equally across geographical range, for example, the high proportion of diversity of the lentil, *Lens culinaris*, held in the western part of its range, cited in Maxted et al. (2011). In practice, the CWR Project set as an initial objective the sampling of ten new populations per target taxon within each region of collection. If sufficient well-identified and georeferenced specimens exist, a map of potential distribution can be prepared for each species such as shown in the Brazil CWR Seed Collecting Guide (Embrapa and Kew 2016). Where specimens' predate use of GPS or taxonomic identifications are in doubt then additional field survey should attempt to both relocate and encounter new populations, such as that achieved for wild barley, *Hordeum chilense*, shown in Fig. 1. Sampling of genetic diversity from populations is explored further in Section 7.

As updates were compiled in 2011 to the Collecting Plant Genetic Diversity technical guidelines, the use of digital data sources was already becoming essential for the planning of germplasm collection (Garruccio 2011). Access to geographical and botanical information through online databases, maps, image libraries, citizen science portals and social media has continued to expand, although care needs to be taken with the quality of georeferencing and of naming which is particularly variable from sources that aggregate massive datasets, for example, the GBIF (https://www.gbif.org/). The checking of raw specimen data during compilation of a research database is therefore essential. In general, one should prioritise georeferenced records that are supported by herbarium voucher specimens determined by a competent botanist. For fieldwork planning, probably the most helpful additional online tool for collectors has been Google Earth which has proved consistently valuable to flag locations of existing collections and to identify remnant vegetation patches and potential access points to habitats for sampling.

Collaboration and consultation at several levels are essential pre-requisites for collecting of CWR. Although the growing breadth of digital data available to collectors through desktop and mobile devices is compelling, this is not a substitute for strengthening cooperation with botanists and communities

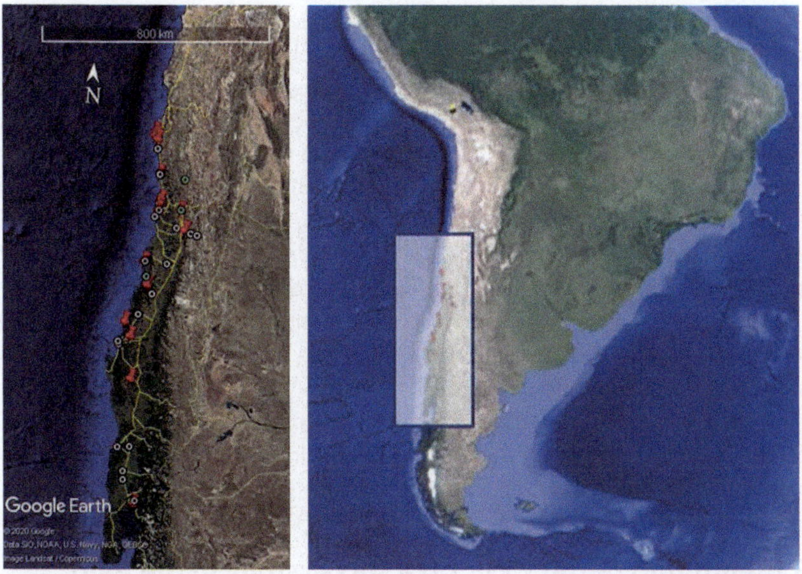

Figure 1 Locations of *Hordeum chilense* specimens compiled by the Chile CWR Project (RBG Kew 2016). Not all historic populations could be relocated for seed collecting, although some new populations were discovered and seed sampling has extended widely across the known distribution of *Hordeum chilense* in Chile. **Grey circles**: Geo-referenced specimens provided by the SGO herbarium, prior to the CWR Project. Collection date ranges from 1854 to 1996 (median 1943). **Green circles**: herbarium specimens from new populations located by the CWR Project. **Red pins**: New seed collections made by the CWR Project.

from the collecting region who can offer local advice and participate in fieldwork. Indeed, they may be able to offer a logistical or staging base for holding botanical collections safely during longer expeditions and could be collaborators for the in situ conservation of plant genetic resources

National legislation implementing the access and benefit-sharing obligations of the Convention on Biological Diversity (CBD) and specifically the Nagoya Protocol usually requires collectors to obtain research and collection permits from national agencies, permit authorities and protected area managers, in consultation with local communities. Specific prior-informed consent may need to be obtained from local providers of genetic resources or for access to traditional knowledge. For ambitious projects operating across multiple administrative boundaries, this consultation and application process is a significant task which needs to be started well in advance of the field season. Projects also need to allow sufficient resources to deliver on the many conditions that attach to collection permits including the deposit of labelled voucher specimens in national facilities, and distribution of reports and publications.

Assuming that permit conditions have been complied with, a set of collections can often be duplicated at research partner facilities in another country under one of the following access routes, which should be determined at the start of a project. Collections of wild species required for biodiversity research (including herbarium vouchers of CWR collections) are often transferred under the terms of the CBD through a bilateral agreement or access contract. In contrast, the supply of germplasm listed within Annex 1 of the International Treaty (FAO 2001) and intended for 'Food and Agriculture' purposes would normally be under the standard Material Transfer Agreement (sMTA) of the IT, following advice from the national focal point for the IT via fao.org/plant-treaty/ (see Louafi, this volume). In addition, the transfer of CWR collections between countries needs to follow plant health regulations and may require inspection of the collections for pest and diseases, a separate phytosanitary certificate and additional packaging for export, and a period of quarantine for germplasm on arrival at the destination.

3 Identifying target CWR taxa successfully in the field

Without up-to-date and accurate botanical keys and descriptions for target taxa, it will be difficult for a field team to be confident that the collecting effort in the field is focussed on the priority species. Even where a team is equipped with field computers and reliable power source, a challenge for collecting CWR taxa is often the scarcity of geolocated, fully identified, digitised collections with accompanying high-quality images either from field or from herbarium, and thus the CWR seed-collecting guides described below provide a practical and consistent way for collectors to locate and identify target species. The content and format of seed-collecting guides continually evolves, but through the use of Brahms database (© University of Oxford) it has been possible to compile essential information to support successful exploration and collecting of seed within the CWR Project. The current format summarises each target taxon on a pair of laminated pages: the first page includes the botanical description, identification notes, and potential distribution maps overlaid with existing specimens, together with habitat and locality information from herbarium labels. The second page, as illustrated in Fig. 2 contains selected images of representative herbarium specimens, and flower, fruit, and mature plant of *Hordeum chilense,* compiled for the Chile seed collecting guide. Additional examples of this approach for CWR Project taxa *Oryza, Solanum, Ipomoea* and *Eleusine* can be reviewed in *Brazil Seed Collecting Guide* (Embrapa and RBG Kew 2016).

Although such guides take several weeks to prepare using all available references and knowledge available to a research team, it is an invaluable resource to collectors in the field and provides all key information required

Figure 2 *Hordeum chilense* excerpt modified from Chile Collecting Guide (RBG Kew 2016). Image top left: Crop Trust - LM Salazar. Other main images: RBG Kew. Inset images at foot: from L to R show: seed; seed collecting method; habit and approximate size of plant; conservation status; phenology information derived from specimens. RBG Kew.

to make collections for a germplasm bank or research project, irrespective of digital connectivity in the field.

In practice, the provision of seed collecting guides has helped facilitate the collection of 94% of the CWR Project's 3713 priority collection targets (Global Crop Diversity Trust 2019). Continual innovation of this approach is required as digital data integration improves and as new mobile devices come to market. Already the capacity to post a photograph with provisional name and receive a confirmed identification within minutes from specialists on platforms such as *iNaturalist* https://www.inaturalist.org/ enables collectors to respond to opportunities during fieldwork and to be confident that they have identified the target taxon. In turn the most reliable records submitted to *iNaturalist* are accepted as 'research grade' records by GBIF for wider use, including for Gap Analysis.

4 Understanding seed development, ripening and dispersal

Knowledge of the development of seeds in key families is helpful in order to plan seed collecting expeditions at the correct time and to make good decisions about seed harvesting.

4.1 Biology of seed development

After fertilisation of the ovule, seeds form within the fruit and develop to a point of mass maturity at which the storage reserves of protein, lipid and carbohydrate reach a maximum level. The four stages of seed formation; reserve accumulation; ripening and dispersal are illustrated in the graph (Fig. 3) below. At seed mass maturity, abscission stops the transfer of fluid or nutrients from the vascular tissue of the parent plant and the fruit then undergoes ripening to prepare the seed for dispersal. Ripening is characterised by texture and colour changes, both of the fruit and of the seed tissues. The ripening of dry fruits is accompanied by loss of moisture and hardening of seeds before dispersal. See Section 5 for an overview of common ripening cues.

In terms of seed physiology, a seed collection capable of use following storage in conventional seed banks needs to have three quality traits:

1. high initial viability (i.e. high % germination of fresh seed, excluding empty or infested seeds);
2. full desiccation tolerance (i.e. ability to germinate following drying to 15% equilibrium relative humidity); and
3. maximum potential longevity under dry cold storage conditions.

In Fig. 3, these three traits are typically achieved successively in wild species with orthodox seeds, although it should be noted that seed of domesticated cereals like wheat, *Triticum aestivum*; barley, *Hordeum vulgare*; and rice, *Oryza sativa* do not follow this pattern (Hay and Smith 2003).

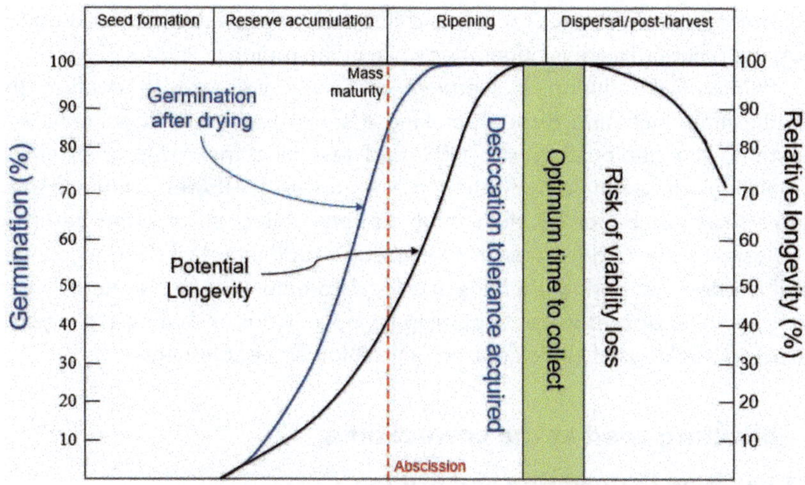

Figure 3 The acquisition of seed quality traits including desiccation tolerance and potential longevity in storage, across four seed development stages, adapted from Hay and Probert (2013) and Way and Gold (2014a). The optimum time to collect seed from wild plant populations for long-term conservation is illustrated with a green column at the boundary of the 'ripening' and 'dispersal' phases, that is, at the onset of natural seed dispersal. Image: RBG Kew. During the final dispersal/post-harvest stage, there is a risk of viability loss if seed is left open to hot and/or humid conditions, and seeds could also be subject to additional predation in this stage especially if within fleshy fruits or dispersed on the ground (see Section 8). For these reasons, the correct timing of seed harvesting is a key factor determining initial seed viability of a collection.

4.2 Predicting timing of fieldwork for seed collecting

In practice, there are two steps that collectors can take to predict optimum time for fieldwork. First, well in advance, herbarium specimen data and literature can be analysed to provide an indicative flowering and fruiting window as shown in the seed collecting guide extract at the foot of Fig. 2. Secondly, it is recommended that one member of the collecting team visit the region for a scouting or survey visit, ideally during the flowering stage of the target species.

To assess phenology of multiple taxa, a review of 111 European species estimated the interval between flowering and seed dispersal as a mean of 3.5 months (range 0.5–12 months) for woody species, and as a mean of 1.8 months (range of 1.0–8.5 months) for herbaceous species. This data is drawn from literature and botanic garden observations (Appendix 3 in ENSCONET 2009) and is an indicative starting point for fieldwork planning. The speed of seed development and ripening in a population depends primarily on plant family and fruit type, described in Section 5 but the altitude, relative humidity and exposure of the population to prevailing weather conditions are also important factors. Weather reports and forecasts can now be accessed during most expeditions to adjust itineraries while in the region.

Priority tasks during a scouting or survey visit are to confirm the identification, including through making a set of herbarium specimens and recording the geolocation with GPS, and assessing the size and extent of populations. If taxon identification mainly relies on flower features, small markers such as paper jewellers' tags discretely attached to stems can help in relocating the targets later in the season. Additional tasks are to contact local landowners, park authorities and local communities to update them on plans and offer opportunities to participate in fieldwork, and complete logistics including bookings of vehicles, accommodation and equipment.

5 Collecting seed at the correct time

5.1 Variation in flowering and fruiting

Flowering of wild plants in a population is not completely synchronous, so collectors should expect variations between individual plants in the onset, peak and completion of flowering and fruiting. Examples of the variation encountered in flowering and seed development of wild plants are detailed in Hay and Smith (2003), and in the same volume Dickie and Stuppy (2003) provide a helpful review of fruit and seed types. For taxa with indeterminate inflorescences such as *Solanum*, a single individual may bear flowers and fruit at various development stages, for example, as shown for *Solanum asperolanatum* in Fig. 4, and CWR collectors must decide which of the ripening fruit can be successfully harvested. Variation is found across all wild plant groups, for

Figure 4 *Solanum asperolanatum* (Ruiz and Pav) showing flower, ripe fruit and abscission scars from previously dispersed fruit. Image: INIAP Ecuador, Edwin J. Naranjo.

example in wild grasses Poaceae; the individual tillers (reproductive units) can exhibit variations in development between inflorescences, between flowering and dispersal (shattering) within the spike.

In contrast to crops, the development and ripening of seed from herbaceous wild plants is rarely studied in detail, but collecting teams develop experience in estimating the date to return for collecting of ripe seed. An exhaustive study reported by Ison and Wagenius (2014) showed that in a restored population of *Echinacaea angustifolia*, a common insect-pollinated prairie perennial within the Asteraceae tribe Heliantheae, the onset, peak and end of flowering varied markedly between individuals both within and between seasons. For example, the authors showed that although position within the population affected seed production, the timing of flowering had a larger and more consistent effect on output. Flowering was also variable between years, for example, the total number of individuals plants recorded in flower varied successively from 204, 572 to 529 over the three years, and the median flowering date differed by around 10 days between years.

As a consequence of this natural variation, collectors should prepare to collect and handle a range of seed ripeness encountered, and they should use cut tests of a random sample of fruits from the population (which primarily assess physical seed quality, detailed in Section 6), to also confirm the seed development, ripening and dispersal stage.

5.2 Cues for natural seed dispersal

Collectors need to use indicators for fruit ripeness and seed dispersal in order to cope with a large diversity of species in the field. According to Spjut (1994) there are actually as many as 95 botanical types of fruits, but for the collection of CWR the three main categories of fruit (fleshy; dry dehiscent; or dry indehiscent) can guide essential seed collecting and handling in the field.

- **Fleshy fruits** include the berry (*Vitis, Solanum*), drupe (*Prunus*) and hesperidium (citrus, Rutaceae). The relatives of potato, tomato and eggplant, *Solanum* spp., produce berries that are attractive to birds and other fauna that may also play a role in dispersal. Fruit traits that consistently indicate ripening in crops and in wild species include fruit colour and texture. However, fruit size and seed coat colour are less reliable as predictors in wild species (Hay and Smith 2003). Entire, whole, fleshy fruits may need to be collected a few days ahead of natural dispersal in order to harvest before seed is predated or damaged, but they will require careful handling and monitoring as they ripen (see Section 8).
- **Simple dehiscent fruits** include follicles (*Asclepias*), many legumes (Fabaceae), capsules (some Iridaceae) and siliqua (Brassicaceae). As

these fruits progressively ripen, they dry and open, sometimes explosively along weak margins or sutures, often with an additional cue that seeds are audibly free and rattling in their fruits. If the team have easy access to the target population, it is possible to tie small mesh or gauze bags around a sample of the ripening dehiscent fruit, on completion of the population and physical seed quality assessments, ready for return as seed is dispersing. Such bags need to be monitored frequently and seed removed promptly from dehisced fruits before it is subject to predation or ageing.

- **Dry indehiscent fruits** include achenes (*Carex, Ranunculus*), cypsela (*Helianthus*) and caryopsis (Poaceae including CWR of *Oryza, Hordeum, Aegilops, Avena, Secale* and *Triticum*). Note that *Arachis* and some *Medicago* (Fabaceae) have indehiscent pods. At dispersal, wild grasses tend to shatter (i.e. disperse seed progressively by abscission and disarticulation of individual seed dispersal units) from the spike, and gentle pressure from a gloved hand will release the proportion of seed that is at natural dispersal. Ideally, collectors would return on several occasions through the season, but if that is not possible, an approach to consider is to collect a sample (maximum of 10%) of entire grass spikes, complete with a short length of stem, checking that most seeds have reached abscission point and are close to dispersal. If the spikes are protected from fast drying for a few days in a ventilated and shaded tray or box, it is likely that the seed will progressively reach dispersal point and acquire the desiccation tolerance and longevity required for transport and preservation.

In addition to these visual cues for dispersal, the moisture levels of seed within dry fruit types can be assessed by use of a portable hygrometer in the field. By measuring the relative humidity of air that is in equilibrium with seed which almost completely fills a hermetic container as shown in Fig. 5, the equilibrium Relative Humidity (eRH) can be quickly determined (Gold and Manger 2014). Before abscission, seeds are moist, that is, close to 100% eRH. A lower measured seed eRH, that is close to ambient RH, indicates that seeds are close to dispersal point (Hay and Smith 2003), but there are relatively few studies of the moisture relations of wild species' seed, and its significance for seed quality.

Areas for future research include the better prediction and remote monitoring of fruit-ripening indicators. For example, Elliot (2019) reported experiments to identify seed-ripening cues in tree crowns via unmanned aerial vehicles (drones). There is a need to better document indicators of ripening in wild species and much scope to explore the use of cameras, loggers, hygrometers and either drones or remote sensing to monitor seed ripening at the population level, especially within forest habitats.

Figure 5 Use of a portable hygrometer to assess the moisture status of a fully ripe seed collection at harvest. Note that at 62% equilibrium relative humidity, this seed collection can be moved to drying conditions to further reduce viability loss. Image: RBG Kew.

6 Assessing seed quality and quantity

Wild plants do not consistently produce healthy, filled seed available for sampling, indeed the Poaceae and Asteraceae are notorious for the high frequency (sometimes over 50%) of seed that at dispersal is found to be either predated or without embryo (the latter is categorised as 'empty'). Evidence at family level from European seed collections with empty or infested seeds are in the Appendix of the ENSCONET (2009) Manual. In the seed lab a quantitative assessment of seed quality can be made precisely using X-ray (Fig. 6), estimating the % of seeds not expected to be viable, and to use that % to calculate usable seed numbers in a collection.

In the field, collectors need to carry out 'cut tests' of a random sample of around 50 seeds from the target population using secateurs, scissors or other cutting tools, assessed with a minimum of ×10 magnification hand lens, before proceeding with a collection. An example of the cut test is shown in Fig. 7, and this simple step is central to the guidelines in Way and Gold (2014a). In addition to recording the proportion of empty and infested seeds, it also helps to confirm indicators for ripening and dispersal such as abscission points, seed tissue and seed coat colour prior to collecting as described in Section 5. Cut test data from the field is not routinely transferred to gene banks for databasing at present, but there are opportunities to inform

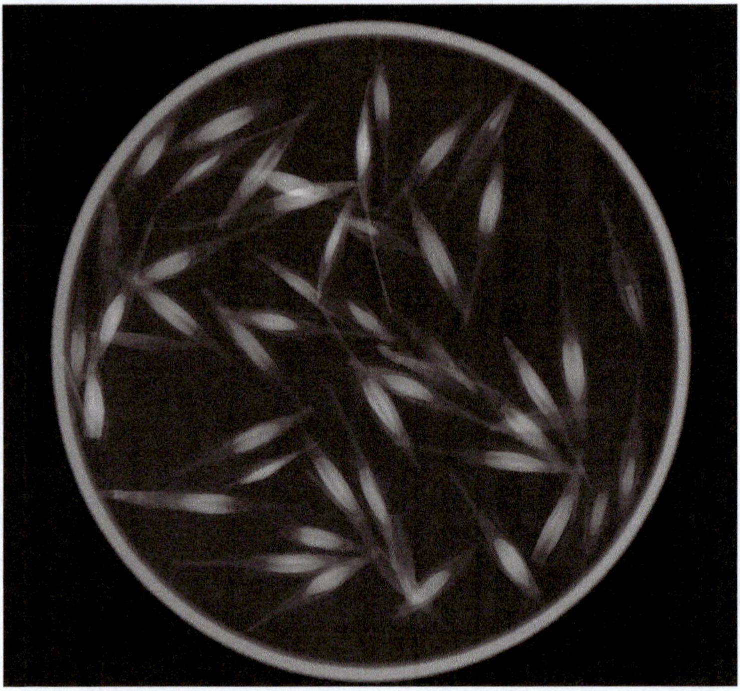

Figure 6 X ray image assessment of 50 seeds from Poaceae, *Hordeum spontaneum*, indicating 41 fully filled, 6 empty and 3 part filled seeds. Image: RBG Kew. Seed provided by Lebanese Agricultural Research Institute.

Figure 7 Cut test through a single cone of Engelmann Spruce *Picea engelmannii*, Oregon 2016, showing at least four filled, hard seeds ready for collecting. Image: RBG Kew.

future seed biology, ecology and plant health studies by publishing these observations.

Cut test data is used to calculate the potentially viable, filled seed available for harvesting from the population, and to ensure that a normal harvesting limit of 20% of seed available on the day of collection is not exceeded (Breman and Way 2018). Although seed production and demographic data is not documented for the majority of wild plants, a review by Menges et al. (2004) is the basis for the comparable CPC Best Practices guideline to generally limit collection to no more than 10% of the whole season's seed crop. In the event that seed availability in a population falls far short of objectives, collectors may choose to make the first of several partial collections from the population, or can continue to explore for more productive populations within the region.

How many seed should be sampled? For two decades the MSB Partnership has recommended that a collection of 10 000 viable seeds is ideal for providing curators, researchers and sister gene banks (i.e. those receiving a duplicate collection for safe storage) with enough seed to meet essential needs for the 50–100 year target lifespan of a typical seed collection (Way and Gold 2014a). If sampled randomly from 50 well-spaced individuals across the full extent of the population, it would also be likely to far exceed the Brown and Marshall (1995) genetic sampling objectives. It is important that any specific demographic or reproductive biology information on the species be used to modify the general approach and to reset the safe harvesting limit if appropriate. With these caveats, this general guideline has been adopted (10 000+ seed objective set by Bureau of Land Management 2015) or adapted for other programmes (e.g. the 5000 seed objective set by ENSCONET 2009).

Allowing for empty and infested seeds in a collection, a well-resourced collection programme targeting widespread species for restoration can routinely achieve 10 000 viable seeds (Carvey and Way 2016), but for species of restricted distribution or under threat these ideal quantities will not often be met. For example, the global median collection size within the CWR Project is approximately 1000 seed (RBG Kew unpublished) that could be due to lower seed quality and smaller populations of many CWR taxa.

7 Sampling genetic diversity effectively from populations

7.1 Development of the guidance for population genetic sampling

The sampling of plant genetic resources in recent decades has been guided by research aiming to collect samples of high genetic diversity, representative of the target population, that are intended primarily for research, breeding and preservation in gene banks (Brown and Marshall 1995), as updated by Parra-Quijano et al. (2011) (Crossa and Vencovsky 2011). To achieve capture of 95%

of the non-rare alleles (i.e. those at a frequency of more than 5%) in a typical population, key tenets have been the random sampling across a population from at least 25 individuals (in a predominantly outcrossing species) and from all populations or sub-populations in which there is any evidence of genetic structure. A particular interest has been to capture any possible local adaptations (i.e. locally abundant at population level) that are of potential value for breeders.

Progressively the guidelines from Guerrant et al. (2004) and the CPC Rare Plant Academy (https://academy.saveplants.org) have continued to refine the genetic sampling recommendations for small or rare plant populations. For example, to enable evaluation, propagation and re-introduction to the wild as part of conservation strategies, it is recommended to keep seed from each maternal line separate when sampling from threatened plants. This approach has also been adopted for programmes confronting specific pest and disease threats, for example, the UK National Tree Seed project (Kallow 2014), and if accompanied by DNA tissue samples collected from the population, provides high resolution for studies of genetic diversity, geneflow and presence of traits such as disease resistance in the population.

In future, research on optimising population sampling must continue rapidly, as it has major resource implications for large-scale ex situ collection programmes supporting food security, species conservation and ecological restoration worldwide. Certainly any knowledge of the reproductive biology and connectivity of populations are central to determining sampling guidance for a species, but as shown by Hoban (2019) it is also important to first agree at programme level the relative value of capturing the rarest of alleles, that is, those present at less than 5% frequency in a population, and how many copies of these alleles are desired in the final seed collection. For collecting of CWR that are thought to have traits of interest to crop breeders, it is especially important to capture locally common alleles by ensuring that populations are sampled across all of the eco-geographic diversity known for the species.

7.2 Determining a plan for genetic sampling within a population

Hoban and Strand (2015) reviewed 15 sets of published seed sampling guidelines, mostly drawing on the underlying work of Brown and Marshall (1995), and drew attention to two factors in particular which were not given weight in the 1995 guidelines: first, the significance of spatial clustering of related individuals in a population, as a result of dispersal and recruitment patterns, and secondly the relatedness of seeds from a single maternal plant as a result of a limited number of pollen donors, especially where pollinated by insects.

In practice, how should collectors plan to sample, given the often spatially heterogeneous (patchy) nature of individuals within a population? Taking into account the best collecting technique (Section 8) and observations on ripeness (Section 5) and physical seed quality (Section 6), collectors need to use a random sampling approach from the full accessible extent of the population. Hoban and Strand (2015) illustrate the likely superior genetic capture modelled from a truly random or grid pattern of sampling, compared to collecting either by transect or from one location on the perimeter of the population (the latter is termed 'points' sampling, evaluated as the worst of all the sampling options). The separation distance between sampled individuals is also critical to minimise relatedness of seed, and typical recommendations would be a minimum of 50 m between mature trees (Kallow 2014), or 10 m between herbaceous plants as shown in Fig. 8.

The research by Ison and Wagenius (2014) has revealed that in the case of *Echinacea angustifolia* a combination of spatial and flowering variation can significantly affect the distance that pollinators travel, and therefore the seed produced from longer-distance pollination events at the beginning and end of flowering could be especially valuable for sampling. For insect-pollinated species in particular, collectors do not need to therefore concentrate effort entirely at the peak of seed production, but may be able to capture higher genetic diversity by sampling randomly on two or more dates during the season and pooling ('bulking') the samples.

Figure 8 A team collecting seed from a large population of *Hordeum chilense* near a windfarm at Talinay, Chile. Collectors are maintaining an agreed separation distance between harvested individuals, using multiple transects as the sampling approach. Image: Crop Trust – LM Salazar.

8 Techniques for collecting and post-harvest handling

A high-quality collection will comprise sufficient ripe seed of high viability, handled with care in the field and accompanied by a set of herbarium vouchers, associated botanical collections and comprehensive field data. The possible transfer of pests or pathogens between sites and collections will be minimised by good plant health practices in the field.

8.1 Seed collecting

Seed collecting from CWR in the field is usually by hand, selecting the most effective of the techniques summarised in Way and Gold (2014b) that will successfully harvest agreed ripeness and quality of seed from randomly selected individuals as described above. In brief, techniques such as plucking individual indehiscent or fleshy fruit, or gently shaking stems with dehiscent fruit or wind dispersed seed, are more likely to harvest a single cohort of seed ripeness, and accordingly post-harvest handling will be straightforward. In contrast, the pruning of clusters of mixed ripeness fruits or entire spikes or racemes may introduce a wider range of seed ripeness into the collection. Ventilated cloth bags are the material of choice for collectors but paper envelopes are preferable for awned seeds or hooked fruits, and plastic is required for handling fleshy fruits. All containers are labelled inside and out with the unique field collection number that is also used for associated botanical collections and data. Seed samples are bulked together from the population unless individual maternal lines have been specifically requested for multiplication, genetic analysis or trait screening.

Compared to some botanical inventory in which dozens of specimens can be quickly collected at a single location, collection of seed of high genetic diversity and quality usually requires more time to be carried out correctly. For example, summary data from the CWR Project (Global Crop Diversity Trust 2019) indicates that a global average of 1.5 seed collections was made per field day (n = 4644 samples) across the 371 taxa sampled from four continents. Allowing for travel time, taxonomic identification and population assessment, this is consistent with a long-term average that from one to two hours is often needed for the practical collection of seed of wild species with accompanying herbarium vouchers and field data (e.g. Way 2003).

Note that although mechanised seed harvesting from wild populations is often attempted in large-scale restoration programmes for immediate sowing of seed, it is not recommended for CWR because of the likelihood of harvesting some seed before abscission and also for the additional physical damage likely to the seed coat (described by Dickie and Stuppy 2003 after seed scarification) which could affect viability and longevity of CWR collections. Such techniques also increase the risk of exceeding the conservation limits to seed collecting from rare or threatened taxa, and could increase collateral damage to other species in the community.

8.2 Post-harvest handling of seed collections

The post-harvest seed handling recommendations of Gold (2014) rely on management of ripeness and speed of drying of a collection, while always avoiding hot, humid and anoxic conditions which would reduce seed viability and longevity. Measurements of seed equilibrium Relative Humidity (eRH) using a hygrometer as described in Section 5 can continue to guide post-harvest handling with reference to safe moisture thresholds, that is, above 85% eRH, seeds are thought to be capable of repair, and below about 60% eRH, seed ageing is significantly reduced (Hay and Smith 2003). For example, Fig. 5 shows a recently sampled seed collection; if the surrounding air is at 40% RH, the reading of 62% eRH indicates that the seed is in the ripening phase just prior to natural dispersal and should be allowed to continue gently drying in a ventilated, shaded tray or box with frequent monitoring.

If the ambient conditions are humid (i.e. the daytime relative humidity is over 50%) then fully ripe, dry collections at risk of re-absorbing moisture can be placed in a hermetic barrel with a little dried silica gel to absorb excess moisture, as shown in Fig. 9. As ambient humidity rises at night, collections should preferably be placed on open newspaper in an air-conditioned room for checking and for ventilation. If this is not possible during remote expeditions, then seed collections can be placed in polythene bags overnight, but must be ventilated as soon as possible in the morning. Under no circumstances should freshly harvested seed, especially if over 50% eRH moisture status, be left in an unventilated or stationary vehicle in the sun – the high temperatures and humidity could be lethal to any fresh germplasm collection.

Mixed ripeness collections present a challenge. If the ripe and unripe portions of the collection cannot be separated for individual handling, then the mixed ripeness collection should be handled as though unripe, that is, not yet desiccation tolerant. Examples are given in Gold (2014) of fruit colour indicators used in practice to make decisions about handling of this material. **Recalcitrant seeds** will not acquire desiccation tolerance during handling in the field and will need to be held in moist, ventilated conditions during fieldwork and transferred promptly to the nursery or gene bank for the appropriate next ex situ preservation step.

8.3 Herbarium vouchers and associated botanical collections

The set of pressed plant (herbarium) specimens taken at the time of seed collection will be used to provide an updatable botanical name for the seed

Figure 9 A hermetically sealed barrel with dried silica gel in the base can be used in the laboratory to actively dry incoming seed collections before processing or storage. In the field, partially dried collections can be held and transported in the barrel to ensure that they do not absorb moisture and enter the 60–85% eRH moisture zone which will be especially damaging. For recently harvested seed collections, significant drying can be achieved by placing an equal mass of silica gel to seed in the barrel. Image: RBG Kew.

collection, and are the first point of reference for any concerns about the identity of the collection. It is important that these 'vouchers' are truly representative of the population and individuals that were harvested. For perennials, it may be possible to take a set of four specimens from an individual that has also been sampled for seed. Annuals that have similar appearance and phenology to the nearby sampled individuals may be collected as vouchers, in all cases sharing the unique field collection number which may follow an individual botanist's number series, or a team's series created for a series of expeditions under a single project or organisation.

The minimum requirements for a quality herbarium specimen, as well as family-specific guidance on material to harvest and prepare have been well documented, for example, in the Herbarium Handbook (Bridson and Forman 2000, 3rd edition) and in the well-illustrated Southern African herbarium user manual (Victor et al. 2004). Rapid drying of herbarium specimens is needed within 48 h of collection to preserve flower and leaf characters and ideally this can be done in ovens or driers at a collaborating facility while the main collecting is underway. If not, portable driers can be used with care on the expedition, or the bundles of specimens can be preserved with ethanol in strong plastic bags to prevent deterioration in transit. The presence of ripe fruit on specimens is usually of interest to botanists, but there is a high chance that the physical pressing and heat of drying will cause any dry fruits to be disarticulated and separated, if not lost, from the herbarium material. Fruits and/or seeds should be placed in a separate envelope, labelled and inserted within the main specimens.

The most common additional material to also sample on a CWR collecting expedition is tissue for DNA analysis. The protocol provided in the CPC Best Plant Conservation Practices (CPC undated) provides advice on material selection and rapid drying with silica gel. Once in silica, the individual labelled packets can be stacked in a rigid box and are quite robust for transport to a herbarium or gene bank. Other germplasm such as tubers, or cuttings, are very vulnerable to transport and storage conditions and should be handled as carefully as recalcitrant seeds, maybe using an insulated coolbox for travel in the hottest part of the day and keeping in shade at all times.

8.4 Documenting the CWR collection in the field

The essential passport data common to germplasm collection programmes is largely defined by Darwin Core (Darwin Core Task Group 2009) with extensions of particular interest and need for programmes. Most collecting organisations have explored the use of tablets and field computers for field data capture on germplasm expeditions. Although mobile applications with positioning and data capture are widespread, paper data forms are always carried as a backup in the event that devices fail or power is lost during the collecting trip. The hoped-for benefits of access in the field to full botanical catalogues and help-lists to speed up and streamline data entry has not yet been demonstrated with resilience in a field setting. However, other sectors are consolidating GIS, GPS and data-form entry for their applications, so collectors need to continue developing workflows and testing systems for field botany and germplasm collecting. Use of pre-printed or locally generated barcodes or QR codes which can speed collection data transfer and specimen sorting and increase the accuracy of record keeping would also be valuable.

There have been advances in integration of GIS base maps, GPS positioning and geolocated image capture successfully in the field, for example, the value of GPS-enabled data recording during seed collecting was reported by Mayzlish Gati et al. (2018). Nowak et al. (2020) review the leading mobile applications available for field mapping based on various workflows, including the ability of the Open Data Kit software to collect data offline from remote sites, followed by aggregation of data centrally once connections are restored. Also within their scope is the interaction with specialist identifiers on citizen science platforms such as iNaturalist and PlantNet (https://plantnet.org/) which could potentially expand the community of people that may be willing to send in records of distribution, flowering or fruiting of target species to support a collecting programme.

Field data that is comprehensive and accurate will be of immediate value to the gene banks receiving the associated material, but as soon as possible data will be accessible to a wide range of users through online portals such as the GBIF (https://www.gbif.org/), Genesys-PGR (https://www.genesys-pgr.org/) and the MSB Partnership Data Warehouse (http://brahmsonline.kew.org/msbp/SeedData/DW). Samples from collections that meet minimum seed and data quality requirements will be promoted to user groups and partners subject to access agreements. For example, the CWR Project has sent samples of 3218 collections to eight international gene banks. They will be used to identify traits of value in crop breeding such as tolerance of heat, drought, salinity and waterlogging, resistance to pests and diseases, resistance to root rot and improved yield. For this reason field data on geology, soil type, exposure and habitat will be of value for selection of potential breeding material.

8.5 Plant health practices

Many CWR genera are listed in national or regional plant health regulations because of the possible risk to agriculture, horticulture and forestry from inadvertent introduction of pests and diseases through plant material movement. This places a duty on programmes to be well informed in advance of fieldwork, using sources cited in Macfarlane et al. (2011), and to maintain cleanliness of clothing and equipment between sites and between collections. Kallow (2014) includes disinfection instructions that the collecting team follow to clean equipment between sites, but the knowledge about likely presence of a pest or pathogen on a particular site is often poor; therefore programme risk assessments and mitigating actions are typically very general. Rapid pest and pathogen prediction or assessment tools with capability on and offline would be valuable to manage these risks in the field.

During post-harvest handling and transport of collections to the gene bank, it is equally important for collections to be labelled and not contaminated by nearby collections. For example, bags should not be re-used for other taxa, and newspaper should be discarded after collections have been ventilated overnight. Controlled (quarantine) collections may need to be triple-bagged within a larger shipment in order to control spread of pests or pathogens, and should be handled to quarantine standards until inspected by a plant health specialist.

9 Future research directions

For orthodox-seeded species, the underpinning sciences of seed biology, taxonomy and ecology continue to inform the essential requirements for the collecting, handling and conservation of germplasm from CWR species. These guidelines are widely available in literature such as the FAO gene bank standards (FAO 2014) the MSBP standards (Breman and Way 2018), the ENSCONET (2009) manual and the CPC Best Plant Conservation Practices. Although effective in most cases to deliver viable seed to gene banks the past guidance has not always been optimal in two respects (a) for sampling of genetic diversity and (b) for achieving target longevity in storage of all species.

9.1 How can knowledge of genetic sampling be strengthened?

Current ex situ conservation targeting has limitations. For example, Griffiths et al. (2014) showed that the large ex situ MSBP collection resulting from priority given to threatened and useful species does not represent optimal sampling of phylogenetic diversity in the Leguminosae. Within the botanic garden community, Hoban et al. (2020) showed that for 11 woody taxa across five genera, the ex situ collections were not optimal for preserving genetic diversity of the taxa, and the minimum collection requirements were not consistent within genera, for example, *Zamia decumbens* would require almost twice the number of conserved individuals to maintain diversity compared to its congener *Z. lucayana*. Simulations by Hoban (2019) indicate that more extensive sampling of as many as 1000 individuals across the natural range may be advisable, with an increasing emphasis on locating additional populations and individuals for sampling to maximise the genetic diversity that can be captured. It is time to expand research across ex situ networks to assess the diversity of germplasm held in terms of ecogeography, genotypes and traits. Resulting knowledge will enable gap analysis for future collecting, and lead to a greater coordination of in situ and ex situ conservation efforts, and between collection holders and germplasm users. Within ex situ collection holders the concept of the multi-institutional meta-collection (Griffith et al. 2019) is encouraging curators to think about appropriate, but not excessive duplication of genotypes between organisations curating collections.

Given the threats to CWR wild populations, for example, that reported by Jarvis et al. (2008), and specifically identified for inventories of Mexican CWR (Contreras-Toledo et al. 2018) and North American CWR (Frances et al. 2018), we cannot wait for completion of species-specific genetic diversity studies before undertaking conservation actions, rather we have to use the best available evidence to guide sampling before wild populations are lost. Where the need is urgent, there is scope to carry out a rapid 'genetic risk assessment' for seed collection and subsequent use, such as that reported by Gargiulo (2019) for 44 native woody plants of the UK in preparation for seed sampling by the UK National Tree Seed Collecting Project. This could be one way that sampling opportunities can be based on minimum evidence without delay or excessive research cost.

9.2 Which taxa are sensitive to standard handling methods and what more research is needed?

The most recent review of seed desiccation tolerance by Wyse and Dickie (2016) showed that there is a high agreement between desiccation sensitivity status of species within the same genus, and so programmes can predict behaviour in most genera using online sources such as the Seed Information Database (https://data.kew.org/sid/). Seed longevity knowledge is also expanding, as Colville and Pritchard (2019) have reviewed several decades of seed performance in gene banks. They note the underlying importance of phylogeny (e.g. Poaceae grasses and Asteraceae are generally shorter lived than Fabaceae and Malvaceae) to determine lifespan in storage, and compare the affect of different gene bank conditions. Building on this and on research reviewed by Hay and Probert (2013), further work is needed to identify taxa for which collecting and handling protocols are not currently optimal, for example, to determine for wild grasses whether the occasional ex situ ripening of whole spikes, as described in Section 5, affects the longevity of seed in storage.

Gene banks responded quickly to evidence (Ballesteros and Pence 2017) that harvested seed of very small seeded orthodox taxa such as Salicaceae were not maintaining viability under conventional conditions and they introduced rapid transport to gene banks for checking and cryopreservation of part of the sample. Further research will alert collectors to other plant groups that need special handling. Research must also continue into the exceptional species (Pence 2013) for which embryo rescue, cryopreservation, in vitro cultivation and micropropagation may be recommended as the only way to safely preserve germplasm that may be sensitive to dry or cold conditions.

9.3 Deploying research knowledge in the field

Ultimately it is the skills and knowledge of the field team that determine the success of a collecting expedition. The core competencies of a collector continue to rely on botanical identification skills, thorough knowledge of seed development and ripening, and attention to detail in preparation for fieldwork, in data recording and the correct handling of seed and herbarium vouchers. Often this is combined with a talent for exploration and the curiosity to research the ecology and distribution of plants and associated species in the target region, together with an ability to engage local botanists and communities to play an active role in conservation of these genetic resources. Collectors continue to experiment with tablets and other field devices to help with field retrieval of the wealth of digital data now available, and to improve capture of their own collecting data, but new workflows resilient to field conditions need to be developed, tested and published.

In conclusion, our objective is not purely to deliver viable, adequately representative germplasm collections of CWR to gene banks and programmes, but to increasingly demonstrate that the collections are directly meeting recognised sampling gaps and research needs. Through application of new knowledge and technology in the field, we need to collect and handle seed of all target species in ways to maximise the longevity in storage and increase the usefulness of samples for research and conservation.

10 Where to look for further information

10.1 Guidelines

- Collecting Plant Genetic Diversity Technical guidelines, 2011 update, edited by L. Guarino, V. Ramanatha Rao and E. Goldberg, Bioversity International https://cropgenebank.sgrp.cgiar.org/index.php/procedures-mainmenu-242/collecting.
- MSB Partnership standards online http://brahmsonline.kew.org/Content/Projects/msbp/resources/Training/MSBP-Seed-Conservation-Standards.pdf.
- MSBP technical information sheets http://brahmsonline.kew.org/msbp/Training/Resources.
- Center for Plant Conservation Best Plant Conservation Practices to Support Species Survival in the Wild (undated) https://academy.saveplants.org/best-practices/cpc-best-practice-guidelines-table-contents.
- ENSCONET (2009) Seed Collecting Manual for Wild Species ISBN: 978-84-692-3926-1 Citation: http://brahmsonline.kew.org/Content/Projects/msbp/resources/Training/ENSCONET_Collecting_protocol_English.pdf.

10.2 Other useful resources

- Research on Exceptional species: Pence, V.C. (2013) "In Vitro Methods and the Challenge of Exceptional Species for Target 8 of the Global Strategy for Plant Conservation," *Annals of the Missouri Botanical Garden* 99(2):

214–220, (13 December 2013). https://doi.org/10.3417/2011112 www.jstor.org/stable/42703725.

- Review of seed conservation research. Hay F.R.; Probert R.J. (2013) Advances in seed conservation of wild plant species: a review of recent research. *Conservation Physiology* 1. doi:10.1093/conphys/cot030 Open Access.
- Crop Wild Relative global inventory https://www.cwrdiversity.org/checklist/
- Seed Information Database (seed storage behaviour and germination) https://data.kew.org/sid/.
- Genesys-PGR the online platform for access to information on PGR collections https://www.genesys-pgr.org/.

11 Acknowledgements

This chapter draws extensively on research and guidelines developed by the Millennium Seed Bank Partnership, ENSCONET, and the Center for Plant Conservation. Images and data have been kindly provided by RBG Kew, the Global Crop Diversity Trust and CWR Project Partners. Particular thanks are due to Chris Cockel (RBG Kew), Luis M. Salazar (Global Crop Diversity Trust), Pedro León Lobos and Sergio Ibañez Browne (INIA Chile), Edwin J. Naranjo (INIAP Ecuador) and to an anonymous reviewer for comments on an earlier version of the manuscript.

12 References

Ballesteros, D. and Pence, V. C. (2017). Survival and death of seeds during liquid nitrogen storage: a case study on seeds with short lifespans. *Cryo Letters* 38(4): 278–289.

Breman, E. and Way, M. (2018). Safe for the future: seed conservation standards developed for the Millennium Seed Bank Partnership. In: *Proceedings of the EuroGard VII Congress*. 6–10 July 2015. Paris, France. pp. 267–274. Available at: http://brahmsonline.kew.org/Content/Projects/msbp/resources/Training/MSBP-Seed-Conservation-Standards.pdf.

Bridson, D. and Forman, L. (2000). *The Herbarium Handbook* (3rd edn.). RBG Kew, UK.

Brown, A. H. D. and Marshall, D. R. (1995). A basic sampling strategy: theory and practice. In: *Collecting Plant Genetic Diversity: Technical Guidelines. 2011 Update* Guarino, L., Ramanatha Rao, V. and Reid, R. (Eds). CABI Publishing, Wallingford, UK, pp. 75–91.

Bureau of Land Management. (2015). *Seeds of success collecting protocol*. Available at: https://www.blm.gov/sites/blm.gov/files/program_nativeplants_collection_quick%20links_technical%20protocol.pdf.

Carvey, N. and Way, M. (2016). *Exploring millennium seed bank collections from North America- a case study*. Millennium Seed Bank Data Warehouse, RBG Kew. Available at: http://brahmsonline.kew.org/Content/Projects/msbp/resources/Training/MSBP_Data_Warehouse_USA_study.pdf.

Castañeda-Álvarez, N. P., Khoury, C. K., Achicanoy, H. A., Bernau, V., Dempewolf, H., Eastwood, R. J., Guarino, L., Harker, R. H., Jarvis, A., Maxted, N., Müller, J. V., RamirezVillegas, J., Sosa, C. C., Struik, P. C., Vincent, H. and Toll, J. (2016). Global conservation priorities for crop wild relatives. *Nature Plants* 2: 16022. https://doi.org/10.1038/nplants.2016.22.

Center for Plant Conservation. (undated). *CPC Best Plant Conservation Practices to Support Species Survival in the Wild.* Available at: https://academy.saveplants.org/best-practices/cpc-best-practice-guidelines-table-contents.

Colville, L. and Pritchard, H. W. (2019). Seed life span and food security. *New Phytologist* 224(2): 557–562. https://nph.onlinelibrary.wiley.com/doi/epdf/10.1111/nph.16006.

Contreras-Toledo, A. R., Cortés-Cruz, M. A., Costich, D., Rico-Arce, M., Brehm, J. M. and Maxted, N. (2018). A crop wild relative inventory for Mexico. *Crop Science* 58(3): 1292–1305. https://doi.org/10.2135/cropsci2017.07.0452.

Convention on Biological Diversity. (2010). *Global strategy for plant conservation*. The Secretariat of the Convention on Biological Diversity, Montreal. Available at: https://www.cbd.int/undb/media/factsheets/undb-factsheet-gspc-en.pdf.

Crossa, J. and Vencovsky, R. (2011) Chapter 5. Basic sampling strategies: theory and practice. In: *Collecting Plant Genetic Diversity: Technical Guidelines. 2011 Update* Guarino, L., Ramanatha Rao, V. and Goldberg, E. (Eds). Bioversity International. Available at: https://cropgenebank.sgrp.cgiar.org/index.php/procedures-mainmenu-242/collecting.

Darwin Core Task Group. (2009). *Simple Darwin Core*. Biodiversity information standards (TDWG). Available at: http://rs.tdwg.org/dwc/terms/simple/.

Dickie, J. B. and Stuppy, W. H. (2003). Seed & fruit structure. In: *Seed Conservation: Turning Science into Practice* Smith, R. D., Dickie, J. D., Linington, S. H., Pritchard, H. W. and Probert, R. J. (Eds). pp. 253–279. Royal Botanic Gardens, Kew, UK.

Elliot, S. (2019). *Droning for seeds, Samara 33*, p. 4. RBG Kew. Available at: https://www.kew.org/sites/default/files/Samara%2033%20web.pdf.

Embrapa and RBG Kew. (2016). *Brazil Seed Collecting Guide*. Allen, R. and Heiden, G. (Compilers). 46 pp. https://doi.org/10.13140/rg.2.2.34706.25289.

ENSCONET. (2009). *Seed collecting manual for wild species*. Available at: http://brahmsonline.kew.org/Content/Projects/msbp/resources/Training/ENSCONET_Collecting_protocol_English.pdf. ISBN: 978-84-692-3926-1.

Food & Agriculture Organisation. (2001). The international treaty on plant genetic resources for food and agriculture, adopted by the thirty-first session of the conference of the food and agriculture organization of the United Nations, 3 November 2001.

Food & Agriculture Organization. (2010). *The second report on the state of the world's plant genetic resources for food and agriculture*. Rome. Available at: http://www.fao.org/3/i1500e/i1500e.pdf.

Food & Agriculture Organisation. (2014). *GeneBank standards for plant genetic resources for food and agriculture* (rev. edn.). Rome. Available at: http://www.fao.org/3/a-i3704e.pdf.

Frances, A. L., Smith, A. B. and Khoury, C. K. (2018). Conservation status and threat assessments for North American crop wild relatives. In: *North American Crop Wild Relatives (Vol. 1)* Greene, S. L., Williams, K. A., Khoury, C. K., Kantar, M. B. and Marek, L. F. (Eds). Springer, Cham. Available at: https://link.springer.com/chapter/10.1007%2F978-3-319-95101-0_7. https://doi.org/10.1007/978-3-319-95101-0_7.

Gargiulo, R. (2019). *Pyrus Cordata Desv. Genetic Risk Assessment*. Research report. Kew Research Repository. Available at: https://kew.iro.bl.uk/work/278bd662-d8af-4017-992d-5a7d26f84ec4. doi: 10.34885/17.

Garruccio, M. (2011), Chapter 13. Published information resources for plant germplasm collectors. In: *Collecting Plant Genetic Diversity: Technical Guidelines. 2011 Update* Guarino, L., Ramanatha Rao, V. and Goldberg, E. (Eds). Bioversity International. Available at: https://cropgenebank.sgrp.cgiar.org/index.php/procedures-mainmenu-242/collecting.

Genesys (undated). *Global Portal on Plant Genetic Resources*. Available at: https://www.genesys-pgr.org/, Accessed 1 May 2020.

Global Crop Diversity Trust. (2019). *A Global Rescue: Safeguarding the Worlds Crop Wild Relatives*. Available at: https://www.cwrdiversity.org/.

Gold, K. (2014). *Post harvest handling of seed collections*. RBG Kew Technical Information Sheet 4. Available at: http://brahmsonline.kew.org/Content/Projects/msbp/resources/Training/04-Post-harvest-handling.pdf.

Gold, K. and Manger, K. R. (2014). Measuring seed moisture status using a hygrometer. RBG Kew Technical Information Sheet 5,. Available at: http://brahmsonline.kew.org/Content/Projects/msbp/resources/Training/05-eRH-moisture-measurement.pdf.

Griffith, M. P., Beckman, E., Callicrate, T., Clark, J., Clase, T., Deans, S., Dosmann, M., Fant, J., Gratacos, X., Havens, K., Hoban, S., Lobdell, M., Jiménez-Rodriguez, F., Kramer, A., Lacy, R., Magellan, T., Maschinski, J., Meerow, A. W., Meyer, A., Sanchez, V., Spence, E., Toribio, P., Walsh, S., Westwood, M. and Wood, J. (2019). *Toward the Metacollection: Safeguarding Plant Diversity and Coordinating Conservation Collections*. Botanic Gardens Conservation International-US, San Marino, CA. Available at: https://www.bgci.org/resources/bgci-tools-and-resources/toward-the-metacollection-coordinating-conservation-collections-to-safeguard-plant-diversity/.

Griffiths, K. E., Balding, S. T., Dickie, J. D., Lewis, G. P., Pearce, T. R. and Grenyer, R. (2014). Maximising the phylogenetic diversity of seed banks. *Conservation Biology* 29: 370–381. https://doi.org/10.1111/cobi.12390.

Guarino, L., Ramanatha Rao, V. and Goldberg, E. (Eds.) (2011). *Collecting Plant Genetic Diversity: Technical Guidelines. 2011 Update*. Bioversity International, Rome, Italy. ISBN 978-92-9043-922-6. Available at: https://cropgenebank.sgrp.cgiar.org/index.php/procedures-mainmenu-242/collecting.

Guerrant, E. O., Fiedler, P. L., Havens, K. and Maunder, M. (2004). Revised genetic sampling guidelines for conservation collections of rare and endangered plants. In: *Ex Situ Plant Conservation: Supporting Species Survival in the Wild* Guerrant, E. O., Havens, K. and Maunder, M. (Eds). Island Press, London.

Hay, F. R. and Probert, R. J. (2013). Advances in seed conservation of wild plant species: a review of recent research. *Conservation Physiology* 1(1). doi: 10.1093/conphys/cot030 Open Access.

Hay, F. R. and Smith, R. D. (2003). Seed maturity: when to collect seeds from wild plants. In: *Seed Conservation: Turning Science into Practice* Smith, R. D., Dickie, J. D., Linington, S. H., Pritchard, H. W. and Probert, R. J. (Eds). Royal Botanic Gardens, Kew, UK, pp. 97–133.

Hoban, S. (2019). New guidance for ex situ gene conservation: sampling realistic population systems and accounting for collection attrition. *Biological Conservation* 235: 199–208. https://doi.org/10.1016/j.biocon.2019.04.013.

Hoban, S., Callicrate,T., Clark, J., Deans, S., Dosmann, M., Fant, J., Gailing, O., Havens, K., Hipp, A. L., Kadav, P., Kramer, A. T., Lobdell, M., Magellan, T., Meerow, A. W., Meyer, A., Pooler, M., Sanchez, V., Spence, E., Thompson, P., Toppila, R., Walsh, S., Westwood, M., Wood, J. and Griffith, M. P. (2020). Taxonomic similarity does not predict necessary sample size for ex situ conservation: a comparison among five genera. *Proceedings of the Royal Society B* 287(1926). http://doi.org/10.1098/rspb.2020.0102.

Hoban, S. and Strand, A. (2015). Ex situ seed collections will benefit from considering spatial sampling design and species' reproductive biology. *Biological Conservation* 187(July): 182–191. https://doi.org/10.1016/j.biocon.2015.04.023.

Ison, J. L. and Wagenius, S. (2014). Both flowering time and distance to conspecific plants affect reproduction in Echinacea angustifolia, a common prairie perennial. *Journal of Ecology* 102(4): 920–929. https://doi.org/10.1111/1365-2745.12262.

Jarvis, A., Lane, A. and Hijmans, R. J. (2008). The effect of climate change on crop wild relatives. *Agriculture, Ecosystems and Environment* 126(1–2): 13–23. https://doi.org/10.1016/j.agee.2008.01.013.

Kallow, S. (2014). *UK national tree seed project seed collecting manual*. RBG Kew. Available at: http://brahmsonline.kew.org/Content/Projects/msbp/resources/Training/UK-national-tree-seed-project-manual.pdf.

Li, D. Z. and Pritchard, H. W. (2009). The science and economics of ex situ plant conservation. *Trends in Plant Science* 14(11): 614–621. https://doi.org/10.1016/j.tplants.2009.09.005.

Macfarlane, R., Jackson, G. V. H. and Frison, E. A. (2011). Chapter 17. Plant health & germplasm collectors. In: *Collecting Plant Genetic Diversity: Technical Guidelines. 2011 Update* Guarino, L., Ramanatha Rao, V. and Goldberg, E. (Eds). Bioversity International. Available at: https://cropgenebank.sgrp.cgiar.org/index.php/procedures-mainmenu-242/collecting.

Maxted, N., Castañeda Álvarez, N. P. and Vincent, H. A. (2011) Chapter 41. Gap analysis: A tool for genetic conservation. In: *Collecting Plant Genetic Diversity: Technical Guidelines. 2011 Update* Guarino, L., Ramanatha Rao, V. and Goldberg, E. (Eds). Bioversity International. Available at: https://cropgenebank.sgrp.cgiar.org/index.php/procedures-mainmenu-242/collecting.

Maxted, N., Dulloo, E., Ford-Lloyd, B. V., Iriondo, J. M. and Jarvis, A. (2008). Gap analysis: a tool for complementary genetic conservation assessment. *Diversity and Distributions* 14(6): 1018–1030. doi:10.1111/j.1472-4642.2008.00512.

Mayzlish Gati, E., Beerman, A., Singer, A., Faraj, T., Lifshitz, D., Bar, D. and Golan, S. (2018). *A new GPS-based system for seed collection at the Israeli Gene Bank. Samara 33*. RBG Kew. Available at: https://www.kew.org/sites/default/files/Samara%2033%20web.pdf.

Menges, E. S., Guerrant, Jr., E. O. and Hamze, S. (2004). Effects of seed collection on the extinction risk of perennial plants. In: *Ex Situ Plant Conservation: Supporting Species Survival in the Wild* Guerrant, E. O., Havens, K. and Maunder, M. (Eds). Island Press, London.

Nowak, M. M., Dziob, K., Ludwisiak, Ł and Chmiel, J. (2020). Mobile GIS applications for environmental field surveys: a state of the art. *Global Ecology and Conservation* 23. https://doi.org/10.1016/j.gecco.2020.e01089.

Parra-Quijano, M., Iriondo, J. M. and Torres Lamas, E. (2011). Chapter 6. Strategies for the collecting of wild species. In: *Collecting Plant Genetic Diversity: Technical Guidelines. 2011 Update* Guarino, L., Ramanatha Rao, V. and Goldberg, E. (Eds).

Bioversity International. Available at: https://cropgenebank.sgrp.cgiar.org/index.php/procedures-mainmenu-242/collecting.

Pence, V. C. (2013). In vitro methods and the challenge of exceptional species for Target 8 of the global strategy for plant conservation 1. *Annals of the Missouri Botanical Garden* 99(2): 214–220. Available at: www.jstor.org/stable/42703725. doi: 10.3417/2011112.

Ramírez-Villegas, J., Khoury, C., Jarvis, A., Debouck, D. G. and Guarino, L. (2010). A gap analysis methodology for collecting crop genepools: a case study with phaseolus beans. *PLoS ONE* 5(10): e13497. https://doi.org/10.1371/journal.pone.0013497.

RBG Kew. (2016) *Chile Seed Collecting Guide*, Allen, R. (Compiler). 57 pp. Unpublished.

Spjut, R. W. (1994). A systematic treatment of fruit types. *Memoirs of the New York Botanical Garden* 70: 182.

Tanksley, S. D. and McCouch, S. R. (1997). Seed banks and molecular maps: unlocking genetic potential from the wild. Science 277 (5329): 1063–1066. https://doi.org/10.1126/science.277.5329.1063.

Victor, J. E., Koekemoer, M., Fish, L., Smithies, S. J. and Mössmer, M. (2004). *Herbarium essentials: the Southern African herbarium user manual*. Southern African Botanical Diversity Network Report No. 25. SABONET, Pretoria. Available at: http://www.sanbi.org/sites/default/files/documents/documents/sabonet-report-no-25-herbarium-essentials-southern-african-herbarium-user-manual.pdf.

Vincent, H., Wiersema, J., Kell, S., Fielder, H., Dobbie, S., Castañeda-Álvarez, N. P., Guarino, L., Eastwood, R., León, B. and Maxted, N. (2013). A prioritized crop wild relative inventory to help underpin global food security. *Biological Conservation* 167: 265–275. https://doi.org/10.1016/j.biocon.2013.08.011.

Way, M. J. (2003). Collecting seed from non-domesticated plants for long term conservation. In: *Seed Conservation: Turning Science into Practice* Smith, R. D., Dickie, J. D., Linington, S. H., Pritchard, H. W. and Probert, R. J. (Eds). Royal Botanic Gardens, Kew, UK, pp. 163–201.

Way, M. J. and Gold, K. (2014a). *Assessing a population for seed collection*. Technical Information Sheet_02. RBG Kew. Available at: http://brahmsonline.kew.org/Content/Projects/msbp/resources/Training/02-Assessing-population.pdf.

Way, M. J. and Gold, K. (2014b). *Seed collecting techniques*. Technical Information Sheet_03. RBG Kew. Available at: http://brahmsonline.kew.org/Content/Projects/msbp/resources/Training/03-Collecting-techniques.pdf.

Wyse, S. V. and Dickie, J. B. (2016). Predicting the global incidence of seed desiccation sensitivity. *Journal of Ecology* 105(4): 1082–1093. https://doi.org/10.1111/1365-2745.12725.

Wyse, S. V., Dickie, J. B. and Willis, K. J. (2018). Seed banking not an option for many threatened plants. *Nature Plants* 4(11): 848–850. https://doi.org/10.1038/s41477-018-0298-3.

Chapter 8

New technologies to improve the ex situ conservation of plant genetic resources

Fiona R. Hay, Aarhus University, Denmark; and Sershen, University of the Western Cape & Institute of Natural Resources, South Africa

1 Introduction

Plant genetic resources, which include traditional and modern varieties, crop wild relatives, genetic stocks, breeding lines and weedy species, form the genetic basis for the improvement and selection of crops through breeding (IPGRI, 1994). While the ex situ conservation of crop genetic resources will contribute towards addressing the challenge of food insecurity, as set out in the United Nations Sustainable Development Goals (www.sustainable development.un.org/), the ex situ conservation of wild species can facilitate strategies (e.g. habitat restoration and species reintroduction) that improve ecosystem resilience under various climate change scenarios (Frankel, 1990; McFerson, 1998; Abbas and Qaiser, 2011). Biotechnology, that is, the use of living organisms or parts thereof to manufacture or modify a product, develop microorganisms for specific uses, or improve plants or animals (Uyoh et al., 2003), has also made plant genetic resource conservation an industry priority in many developed parts of the world.

In this regard, seed storage is the most efficient and cost-effective means of ex situ plant germplasm conservation, and has been used to conserve a sizeable amount of plant biodiversity worldwide for decades (IBPGR, 1976; Engels and Engelmann, 1998). The science of seed genebanking - of storing orthodox seeds under dry, cold conditions (orthodox seeds being those that tolerate such conditions) - is still relatively basic and, given the lengths of time that the seeds

http://dx.doi.org/10.19103/AS.2020.0085.14

are expected to remain alive compared with the age of many well-documented collections, could still be considered somewhat experimental. Nonetheless, over recent years, we have gained greater insight into the effectiveness of the basic processes that most genebanks follow in managing their collections of orthodox seeds and, since budgets are often limited, advanced technologies are being introduced, with the tag line of improving efficiency.

However, seed storage cannot be used to conserve genetic variation of all species, due to fundamental differences in innate reproductive characteristics (specifically the ability of seeds to survive long periods of dry storage) or the need to fix and multiply favourable genetic combinations. In such cases, seed-derived embryonic tissues, somatic embryos or vegetative tissues (e.g. shoot tips) are the materials of choice for storage. Collections housed in both the developed and developing world (Thailand, Australia, United States of America, Italy, Belgium, India and China, among others) include germplasm of vegetatively propagated fruit, commercial and speciality crops, ornamentals, nuts, vegetables and wild relatives, from temperate, subtropical and tropical zones (Jenderek and Reed, 2017). Where these are clonal collections the plant material is maintained in an actively growing state using a variety of techniques for one or more of the following purposes: (1) to preserve the selected genotypes; (2) to maintain their sterile nature; (3) to produce seed unamenable to storage. While the development of cryopreservation techniques for clonal crops began as far back as the 1970s, involving controlled-rate cooling and subsequent plunging in liquid nitrogen (LN) of shoot tips (Grout et al., 1978) and buds (Sakai and Nishiyama, 1978), the last two decades has seen the development of a range of cryopreservation techniques that have been applied across numerous genera and species (Benson, 1999; Reed, 2008; González-Arnao et al., 2014), including both orthodox-seeded (desiccation tolerant) and recalcitrant-seeded (desiccation sensitive) species (Berjak et al., 2011a; Jenderek and Reed, 2017).

In this chapter, we describe some of the recent applications of science and technology to improve the management of genebank collections of orthodox seeds, not least through the introduction of automation. In addition, we look at how routine cryopreservation procedures are now possible for many species, and how storage procedures for recalcitrant-seeded and vegetatively propagated species continue to evolve, in light of the challenges they present in terms of long-term ex situ germplasm conservation.

2 Improving the management of orthodox seeds

2.1 Routine operations

Differences in seed post-harvest physiology are reflected in the categorisation of seeds as orthodox, recalcitrant (Roberts, 1973) and intermediate (Ellis et al.,

1990), based on their responses to loss of moisture and to storage at different temperatures. Orthodox seeds can tolerate drying without damage to low-moisture contents (2–6%; Roberts, 1973; Chin and Roberts, 1980; Hong and Ellis, 1996) and their longevity, spanning perhaps many decades, is increased as the moisture content and temperature at which they are stored (within limits) are reduced (Ellis and Roberts, 1981). The core operations of a genebank with the mandate to conserve seeds of orthodox species comprise seed production, drying, cleaning, viability and health testing, packing, storage and distribution; all of these activities are documented and driven by the genebank information system (Fig. 1).

Most genebanks store seeds of every accession in both an active collection and a base collection. The active collection is generally stored at 2–4°C (medium-term storage; MTS) and it is from this collection that samples are taken for distribution. This is therefore where the bulk of the seeds are stored. The base collection is stored at a lower temperature, typically −20°C (long-term storage; LTS). Samples in the base collection are usually much smaller than the samples in the active collection, and the purpose is long-term conservation. The point of having all the accessions in two environments is to reduce costs. In theory, the cost of storing a sufficient quantity of seeds for distribution at the lower temperature is too high. This is particularly true for agricultural genebanks with high rates of distribution. In contrast, at, for example, the Millennium Seed Bank (MSB) of the Royal Botanic Gardens Kew, whose main purpose is long-term conservation of species and which has relatively few sample requests, all seeds are under LTS conditions. Other genebanks may also maintain only one storage environment, based on operational needs and costs. The United States Department of Agriculture has a network of genebanks holding the active collections for different crops, with one main base collection serving all the regional genebanks (National Research Council, 1991). Some genebanks additionally cryopreserve samples of orthodox seeds, for example, of orthodox seeds which are expected to be very short-lived, even under 'conventional' genebank storage conditions of −20°C (Davies et al., 2018; Ballesteros et al., 2021).

Testing the viability of seeds before they are placed into storage, and at regular intervals during storage, is a key 'quality control' activity that genebanks are expected to do as a matter of routine (FAO, 2014). For genebanks with accessions in both an active and a base collection, if the sample in both comes from the same seed lot, it is not necessary to check the viability of the seeds in both collections. Since the longevity of orthodox seeds improves in a predictable manner as the temperature is reduced, the seeds in the base collection should always have higher viability than the seeds from the same seed lot stored in the active collection after the same period of time. In other words, the viability of the seeds in the active collection

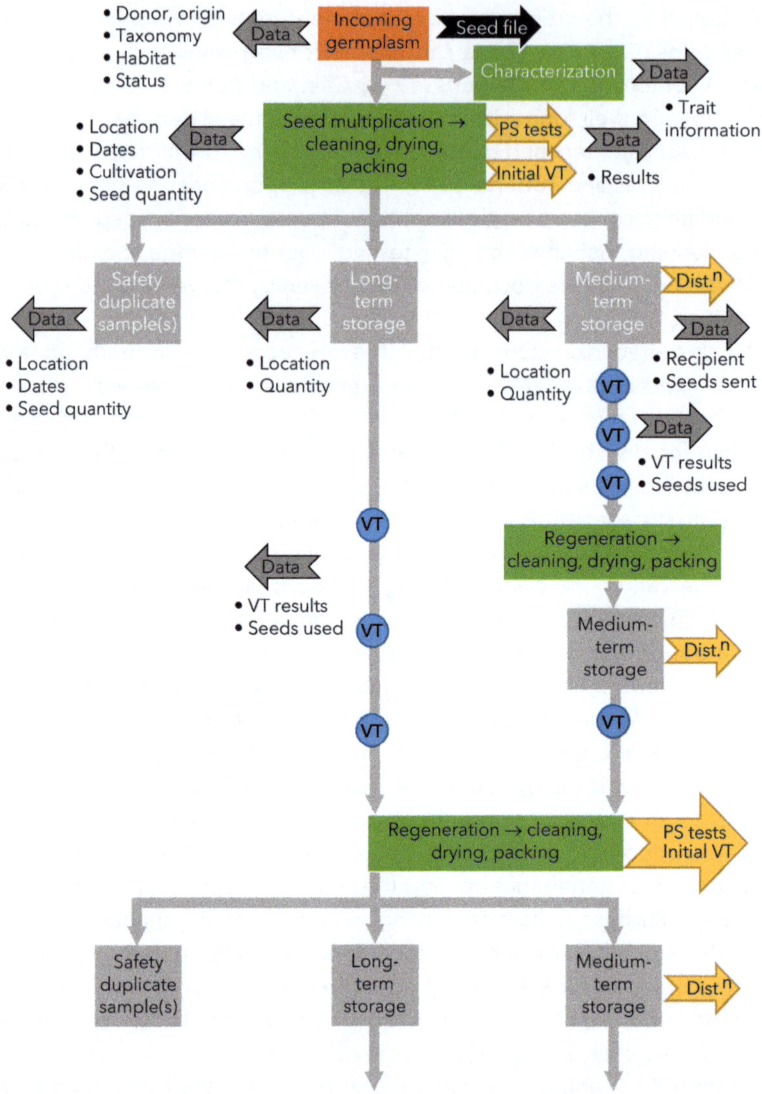

Figure 1 Overview of operations in a typical seed genebank. When a sample first arrives at the genebank, it will be assessed for uniqueness and initially grown under quarantine conditions to ensure there is no disease. If there are enough seeds, a small sample (e.g. 20–30 seeds) of the incoming seeds will also be added to the seed file, for future reference and variety verification. The accession will then be multiplied to produce sufficient seeds for storage. Characterisation of the accession for morpho-agronomic traits according to defined crop descriptors may be carried out on plants used for seed multiplication, or may be an independent operation. Characterisation also includes collection of data on features of the seeds. After harvest, seeds will be cleaned and dried. They may also undergo a pest control treatment. Samples will be taken for phytosanitary (PS) testing and for an initial viability test (VT). After final equilibrium drying, seeds will

Figure 1 (*Continued*)
be packed for storage in the long-term (base) and medium-term (active) collections (if appropriate). Some genebanks may additionally cryopreserve a sample of seeds. Safety duplicate samples will be sent for long-term conservation, without any management interventions, at another genebank and/or in the Svalbard Global Seed Vault. Seeds in the active collection are used for distribution (Dist.^{n}). They are also regularly tested for viability (viability test; VT). Once the viability falls below the threshold level or the quantity of seeds remaining gets low, the accession will be regenerated and the seeds produced are processed for storage in the active collection again. A few cycles of regeneration using seeds from the active collection may occur, before seeds from the base collection should be used. After replacing the seeds in the active collection, meaning that the seeds in the two collections are not from the same seed lot, viability testing of seeds in the base collection should commence. When the viability of the seeds in the base collection falls below the threshold level (or, but theoretically unusually, the quantity of seeds remaining gets low), a new sample should be produced for the base collection. A sample of seeds from this seed lot should also be sent as a safety duplicate sample, because it is expected that the samples in the base collection and in long-term storage in another location would lose viability at a similar rate. Seeds from this same seed lot may also be used to replenish the active collection and the 'cycle' begins again. Across all the operations, data is collected and entered into the genebank information system. This data includes information on the processes themselves, such as number of seeds sown, number of plants harvested, viability test date and so on. The data examples shown represent only a fraction of all the data that would, in reality, be recorded.

will fall faster than the seeds stored in the base collection – unless there has been any divergence in the handling that compromises the initial quality or moisture content of the seeds intended for storage in the base collection. It is only necessary to start testing the viability of seeds in the base collection when the seeds in the active collection are replaced by a new seed lot (Whitehouse et al., 2020). Similarly, if seed samples are taken from the same seed lot and prepared in the same way for storage in the base collection and for sending to another genebank as a safety duplicate, it is not necessary to monitor the viability of that safety duplicate sample. Safety duplicate samples are only intended to be returned to the originating genebank, and only if that genebank cannot recover the accessions from their own samples. When the viability of the seeds in the base collection declines, the safety duplicate sample may be replaced Until now, this has been happening at low rates, although it is likely to increase, particularly for species with relatively short-lived seeds, as the viability of more samples in the base collection falls and the samples are replaced. For many genebanks, in addition to having every accession in the two collections, a very small sample of the 'most original seeds' of each accession is also placed in the seed file. The seed file is the reference sample against which comparisons can be made, for example, of regenerated material or of new material that is submitted to the genebank. Ideally, the most original sample would comprise some of the seeds originally donated or arriving at the genebank, but in some cases it will be seeds from the first round of multiplication.

Given the interdependency of operations, a robust information system is crucial to record all the data collected for each process and to ensure timely and efficient sequencing of operations. Unfortunately, data management may be overlooked as key 'infrastructure', particularly in national genebanks, where resources may be limited. Genebank information systems also store passport data and, increasingly, characterisation and evaluation data, and links to genomic data, where available. More genebanks are genotyping accessions or even, whole collections, and/or are handling genetic stocks such as mapping populations, mutant populations and recombinant inbred lines. Genebank information systems such as GRIN-Global (www.grin-global.org) also allow users to search collections and order accessions online.

2.2 Changing procedures

Recommendations on how to handle seeds in genebanks were first published in 1975 (Cromarty et al., 1990) and, while there has been some evolution, the essence of those recommendations has barely changed. For example, Cromarty et al. (1990) recommended drying seeds in a drying room at 'about 15°C and 10–15% relative humidity with good air circulation'; in the IPGRI/FAO genebank standards published in 1994, the recommendations were to dry at 10-25°C and 10–15% relative humidity (IPGRI/FAO, 1994); and in the latest version of the standards, the recommended drying environment is 5-20°C and 10-25% relative humidity (FAO, 2014). Similarly, storage temperatures and monitoring intervals remain largely unchanged, even after decades of storing seeds in genebanks. Unfortunately, because some of those original recommendations are now relatively old, it may be difficult to know how they came about; old data sets might not be archived and knowledge of the hows and wherefores not passed on to younger genebank managers/scientists. Nonetheless, we should still be considering whether genebank standards are optimal and in particular, given that the standards are intended to cover such a broad range of species, whether they are optimal for particular species.

2.2.1 Seed drying

The conditions under which orthodox seeds are dried are intended to result in seeds being at a moisture content where seed longevity is optimal (regardless of whether the seeds are subsequently stored at 2-4°C or −20°C). Various combinations of temperature and humidity would allow seeds to reach the same moisture content, however, the drying treatment itself should not be detrimental to the quality of the seeds. In theory, if very wet seeds are placed at a high temperature there could be a high rate of seed ageing. Thus, the genebank standards recommend drying at a relatively cool temperature, in contrast with the higher temperatures used for drying grain. Many genebanks have a drying room that runs at 15°C and 15% relative humidity. In the case of rice, however, it has been found that drying at 15°C is not optimal for the subsequent longevity of the seeds (Crisostomo et al., 2011; Whitehouse et al., 2015, 2018a). If the rice seeds, due to the typically humid environment of rice fields, are not able to dry in situ and have a high moisture content at the time of harvest (> 16.5%), initial drying at 40–45°C followed by final equilibrium drying at 15°C and 15% relative humidity improves the subsequent longevity of the seeds significantly compared with only drying at 15°C and 15% relative humidity (Whitehouse et al., 2018a). This response is perhaps not surprising for rice: in the tropics, ambient temperature is, of course, relatively high, and rice (and other grains) is often dried on roads or flat cement areas, where temperatures in the middle of the day under bright sunlight can get quite high. A similar response was also seen for seeds of four wild rice species (Timple and Hay, 2018). As a consequence of this research, the rice genebank at the International Rice Research Institute now routinely uses two-stage drying for all the collection. Freshly harvested seeds are initially dried for 3 days in a drying room at 40°C and 30% relative humidity, after which, they are transferred for final equilibrium drying in the drying room at 15°C and 15% relative humidity. There is some evidence that drying seeds of some other species at a higher temperature than 15°C may similarly be better for the subsequent longevity of the seeds (Whitehouse et al., 2018b).

2.2.2 Revising monitoring intervals

As indicated above, many genebanks still adopt the default viability testing intervals of 5 or 10 years, depending on the expected longevity in storage (FAO, 2014). Otherwise, '*Viability monitoring test intervals should be set at one-third of the time predicted for viability to fall to 85% of initial viability or lower depending on the species or specific accessions, but no longer than 40 years*' (FAO, 2014). The recommended way to predict the viability period is to use the Ellis and Roberts (1980) viability equation, using parameters determined from experiments in which seeds are stored at a range of temperatures and moisture contents (e.g. Hay et al., 2003). This can be done using the seed viability constants menu in the Seed Information Database (SID; Royal Botanic Gardens Kew, 2020), but is only possible for a small number of species. Furthermore, as discussed by Hay and Whitehouse (2017), the time for viability to fall to 85% of the initial viability is perhaps not an ideal standard, since it depends on an estimate of initial viability, and because it means that the absolute threshold value could potentially vary among seed lots (i.e. of seed lots of the same or different accessions). For example, if the initial viability ranges between 85% and 100%, the monitoring intervals would be calculated based on the time for viability to fall anywhere between 72% and 85%. Further, it is known, based on the typical shape of the seed survival curve (the curve describing the loss of viability during storage), that once the viability of a seed lot has fallen to 72%, the rate of percentage viability loss is nearing its peak. In practice, most genebanks still use set-standard monitoring intervals, at the species level at least, and have a fixed viability threshold (e.g. 85%), rather than setting seed lot-specific test intervals and thresholds. Nonetheless, in future, there could be a more dynamic approach for setting viability-monitoring intervals (Whitehouse et al., 2020).

A more dynamic, flexible approach will probably only emerge as we understand more about the real performance of seed lots in genebanks. Some genebanks have already analysed their historical viability monitoring data, using a variety of approaches to model the longevity of seed lots, within species or species groups and in some cases, to formulate more efficient monitoring intervals (e.g. Walters et al., 2005; van Treuren et al., 2013; Hay et al., 2015; Ellis et al., 2018, 2019; Yamasaki et al., 2020). In some cases, the conclusions drawn from such studies, while contributing to scientific knowledge, are only really relevant to the genebank where the data was collected, due to crop focus or more likely, regeneration and processing protocols, and storage conditions that are perhaps unique to that genebank, even if they are more-or-less consistent with the Genebank Standards (which cannot be species-prescriptive because they are meant to cover so many species). In future, there may be more meta-analyses, using viability data for the same crop from different genebanks. In particular, this could help verify the applicability of the Ellis-Roberts viability equations, particularly at low temperatures for which data is still lacking (Pritchard and Dickie, 2003).

There have also been some advances in terms of understanding the molecular basis of seed longevity for genebank accessions, which raises the possibility of using genotype data to predict the relative longevity of seeds in genebank storage and hence set appropriate monitoring intervals (Lee et al., 2019). Screening seed lots for longevity under experimental conditions to get an initial measure of relative longevity could also help identify which seed lots to test first (Davies et al., 2016; Hay and Whitehouse, 2017), perhaps in particular if used in conjunction with genotype data. Taking this idea further, it has been suggested that a more efficient strategy for genebanks would be to preclude the need to do viability monitoring tests, by taking other measurements (ideally simple, cheap and non-destructive tests) that are predictive of the extent to which a seed lot has aged (Fu et al., 2015). However, this still seems to be somewhat futuristic in practice, and may only ever be feasible for large agricultural genebanks with many accessions of a single or few crops.

Figure 2 Barcodes are used to track samples through genebank operations: seed drying (a) and transplanting (b) at the International Center for Tropical Agriculture (CIAT) in Colombia; and germination testing and scoring (c, d) at the International Institute of Tropical Agriculture (IITA) in Nigeria. All photos taken by the first author (Fiona R. Hay).

2.2.3 Electronic data collection

The use of barcodes to label the packets of seeds going into genebank storage has been a common practice for some years now (Rao et al., 2006). Indeed, barcodes are now used to track samples through many, if not all, genebank operations, from the field to storage, including during characterisation/evaluation activities and distribution (Fig. 2). This has also made it possible to collect data electronically, even in the field: the barcode can be scanned and then the relevant data entered. To further minimise the risk of wrong data entry, genebank staff may use a portable tablet or similar device with software that is customised to show, for example, value options with pictures for a particular characterisation trait. In theory, not only is this more accurate, but also improves efficiency since staff do not have to manually enter data from paper score sheets after data collection.

2.2.4 Improving viability testing

Modifying and improving germination procedures used in viability monitoring should be a constant process in genebanks, particularly those conserving wild species diversity. Such species, compared with major crop species, are likely to have some sort of dormancy mechanism which prevents germination when moisture, temperature and light requirements are met. It is also more likely that dormancy-breaking procedures have not been documented and/or widely validated for such species. Despite the need for reliability in viability testing and to provide germplasm users with seed germination advice, research in this area has been under-resourced, particularly at crop genebanks where wild species have historically perhaps received less focus in the context of managing germplasm. Seed banks entirely focussed on conserving wild species, in contrast, have invested more resources in determining optimum germination protocols, as documented, for example, in the Seed Information Database (Royal Botanic Gardens Kew, 2020). There are other comprehensive resources that would be helpful to any genebank needing to improve their germination protocols (e.g. Baskin and Baskin, 2014; ISTA, 2020).

The preferred germination-test sample size in official testing is 400 seeds, sown as four replicates of 100 (ISTA, 2020). This is far too many seeds than most genebanks would like to use in a viability monitoring test, though the actual number tested varies among genebanks depending on species, quantity of seeds available, facilities and other logistical constraints. As also discussed elsewhere (Hay and Whitehouse, 2017), there are other ways in which the number of seeds used overall for viability monitoring might be reduced, not just by having more knowledge of longevity in storage, but also, for example, by following a sequential monitoring scheme (Ellis et al., 1985). A recent study across three international genebanks, involving 111 accessions from 11 species, confirmed that sequential sampling is an efficient alternative to the fixed size sampling method for making decisions based on viability for tropical forage species, although for three *Leucaena* species, there was no advantage in using sequential sampling to save seeds in viability monitoring (A. Sartie, L. Santos and Z. Kinyanjui, *pers. comm.*).

While 'viability' is most often assessed using a germination test, there has long been an interest in developing a non-destructive test to predict the viability of a seed lot. Research in this area has considered, for example, the nature and quantity of volatiles released by seeds during storage (Colville et al., 2012) and molecular markers (Boniecka et al., 2019; Fleming et al., 2019). However, while these studies certainly cast light on the process of seed deterioration in storage, there is still a lot of research needed before we get close to a practical non-destructive predictive test of seed viability during genebank storage (reviewed by Fu et al., 2015).

2.3 Automating processes

Some genebanks, notably those of the international genebanks of the CGIAR and a few national genebanks, have very large collections. For example, the genebank at the International Rice Research Institute (IRRI) in the Philippines is the largest single-crop collection with more than 132 000 accessions (https://www.genesys-pgr.org/). Staff at such genebanks may work as if they were

Figure 3 Images of the robotic storage and retrieval system in the long-term storage room of the National Agrobiodiversity Center, Rural Development Administration, Republic of Korea. Images provided by the National Agrobiodiversity Center.

working in a factory line, with groups of staff focussing on one particular activity at any one time and often for quite long periods of time. In such genebanks, that process thousands of seed samples of one or just a few crops each year, automating some of those processes is more likely to be feasible, compared with a genebank that is perhaps processing one species for one week, and a completely unrelated species, the next. The genebank storage rooms can thence be considered the warehouses for the factory and also be managed in a similar way.

2.3.1 Robotic storage and retrieval

A few genebanks around the world, for example, the Genetic Resources Center of Japan's National Agriculture and Food Research Organization and the National Agrobiodiversity Center in the Republic of Korea (Fig. 3), have incorporated warehouse management-type technology to manage their collections. All the placing and retrieval of samples is handled by a robot, so staff rarely enter the stores themselves. There are a number of reasons why this could be considered advantageous. The long-term storage environment, at −20°C, is not a pleasant working environment for staff and could even present a health hazard to some people, even if they wear insulated clothing. Therefore, restricting access and limiting the length of time anyone is in that room, might reduce any risk associated with working at such a low temperature. On the other hand, it should not be necessary to enter this storage room very often, as its intended use is long-term conservation. The active collection is stored at a higher temperature (2-4°C), which is a more tolerable temperature, nonetheless, a robotic system will increase security and ensure correct placement of samples. It may also mean that there is less exchange of air so it is easier to maintain the correct temperature and energy is not being wasted. This is perhaps particularly relevant in hot and humid environments. Of course, genebanks that do not install a robot system, will still make sure that the location of accessions in both stores (e.g. shelf number, tray number) is recorded in the genebank information system.

2.3.2 Seed phenotyping

Analysis of images captured by different types of cameras has many potential applications in seed testing (Dell'Aquila, 2009; Boelt et al., 2018). One way in

Figure 4 (a) Samples from the genebank's physical seed file at the International Rice Research Institute (IRRI); (b) selecting the training-set seeds from the bulk prior to automated sorting (image used with permission of N. Ruaraidh Sackville Hamilton); (c) the custom seed sorter at IRRI. (a) and (c) were taken by the first author (Fiona R. Hay).

which image analysis could be applied to improve genebank operations is by using it as a virtual seed file (Edberg Hansen et al., 2016). Most agricultural genebanks keep a small sample of seeds of perhaps the material originally received at the genebank, or otherwise, of the first cycle of seed multiplication, for use as a reference against which new harvests can be compared. This is called the seed file (Fig. 4A). This seed file does not have a physical backup, and if any seed file samples are lost, they would have to be replaced by the next available 'most original sample' (i.e. from the first regeneration cycle, if available, rather than the second or third). The seed file is not just used to verify that harvested seeds appear to be of the same accession, but they are also used, for example, to check incoming material to see whether the sample is a potential varietal mixture with different seed phenotypes and/or to determine whether it might be a variety that is already represented in the genebank. However, use of the seed file relies on having trained personnel who are able to detect subtle differences between seeds. Some genebanks already capture images of the seeds of different accessions (as well as other parts of the plants), to use internally and/or to make available to germplasm browsers on their online ordering portal, and there is a lot of potential to analyse such images and make the validation/checking process more objective. Edberg Hansen et al. (2016) proposed the use of multispectral imaging since it offers the possibility of identifying and measuring some of the more subtle characteristics of seeds, which might not be apparent in images from an ordinary digital RGB camera.

2.3.3 Seed sorting

Most seed lots intended for genebank storage will go through a cleaning process, though the extent and method of cleaning may vary. There are various types of seed cleaning equipment developed for the seed industry, including seed blowers, sieves, brushes and colour sorters, which can be used to remove non-seed material such as plant stems or dispersal structures. The choice of method will very much depend on the seeds being cleaned and the quantity of seeds to be cleaned. For very small seed lots, the amount of time it takes to clean the equipment between seed lots may mean that it is not practical to use such machines. Some genebanks do a considerable amount of hand sorting. For many years, all the seeds intended for storage in the rice genebank at the International Rice Research Institute (IRRI) were entirely sorted by hand, removing off-types, immature (green), damaged or diseased seeds. This used to be a year-round activity and as a consequence, for some seed lots, the length of time between harvest and final storage was many months. Although the seeds were kept in the drying room until they could be cleaned, this could nonetheless impact the initial viability and storage potential of the seeds when put into the genebank stores. The IRRI genebank now has a unique, purpose-built seed sorter that selects seeds by individual-seed image analysis, making sure the colour and shape parameters of each seed matches that of a training set (Fig. 4B, C). This automated seed sorting is considered a first-step in the cleaning process, and the 'selected' seeds are still manually sorted, but introducing this automation has increased the throughput of the seed cleaning process.

2.3.4 Germination scoring

Automated scoring of the germination process has attracted quite a lot of interest in the seed testing community over the last decade or so, to increase the number of samples that can be tested for germination in a short period of time and/or to get more information on the rate of germination as a measure of seed lot vigour (Joosen et al., 2010; Harper and Long, 2011; Demilly et al., 2015). Some genebanks test the germination of thousands of samples every year, as part of the routine viability monitoring. For genebanks handling seeds of many different species, germination protocols (pretreatment, germination method and germination temperature) are likely to vary from one species to the next, and there is little opportunity to automate the scoring process. However, for large crop genebanks, testing one or a few species following the same protocols for each accession within a crop, automated scoring of germination may be feasible. Various systems are available, from fully automated robotic systems in a controlled-environment room with the capacity for hundreds of tests at any one time, to semi-automated systems in which tests are manually moved from the germination environment to the camera. The software for these systems use image analysis to determine how many seeds have germinated at each observation time and, once the test is finished, calculates various parameters, for example, final germination percentage and mean germination time. From the genebank perspective, the parameter of most interest is the final germination percentage at the end of the germination test, but other parameters may provide insight into the seed-ageing process (since vigour is expected to decline before there is loss of viability) and/or may be of interest to breeders who are trying to enhance the seed vigour of varieties, so that they can, for example, still give good crop establishment in marginal environments.

3 Improving the management of recalcitrant-seeded and vegetatively propagated species

3.1 Routine operations: recalcitrant seeds and vegetatively propagated species

Recalcitrant seeds are sensitive to desiccation (Roberts, 1973), freezing (reviewed by Walters et al., 2008) and very often chilling, in some cases at temperatures as high as 15°C (Berjak et al., 1995). This precludes their maintenance under conventional orthodox seed storage conditions (e.g. King and Roberts, 1980; Farrant et al., 1989). Short- to medium-term storage methods and long-term storage via cryopreservation of seed-derived explants represent some of the storage options employed for such species over the last few decades. Similarly, alternative germplasm conservation strategies, such as storage of actively growing cultures, minimal-growth storage and cryopreservation, have been adopted for species that do not produce seeds (e.g. banana), root and tuber vegetables, clonally propagated crops, and where the unique true-to-type genomic constitution of a cultivar needs to be maintained.

For recalcitrant seeds, short- to medium-term storage has involved maintenance at water contents as close to that at shedding, and at ambient or slightly lower temperatures, and this is defined as 'hydrated-storage' (e.g. Berjak et al., 1989). However, under these conditions, recalcitrant seeds still initiate germination-associated events, culminating in seed death (Berjak et al., 1989; Pammenter et al., 1994; Chandra et al., 2019). Storage at temperatures lower than ambient has been shown to postpone the onset of germination in recalcitrant seeds by slowing down the metabolic rate (e.g. Pritchard et al., 1995), provided the seeds are not chilling-sensitive. However, even then, storage longevity generally ranges from a few weeks to months e.g. for *Scadoxus membranaceus* and *Landolphia kirkii* (Farrant et al., 1989), various amaryllid species (Sershen et al., 2008) and *Madhuca latifolia* (Chandra et al., 2019). The lifespan of recalcitrant seeds in hydrated storage is often further curtailed by the proliferation of a spectrum of fungi (Mycock and Berjak, 1990; Sutherland et al., 2002), even when seeds are treated with fungicidal agents (Mycock and Berjak, 1995; Moothoo-Padayachie et al., 2018).

In vitro storage, that is, the use of tissue culture, by which cells, tissues or organs are excised from parent plants, decontaminated and then transferred to artificial growth media in vitro (Krøgstrup et al., 1992; George, 1993; Mandal et al., 2000), represents the major short- to medium-term germplasm storage option for vegetatively propagated species. A wide range of explants have been used (e.g. buds, cuttings, seeds, shoot apices and leaves) based on the fact that media can be manipulated to produce different cultures such as unorganised, undifferentiated callus or organised tissues and organs that can be converted into plantlets in the regeneration phase. Variations to impose minimal growth include reduction in nutrient (e.g. Schnapp and Preece, 1986) and/or sucrose concentration (e.g. Kartha and Engelmann, 1994) in the growth medium, alterations to culture medium osmotic potential (using osmotica such as sucrose, sorbitol, ribose and mannitol (e.g. Zandvoort et al., 1994)), the addition of growth retardants such as abscisic acid (ABA) (e.g. Jarret and Gawel, 1991; Taylor et al., 1996), and placing explants in a controlled atmosphere or beneath a liquid medium or mineral oil (Paunescu, 2009). Lowering the partial pressure of oxygen with temperature (between 0°C and 10°C (e.g. Blakesley et al., 1996)) or light (e.g. Grout, 1995) below optimum in the culture environment can also limit in vitro growth. However, irrespective of whether germplasm is stored in vitro as actively growing cultures (e.g. Krøgstrup et al., 1992) or as minimal-growth cultures (e.g. Schnapp and Preece, 1986), such storage will at some stage require transfer of material onto new media (Krøgstrup et al., 1992; Mandal et al., 2000; Mycock et al., 2004), introducing the risk of contamination. Surviving, uncontaminated material can potentially be rapidly micropropagated to bulk up reserves (Mandal et al., 2000) but such clonal propagation, apart from being labour intensive and expensive, limits biological diversity within the collection and can impose selection pressures and environmental stresses, resulting in plants with genetic modifications (Staritsky, 1997). This variation arises from somaclonal variation during culture (Panis and Lambardi, 2006), but the use of organised systems such as embryos, meristems and shoot tips can reduce this risk (Engelmann, 1997; Mandal et al., 2000).

Long-term germplasm storage of recalcitrant-seeded and vegetatively propagated species is achieved via cryopreservation which involves the cooling of biological material to, and subsequent storage at, cryogenic temperatures, typically in liquid nitrogen (LN; −196°C), or its vapour (at approximately -150°C). Less ideally, material can be stored at some other temperature below −80°C (Finkel and Ulrich, 1983; Berjak et al., 1999a). Cryopreservation is regarded as the ultimate long-term storage approach since it is believed to arrest metabolic activity and deterioration, thus minimising, if not precluding, genetic changes (Krøgstrup et al., 1992; reviewed by Harding, 2004). With the exception of mature orthodox seeds (Pritchard and Nadarajan, 2008) and certain varieties of pollen (Ganeshan et al., 2008) and spores (Ingram and Bartels, 1996), biological tissues almost always contain considerable cellular water. This implies that successful cryopreservation of hydrated explants from recalcitrant seeds and vegetatively propagated species is best achieved when lethal intracellular ice-crystal formation is mitigated, as this can cause irreversible intracellular damage (Wesley-Smith et al., 1992). While cryopreservation is generally considered in terms of water's liquid and solid (ice) phases, it is also possible to cryopreserve plant material by inducing the process of vitrification, that is, the transition from liquid to glass phase without ice-crystal formation (Sakai, 2004). The phenomenon of vitrification has allowed for the development of 'ice-free' cryopreservation, which has been extensively applied to plant tissues (Sakai et al., 1992) of a number of vegetatively propagated species. However, freezing of plant tissues that inherently have high water content will inevitably involve the conversion of at least some of the water to ice and this is what precludes this approach in recalcitrant seed-derived explants of a number of species (Wesley-Smith et al., 1992).

Even though cooling and dehydration are the greatest sources of failure, and under other circumstances, the greatest contributors to post-cryo preservation survival, the success or failure of any plant cryopreservation protocol is a consequence of optimisation of all the manipulations involved in the preparation of the tissue for cooling, and all the steps involved in the recovery of that tissue after cooling (Berjak et al., 1999b; Pammenter and Berjak, 2014). These manipulations generally include variable combinations of the following (Funnekotter et al., 2017; Kaczmarczyk et al., 2012):

1 excision of explants such as seed embryos/embryonic axes or shoot tips to obtain sufficiently small samples;
2 partial physical desiccation or pre-growth on osmotic media to reduce water content;
3 cryoprotection through exposure to penetrating or non-penetrating cryoprotective agents to promote vitrification;
4 cryopreservation in LN;
5 re-warming which can be accompanied/followed by unloading of the cryoprotectant(s); and
6 washing (unloading of CPA solutions), recovery and regeneration of germplasm into plantlets (usually in vitro).

Samples that have been used for cryopreservation over the last three decades or so include buds, shoots, meristems, cell cultures, protoplast cultures, anthers, pollen, somatic and zygotic embryos, embryonic axes, callus, and whole seeds, if they are sufficiently small (Benson, 2008a). Success has been achieved in cryopreserving whole seeds (e.g. Hor et al., 1990; Kioko et al., 2003), shoot tips (Varghese et al., 2009), embryonic axes and zygotic embryos (Berjak et al., 2011b; Sershen et al., 2012) and somatic embryos of recalcitrant-seeded species (e.g. Mycock and Berjak, 1993). However, the number of successful protocols developed for embryonic axes and zygotic embryos from recalcitrant seeds is extremely low in the context of the number of species within this seed category (Subbiah et al., 2019).

A number of protocols for vegetative explants incorporate cold acclimation to pre-condition samples for exposure to cryogenic temperatures (reviewed by Reed, 1996, 2008; Benson, 2008b). Cold acclimation is generally induced in the laboratory by exposing explants to low in vitro growth temperatures (1–6°C) generally in combination with shortened day-length and/or high-sugar pre-treatments (Reed, 1996, 2008). Most of these protocols have also been based on what are termed classical methods (reviewed by Engelmann, 1997) which generally involve explant chemical cryoprotection, followed by slow cooling (0.5–2.0°C min^{-1}) down to −30°C to −40°C, or even −60°C (Krøgstrup et al., 1992). This controlled slow cooling (step 1) is said to encourage the formation of extracellular ice, progressively dehydrating the cells, as intracellular water is lost to exterior ice nucleation sites. This step is usually followed by immersion in LN (step 2). When freeze-induced dehydration during step 1 is too intense, various damaging events associated with the concentration of intracellular salts and changes in cellular membranes are possible (Mazur, 1990; Pritchard et al., 1995a). Also, some of the cells may fail to reach the optimum intracellular concentration and upon super-cooling undergo lethal intracellular ice-crystal formation (Mazur, 1990). Thus, while slow cooling may retain the integrity of individual cells, it has been shown to be less efficient at maintaining the tissue integrity required for survival of complex tissues, that is, meristems and embryos (Panis and Lambardi, 2006). Nevertheless, this 'two-step' cooling method, regarded as the first standard protocol developed for hydrated plant tissue (e.g. Withers and King, 1980), is still used for the cooling of undifferentiated culture systems (e.g. cell suspensions and calli (Withers and Engelmann, 1997)), and even differentiated structures such as the shoot apices of cold-tolerant species (e.g. Reed and Chang, 1997) and in isolated cases, excised embryonic axes/zygotic embryos from recalcitrant seeds (Mroginski et al., 2008).

3.2 Changing procedures: recalcitrant seeds and vegetatively propagated species

While minimal-growth storage has been successfully applied to explants from recalcitrant seeds (e.g. Chin, 1996; Sershen et al., 2008), recent reports of its use for the storage of such germplasm are scarce. Its continued use for the storage of somatic embryos (Hassan, 2017) and vegetative tissues is, however, still prevalent in the literature (Chauhan et al., 2019). In this regard, temperature reduction appears to be the most widely applied procedure in slow growth preservation, but it is now clear that temperature requirements vary from species to species, possibly based on the agroclimatic conditions in which a particular species occurs (Thakur et al., 2015). A more recent development in the design of minimal growth storage protocols involves the use of artificial seeds, produced by encapsulating propagules such as shoot buds or somatic embryos in a synthetic matrix, and this approach has been used successfully for the medium-term storage of a variety of plant species and propagule types (Cruz-Cruz et al., 2013).

There is also an increasing emphasis being placed on monitoring explants during storage, for example, changes in sugar content (total and reducing and non-reducing) due to sucrose-imposed stress in order to understand the effect of sucrose concentrations during slow growth conditions on survival and regeneration rate (El-Dawayati et al., 2018). Accommodating the interactive effects of light and ABA, and assessing relative nuclear DNA content in regenerants to ensure retention of ploidy level have also become important considerations (El-Dawayati et al., 2018).

The short- to medium-term storage of recalcitrant seeds via hydrated storage is no longer a common practice, but there are recent reports (Moothoo-Padayachie et al., 2016; Chandra et al., 2019) that the exogenous application of reactive oxygen species blocking agents (e.g. diphenyleneiodonium (DPI) and dimethyl thiourea (DMTU)) can be used to extend the storage lifespan of recalcitrant seeds (by days to weeks). In terms of long-term storage of recalcitrant seeds, in recent years it has become increasingly apparent that the success of cryopreservation protocols for their zygotic germplasm depends on the optimisation of cooling rates in parallel with explant moisture content, to eliminate or at least minimise intracellular ice crystal formation (Wesley-Smith et al., 2004). Success in this regard has, however, been hampered by (a) lethal freezing damage occurring when hydrated seeds/embryos/axes are exposed to LN (Wesley-Smith et al., 1992; Berjak et al., 1999b); and (b) drying to water contents precluding ice formation to the extent of desiccation damage, which generally culminates in loss of viability (Walters et al., 2008). In contrast, pollen (reviewed by Ganeshan et al., 2008) and seeds and somatic embryos of most desiccation-tolerant species (reviewed by Pritchard and Nadarajan, 2008) appear to be highly amenable to cryopreservation, shifting the focus for successful cryopreservation of recalcitrant seed germplasm from freezing tolerance to dehydration tolerance (Panis and Lambardi, 2006). The large-scale, routine application of cryopreservation to recalcitrant-seeded species is very limited globally, with many protocols remaining unpublished, but a recent study has validated the use of embryonic axes for cryopreservation of this group of species by showing that viability can be retained for decades (Ballesteros and Pence, 2019). For some recalcitrant-seeded species, such as cocoa and *Avicennia marina*, however, seed-derived germplasm will never be suitable for cryopreservation based on high levels of microbial contamination and/or the absence of suitably small zygotic explants. In these cases storage of alternate explants (e.g. shoot tips and meristems) generated via tissue culture may be more suitable.

While classical cryopreservation protocols used freeze-induced dehydration, modern protocols are predominantly vitrification-based (e.g. Fu et al., 1990; reviewed by Engelmann, 1999). In the latter case, cell dehydration to increase cytoplasmic viscosity precedes cooling, promotes the formation of glasses and avoids the factors that encourage ice-crystal formation - before exposure to the cryogen (e.g. Stanwood, 1985; Wesley-Smith et al., 1992; reviewed by Walters et al., 2008). Cooling rates typically used to cool embryonic axes/zygotic embryos in these modern protocols, range from 10°C min^{-1} (e.g. Vertucci, 1989; Sershen et al., 2007) to hundreds of °C s^{-1} (e.g. Wesley-Smith et al., 2004; Walters et al., 2002). Additionally, most recent plant cryopreservation studies involve one of the following vitrification-based procedures: pre-growth, dehydration, pre-growth-dehydration, encapsulation-dehydration, vitrification, encapsulation-vitrification, droplet-vitrification (González-Arnao et al., 2008) and, over the last decade or so, the novel cryo-plate procedure (Yamamoto et al., 2011).

Pre-growth involves culturing explants on cryoprotectants, followed by rapid, direct immersion in LN (Engelmann, 2004), while for dehydration explants are usually partially dehydrated in a laminar flow or above a bed of silica gel before immersion in LN. Pre-growth-dehydration involves the combination of both these procedures. These methods are now widely applied for cryopreserving meristems and polyembryonic cultures, small seeds and seed zygotic embryos or embryonic axes (González-Arnao et al., 2008). However, reports on their successful application to recalcitrant seed embryonic explants are rare. The encapsulation-dehydration technique is based on the methods used for artificial seed production, and involves encapsulating explants in alginate beads, followed by pre-growth in sucrose-enriched liquid medium and then partial desiccation in a laminar flow or using silica gel. The explants, often shoot apices (González-Arnao and Engelmann, 2006), are usually dried to a water content around 20% (fresh mass basis), the effects of which are minimised by being encapsulated, and they are then rapidly immersed in LN.

Vitrification techniques have remained relatively unchanged over the last few decades and involve the immersion of explants (most often somatic embryos, apices and cell suspensions) in cryoprotective solution (loading), chemical dehydration in highly concentrated plant vitrification solutions (PVS), followed by rapid cooling and rewarming, and finally removal (unloading) of cryoprotectants before in vitro recovery (Sakai and Engelmann, 2007; Sakai et al., 2008). Encapsulation-dehydration and vitrification procedures are sometimes combined (encapsulation-vitrification) and involve encapsulating explants before cryoprotection and cooling. The droplet vitrification technique (Kartha et al., 1982) is also based on the vitrification procedure and is still in use; samples are loaded, dehydrated with vitrification solutions and then placed within small droplets of these solutions on aluminium foil, which are then immersed with the samples in LN.

The cryo-plate procedure (Yamamoto et al., 2011) is one of the most revolutionary technical developments in plant cryopreservation over the last decade and combines the droplet-vitrification and encapsulation-dehydration techniques. The methods involve using a thin layer of calcium alginate to attach shoot tips to an aluminium cryo-plate. The explants are then loaded, dehydrated with PVS and then cooled in LN by direct immersion of the cryo-plates (Yamamoto et al., 2011, 2012). It is evident that the commonalities between both these relatively recent cryopreservation techniques, droplet-vitrification and cryo-plate, are the rapid cooling and warming rates (achieved by immersion in a sucrose-enriched medium at ambient temperature) compared with other vitrification-based procedures. These rapid cooling and warming rates are a consequence of the fact that the aluminium contact surfaces have a very high thermal conductivity and this increases the probability of vitrification during cooling and the avoidance of devitrification during subsequent warming. This may explain why vitrification-based protocols have been shown to improve post-cryo preservation survival in complex organs (e.g. shoot-tips) of species that responded poorly to classical protocols (Panis, 1995; González-Arnao et al., 2003).

4 Conclusion and future trends

Over the last 50–60 years since many of the first national and international genebanks were established, the science of conventional seed banking has barely changed and it seems unlikely to change drastically in the near future, not least since many genebanks have, to a seemingly large extent, effectively conserved and distributed seed germplasm. Despite that apparent success, it is difficult to know the extent of genetic erosion that might have occurred since the original sample arrived at the genebank, through cycles of regeneration or loss of alleles (due to seed death) during storage (Fu, 2017). There has been concern that seeds of many species, even though desiccation tolerant, are perhaps 'minimally orthodox' in that their lifespan in conventional genebank storage may not extend to many decades. Colville and Pritchard (2019) recently published a meta-analysis of seed longevity data that suggested that many more species have relatively short seed lifespans, compared with the extreme longevity reported for few species.

The number of seed accessions in genebank storage continues to increase, perhaps in response to the identification of gaps in existing collections (as discussed elsewhere in this book), but also due to changing priorities and advances in plant science. Many genebanks now conserve genetic stocks such as recombinant inbred lines, MAGIC populations and similar, and provide direct or indirect access to genomic information. A number of genebanks have also started or plan to genotype all or subsets of their collection (as discussed by McCouch et al., 2012). Genetic markers may also be used to, for example, identify duplicates within and among collections or to track accessions through storage and cycles of regeneration, although the cost and additional logistics required for such tracking is probably unworkable for most genebanks currently (van Treuren and van Hintum, 2014).

Large genebanks, particularly those conserving only one or two crops, process, perhaps, many thousands of new seed lots every year. As such, by necessity, they often adopt a factory-like approach to manage operations, with staff trained for a particular activity and the seed lot then moved to the next position/task. In contrast, smaller genebanks, or seed banks conserving diverse species, particularly wild species, may take a different approach and have the same member(s) of staff follow the whole cycle for a particular seed lot. Whichever strategy is in place, it is beneficial for the genebank and for the wider seed conservation community, for there to be regular evaluation of procedures and active innovation where improvements are needed or offer advantages, for example, in terms of accuracy, consistency and/or efficiency. Related to this, the documentation of standard operating procedures (SOPs) is expected to be revised as procedures are adapted, as part of the continuous improvement principle of a genebank quality management system (as discussed elsewhere in this book).

Some genebanks have a research team responsible for developing new procedures, technologies and understanding, although in general, this is not something that has been prioritised at most genebanks. Nonetheless, it is helpful for genebanks to follow the seed science/testing and plant phenotyping communities to see whether emerging technologies could be applied in the genebank context (Whitehouse et al., 2020). As with many industries, more genebanks may have robots in the future. However, there will always be a need for people with an understanding of the value and uniqueness of germplasm samples, not least since it is for that very uniqueness and the underlying genes, that we conserve our species and agrobiodiversity for future generations.

The last two to three decades have seen significant progress being made in terms of conserving non-orthodox seeded and vegetatively propagated plant species from temperate, tropical and subtropical zones. This has been aided by the development of in vitro seed germination, zygotic embryo and callus culture, somatic embryogenesis and micropropagation systems for many of these species (e.g. a collection of Brazilian species (Pilatti et al., 2011)). The development of these in vitro systems has enabled short- and medium-term conservation for a number of these species, but successful cryopreservation protocols are still elusive for the vast majority. Nevertheless, there are presently collections in cryobanks in close to 15 countries, based on a list published by Cruz-Cruz et al. (2013) and our consultation with various specialists in the field. These collections, which include callus, pollen, shoot tips, dormant buds, seeds and embryogenic cell lines for vegetatively propagated species (Cruz-Cruz et al., 2013), have achieved significant species coverage, with cryopreservation protocols being established for root and tubers, ornamentals, crops and fruit trees of temperate and tropical origin (Engelmann, 2000; Benelli et al., 2013). As these collections expand and the production of clones obtained from elite genotypes, unique/important cell lines and genetically transformed material increases, genotyping collections in cryobanks to avoid duplication is going to become increasingly important. A recent review by Wang et al. (2020), for example, reports on the use of stem disc-bearing adventitious buds, small leaf square-bearing adventitious buds, rhizome buds and microtubers as novel propagule types and hence, possible new alternative explants for cryopreservation. Cryobanks of the future are therefore likely to house much more diverse germplasm than at present.

While the latest two cryopreservation techniques, namely, droplet-vitrification and cryo-plate, have proven to be more beneficial for cryopreserving complex organs (e.g. shoot-tips) than classical protocols, of tropical species in particular, successful cryopreservation of explants from recalcitrant seeds remains challenging for the vast majority of species of tropical and temperate origin (Ballesteros et al., 2021). Large-scale, routine application of cryopreservation is therefore still very limited in comparison with conventional low-temperature seed storage. However, the benefits of cryotherapy in eliminating viruses in explants from vegetatively propagated species (e.g. sugarcane (González-Arnao et al., 2020)) increased production of artificial seeds in breeding programmes for non-orthodox or non-seed producing plants (Ravi and Anand, 2012), and more examples of orthodox species with poor longevity in conventional seed/genebank storage (e.g. Ali et al., 2007; Mondoni et al., 2011; Davies et al., 2018) may encourage practitioners to integrate cryopreservation into existing plant biodiversity conservation procedures, whether or not these facilities focus on vegetatively propagated, orthodox- or recalcitrant-seeded species.

5 Where to look for further information

- CGIAR Genebank Platform (www.genebanks.org).
- Food and Agriculture Organization of the United Nations (FAO) (www.fao.org/genetic-resources/en/).
- International Society for Low Temperature Biology and Medicine (www.societyforcryobiology.org/).
- Millennium Seed Bank (www.kew.org/wakehurst/whats-at-wakehurst/millennium-seed-bank).
- United States Department of Agriculture Germplasm Resources Information Network (GRIN) (www.ars-grin.gov).

6 References

Abbas, H. A. I. D. E. R. and Qaiser, M. (2011). *Ruellia linearibracteolata*: conservation assessment and strategies to avoid extirpation, *Pak. J. Bot.* 43(5), 2351–2357.

Ali, N., Probert, R., Hay, F., Davies, H. and Stuppy, W. (2007). Post-dispersal embryo growth and acquisition of desiccation tolerance in *Anemone nemorosa* L. seeds, *Seed Sci. Res.* 17(3), 155–163.

Ballesteros, D., Fanega-Sleziak, N. and Davies, R. (2021). Cryopreservation of seeds and seed embyros in orthodox-, intermediate, and recalcitrant-seeded species. In: Wolkers, W. F. and Oldenhop, H. (Eds), *Cryopreservation and Freeze-Drying Protocols, Methods in Molecular Biology* (vol. 2180), Springer Science+Business Media, LLC, part of Springer Nature, New York, NY: Humana. doi: 10.1007/978-1-0716-0783-1_36.

Ballesteros, D. and Pence, V. C. (2019). Survival and growth of embryo axes of temperate trees after two decades of cryo-storage, *Cryobiology* 88, 110–113.

Baskin, C. C. and Baskin, J. M. (2014). *Seeds, Ecology, Biogeography, and Evolution of Dormancy and Germination* (2nd edn.), San Diego, CA: Academic Press.

Benelli, C., de Carlo, A. and Engelmann, F. (2013). Recent advances in the cryopreservation of shoot-derived germplasm of economically important fruit trees of *Actinidia*, *Diospyros*, *Malus*, *Olea*, *Prunus*, *Pyrus* and *Vitis*, *Biotechnol. Adv.* 31(2), 175–185.

Benson, E. E. (2008a). Cryopreservation of phytodiversity: a critical appraisal of theory and practice, *Crit. Rev. Plant Sci.* 27(3), 141–219.

Benson, E. E. (2008b). Crypreservation theory. In: Reed, B. M. (Ed), *Plant Cryopreservation: A Practical Guide*, New York: Springer, pp. 15–30.

Benson, E. E. (1999). *Plant Conservation Biotechnology*, London: Taylor and Francis.

Berjak, P., Bartels, P., Benson, E. E., Harding, K., Mycock, D. J., Pammenter, N. W. and Wesley-Smith, J. (2011a). Cryoconservation of South African plant genetic diversity, *In Vitro Cell. Dev.-Pl.* 47(1), 65–81.

Berjak, P., Campbell, G. K., Farrant, J. M., Omondi-Oloo, W. and Pammenter, N. W. (1995). Responses of seeds of *Azadirachta indica* (neem) to short-term storage under ambient or chilled conditions, *Seed Sci. Technol.* 23, 779–792.

Berjak, P., Farrant, J. M. and Pammenter, N. W. (1989). The basis of recalcitrant seed behaviour. In: Taylorson, R. B. (Ed), *Recent Advances in the Development and Germination of Seeds*, New York: Plenum Press, pp. 89–105.

Berjak, P., Kioko, J. I., Walker, M., Mycock, D. J., Wesley-Smith, J., Watt, M. P. and Pammenter, N. W. (1999a). Cryopreservation - an elusive goal. In: Marzalina, M., Khoo, K. C., Jayanthi, N., Tsan, F. Y. and Krishnapillay, B. (Eds), *Recalcitrant Seeds: Proceedings of the IUFRO Seed Symposium*, 1998, Kuala Lumpur: Forest Research Institute Malaysia, pp. 96–109.

Berjak, P., Sershen, Varghese, B. and Pammenter, N. W. (2011b). Cathodic amelioration of the adverse effects of oxidative stress accompanying procedures necessary for cryopreservation of embryonic axes of recalcitrant-seeded species, *Seed Sci. Res.* 21(3), 187–203.

Berjak, P., Walker, M., Watt, M. P. and Mycock, D. J. (1999b). Experimental parameters underlying failure or success in plant germplasm cryopreservation: a case study on zygotic axes of *Quercus robur* L., *CryoLetters* 20, 251–262.

Blakesley, D., Pask, N., Henshaw, G. G. and Fay, M. F. (1996). Biotechnology and the conservation of forest genetic resources; in vitro strategies and cryopreservation, *Plant Growth Regul.* 20(1), 11–16.

Boelt, B., Shrestha, S., Salimi, Z., Jørgensen, J. R., Nicolain, M. and Carstensen, J. M. (2018). Multispectral imaging - a new tool in seed quality assessment?, *Seed Sci. Res.* 28(Special Issue 3), 222–228.

Boniecka, J., Kotowicz, K., Skrzypek, E., Dziurka, K., Rewers, M., Jedrzejczyk, I., Wilmowicz, E., Berdychowska, J. and Dąbrowska, G. B. (2019). Potential biochemical, genetic and molecular markers of deterioration advancement in seeds of oilseed rape (*Brassica napus* L.), *Ind. Crop. Prod.* 130, 478–490.

Chandra, J., Sershen, Varghese, B. and Keshavkant, S. (2019). The potential of ROS inhibitors and hydrated storage in improving the storability of recalcitrant *Madhuca latifolia* seeds, *Seed Sci. Technol.* 47(1), 33–45.

Chauhan, R., Singh, V. and Quraishi, A. (2019). In vitro conservation through slow-growth storage. In: Faisal, M. and Alator, A. A. (Eds), *Synthetic Seeds*, Cham, Switzerland: Springer Nature, pp. 397–416.

Chin, H. F. (1996). Strategies for conservation of recalcitrant species. In: Normah, M. N., Narimah, M. K. and Clyde, M. M. (Eds), *In Vitro Conservation of Plant Genetic Resources*, Kuala Lumpur, Malaysia: Percetakan Watan Sd. Bhd., pp. 203–215.

Chin, H. F. and Roberts, E. H. (1980). *Recalcitrant Crop Seeds*, Kuala Lumpur, Malaysia: Tropical Press.

Colville, L. and Pritchard, H. W. (2019). Seed life span and food security, *New Phytol.* 224(2), 557–562.

Colville, L., Bradley, E. L., Lloyd, A. S., Pritchard, H. W., Castle, L. and Kranner, I. (2012). Volatile fingerprints of seeds of four species indicate the involvement of alcoholic fermentation, lipid peroxidation, and Maillard reactions in seed deterioration during ageing and desiccation stress, *J. Exp. Bot.* 63(18), 6519–6530.

Crisostomo, S., Hay, F. R., Reaño, R. and Borromeo, T. (2011). Are the standards for genebank drying conditions optimal for rice seed quality?, *Seed Sci. Technol.* 39(3), 666–672.

Cromarty, A. S., Ellis, R. H. and Roberts, E. H. (1990). *Design of Seed Storage Facilites for Genetic Conservation, Revised*, Rome: International Board for Plant Genetic Resources.

Cruz-Cruz, C. A., González-Arnao, M. T. and Engelmann, F. (2013). Biotechnology and conservation of plant biodiversity, *Resources* 2(2), 73–95.

Davies, R. M., Newton, R. J., Hay, F. R. and Probert, R. J. (2016). 150-seed comparative longevity protocol – a reduced seed number screening method for identifying short-lived seed conservation collections, *Seed Sci. Technol.* 44(3), 569–584.

Davies, R. M., Dickie, J. B. and Ballesteros, D. (2018). Evaluation of short-lived seeds' cryopreservation as alternative to conventional seed banking, *Cryobiology* 85, 140–141.

Dell'Aquila, A. (2009). Development of novel techniques in conditioning, testing and sorting seed physiological quality, *Seed Sci. Technol.* 37(3), 608–624.

Demilly, D., Ducournau, S., Wagner, M.-H. and Dürr, C. (2015). Digital imaging of seed germination. In: Gupta, S. D. and Ibaraki, Y. (Eds), *Plant Image Analysis: Fundamentals and Applications*, Boca Raton, FL: CRC Press, pp. 147–164.

Edberg Hansen, M. A. E., Hay, F. R. and Carstensen, J. M. (2016). A virtual seed file: the use of multispectral image analysis in the management of genebank seed accessions, *Plant Genet. Resour.* 14(3), 238–241.

El-Dawayati, M., Baki, M. A. A. and Abdelgalil, L. M. (2018). Effect of different conservation period with different sucrose concentrations on conserving somatic embryo clusters of date palm (*Phoenix dactylifera* L.) under minimal growth conditions, *Appl. Sci. Rep.* 21(1), 14–21.

Ellis, R. H., Hong, T. D. and Roberts, E. H. (1985). *Handbook of Seed Technology for Genebanks Volume I: Principles and Methodology*, Rome: International Board for Plant Genetic Resources.

Ellis, R. H., Hong, T. D. and Roberts, E. H. (1990). An intermediate category of seed storage behaviour? I. Coffee, *J. Exp. Bot.* 41(9), 1167–1174.

Ellis, R. H., Nasehzadeh, M., Hanson, J., Ndiwa, N. and Woldemariam, Y. (2019). Medium-term seed storage of diverse genera of forage grasses, evidence-based genebank monitoring intervals, and regeneration standards, *Genet. Resour. Crop Evol.* 66(3), 723–734.

Ellis, R. H., Nasehzadeh, M., Hanson, J. and Woldemariam, Y. (2018). Medium-term seed storage of 50 genera of forage legumes and evidence-based genebank monitoring intervals, *Genet. Resour. Crop Evol.* 65(2), 607–623.

Ellis, R. H. and Roberts, E. H. (1980). Improved equations for the prediction of seed longevity. *Ann. Bot.* 45,13–30.

Ellis, R. H. and Roberts, E. H. (1981). The quantification of ageing and survival in orthodox seeds, *Seed Sci. Technol.* 9, 373–409.

Engelmann, F. (1997). In vitro conservation methods. In: Callow, J. A., Ford-Lloyd, B. V. and Newbury, H. J. (Eds), *Biotechnology and Plant Genetic Resources: Conservation and Use*, Wallingford, UK: CABI Publishing, pp. 159–170.

Engelmann, F. (1999). Alternative methods for the storage of recalcitrant seeds- an update. In: Marzalina, M., Khoo, K. C., Jayanthi, N., Tsan, F. Y. and Krishnapillay, B. (Eds), *Recalcitrant Seeds: Proceedings of the IUFRO Seed Symposium, 1998*, Kuala Lumpur: Forest Research Institute Malaysia, pp. 96–109.

Engelmann, F. (2000). Importance of cryopreservation for the conservation of plant genetic resources. In: Engelmann, F. and Takagi, H. (Eds), *Cryopreservation of Tropical Plant Germplasm: Current Research Progress and Application*, Tsukuba, Japan: Japan International Research Center for Agriculture and Rome, Italy: International Plant Genetic Resources Institute, pp. 8–20.

Engelmann, F. (2004). Plant cryopreservation: progress and prospects, *In Vitro Cell. Dev. Biol. Plant* 40(5), 427–433.

Engels, J. and Engelmann, F. (1998). Introductory statement, *Seed Sci. Res.* 8(1), 1–8.

FAO (2014). *Genebank Standards for Plant Genetic Resources for Food and Agriculture*, Rev Ed., Rome: FAO.

Farrant, J. M., Pammenter, N. W. and Berjak, P. (1989). Germination-associated events and the desiccation sensitivity of recalcitrant seeds – a case study in three unrelated species, *Planta* 178(2), 189–198.

Finkel, B. and Ulrich, J. (1983). Protocols of cryopreservation. In: Evans, D. A., Sharp, W. R., Ammirato, P. V. and Yamada, Y. (Eds), *Handbook of Plant Cell Culture (Vol. 1): Techniques for Propagation and Breeding*, New York: Macmillan Publishing, pp. 806–815.

Fleming, M. B., Hill, L. M. and Walters, C. (2019). The kinetics of ageing in dry-stored seeds: a comparison of viability loss and RNA degradation in unique legacy seed collections, *Ann. Bot.* 123(7), 1133–1146.

Frankel, O. H. (1990). Germplasm conservation and utilization in horticulture. In: *Horticultural Biotechnology*, New York: Wiley Liss Inc., pp. 5–17.

Fu, J., Zhang, B. Z., Wang, X. P., Qiao, Y. Z. and Huang, X. L. (1990). Physiological studies on desiccation, we storage and cryopreservation of recalcitrant seeds of three fruit species and their excised embryonic axes, *Seed Sci. Technol.* 18, 743–754.

Fu, Y. B. (2017). The vulnerability of plant genetic resources conserved ex situ, *Crop Sci.* 57(5), 2314–2328.

Fu, Y. B., Ahmed, Z. and Diederichsen, A. (2015). Towards a better monitoring of seed ageing under ex situ seed conservation, *Conserv. Physiol.* 3(1), cov026.

Funnekotter, B., Mancera, R. L. and Bunn, E. (2017). Advances in understanding the fundamental aspects required for successful cryopreservation of Australian flora, *In Vitro Cell.Dev.Biol. Plant* 53(4), 289–298.

Ganeshan, S., Rajasekharan, S., Shashikumar, S. and Decruze, W. (2008). Cryopreservation of pollen. In: Reed, B. M. (Ed), *Plant Cryopreservation: A Practical Guide*, New York: Springer, pp. 15–30.

George, E. F. (1993). *Plant Propagation by Tissue Culture. Part 1: The Technology*, Westbury, UK: Exegetics Limited.

González-Arnao, M. T., Banasiak, M., Snyman, S. J. and Sershen (2020). The potential of cryotherapy to remove sugarcane mosaic virus from sugarcane (*Saccharum* spp. hybrids) shoot tips, *CryoLetters* 41(5), 266–270.

González-Arnao, M. T. and Engelmann, F. (2006). Cryopreservation of plant germplasm using the encapsulation-dehydration technique: review and case study on sugarcane, *Cryoletters* 27(3), 155–168.

González-Arnao, M. T., Juarez, J., Ortega, C., Navarro, L. and Duran-Vila, N. (2003). Cryopreservation of ovules and somatic embryos of citrus using the encapsulation-dehydration technique, *CryoLetters* 24(2), 85–94.

González-Arnao, M. T., Martinez-Montero, M. E., Cruz-Cruz, C. A. and Engelmann, F. (2014). Advances in cryogenic techniques for the long-term preservation of plant biodiversity. In: Ahuja, M. R. and Ramawat, K. G. (Eds), *Biotechnology and Biodiversity*, Heidelberg, New York, Dordrecht, London: Springer Cham, pp. 129–170.

González-Arnao, M. T., Panta, A., Roca, W. M., Escobar, R. H. and Engelmann, F. (2008). Development and large scale application of cryopreservation techniques for shoot and somatic embryo cultures of tropical crops, *Plant Cell. Tissue Org. Cult.* 92, 1–13.

Grout, B. W. W. (1995). *Genetic Preservation of Plant Cells In Vitro*, Berlin: Springer-Verlag.

Grout, B. W. W., Westcott, R. J. and Henshaw, G. G. (1978). Survival of shoot meristems of tomato seedlings frozen in liquid nitrogen, *Cryobiology* 15(4), 478–483.

Harding, K. (2004). Genetic integrity of cryopreserved plant cells: a review, *CryoLetters* 25(1), 3–22.

Harper, S. and Long, J. (2011). 'Automated analysis of primed seed', presentation at the ISTA Seminar New Developments and Technologies in Seed Testing, 11 June 2012, Venlo, The Netherlands. Available at: https://www.seedtest.org/upload/cms/user/ISTA-June11-Seminar-0935-Harper.pdf [Last accessed 25 May 2020].

Hassan, M. M. (2017). In vitro conservation of date palm somatic embryos using growth-retardant conditions. In: Al-Khayri, J. M., Jain, S. M. and Johnson, D. V. (Eds), *Date Palm Biotechnology Protocols Volume II*, New York: Humana Press, pp. 61–70.

Hay, F. R. and Whitehouse, K. J. (2017). Rethinking the approach to viability monitoring in seed genebanks, *Conserv. Physiol.* 5(1), cox009. doi:10.1093/conphys/cox009.

Hay, F. R., Mead, A., Manger, K. and Wilson, F. J. (2003). One-step analysis of seed storage data and the longevity of Arabidopsis thaliana seeds, *J. Exp. Bot.* 54(384), 993–1011.

Hay, F. R., de Guzman, F. and Sackville Hamilton, N. R. S. (2015). Viability monitoring intervals for genebank samples of *Oryza sativa*, *Seed Sci. Technol.* 43(2), 218–237.

Hong, T. D. and Ellis, R. H. (1996). *A Protocol to Determine Seed Storage Behaviour*, Rome: International Plant Genetic Resources Institute.

Hor, Y. L., Stanwood, P. C. and Chin, F. C. (1990). Effects of dehydration on freezing characteristics and survival in liquid nitrogen of three recalcitrant seeds, *Pertanika* 13(2), 309–314.

IBPGR (1976). *Report on the IBPGR Working Group on Engineering, Design and Cost Aspects of Long-Term Seed Storage Facilities*, Rome: International Board for Plant Genetic Resources.

Ingram, J. and Bartels, D. (1996). The molecular basis of dehydration tolerance in plants, *Annu. Rev. Plant Physiol. Plant Mol. Biol.* 47(1), 377–403.

IPGRI/FAO (1994). *Genebank Standards*, Rome: International Plant Genetic Resources Institute, Food and Agriculture Organization of the United Nations.

ISTA (2020). *International Rules for Seed Testing*, Basserdorf, Switzerland: International Seed Testing Association.

Jarret, R. L. and Gawel, N. (1991). Abscisic acid-induced growth inhibition of sweet potato (*Ipomoea batatas* L.) in vitro, *Plant Cell. Tiss. Org.* 24(1), 13–18.

Jenderek, M. M. and Reed, B. M. (2017). Cryopreserved storage of clonal germplasm in the USDA National Plant Germplasm System, *In Vitro Cell.Dev.Biol.-Plant* 53(4), 299–308.

Joosen, R. V. L., Kodde, J., Willems, L. A. J., Ligterink, W., van der Plas, L. H. W. and Hilhorst, H. W. M. (2010). GERMINATOR: a software package for high-throughput scoring and curve fitting of Arabidopsis seed germination, *Plant J.* 62(1), 148–159.

Kaczmarczyk, A., Funnekotter, B., Menon, A., Phang, P. Y., Al-Hanbali, A., Bunn, E. and Mancera, R. (2012). Current issues in plant cryopreservation. In: Katkov, I. (Ed.), *Current Frontiers in Cryobiology*, Rijeka, Croatia: IntechOpen, pp. 417–438.

Kartha, K. K. and Engelmann, F. (1994). Cryopreservation and germplasm storage. In: Vasil, I. K. and Thorpe, T. A. (Eds), *Plant Cell and Tissue Culture*, Dordrecht, Netherlands: Kluwer Academic Publishers, pp. 195–230.

Kartha, K. K., Leung, N. L. and Mroginski, L. A. (1982). In vitro growth responses and plant regeneration from cryopreserved meristems of cassava (*Manihot esculenta* Crantz), *Z. Pflanzenphysiol.* 107(2), 133–140.

King, M. W. and Roberts, E. H. (1980). Maintenance of recalcitrant seeds in storage. In: Chin, H. F. and Roberts, E. H. (Eds), *Recalcitrant Crop Seeds*, Kuala Lumpur, Malaysia: Tropical Press, pp. 53–89.

Kioko, J. I., Berjak, P., Pammenter, N. W., Lamb, J. and Gumede, Z. (2003). Is genetic fidelity maintained following the cryopreservation of the seeds of the endangered African Pepper Bark (*Warburgia salutaris*)? In: Nicolas, G., Bradford, K. J., Côme, D. and Pritchard, H. W. (Eds), *The Biology of Seeds: Recent Research Advances*, Wallingford, UK: CABI Publishing, pp. 327–335.

Krøgstrup, P., Baldursson, S. and Norg, J. V. (1992). Ex situ genetic conservation by tissue culture, *Opera Bot. Belg.* 113, 49–53.

Lee, J. S., Velasco-Punzalan, M., Pacleb, M., Valdez, R., Kretzschmar, T., McNally, K. L., Ismail, A. M., Cruz, P. C. S., Sackville Hamilton, N. R. and Hay, F. R. (2019). Variation in seed longevity among diverse Indica rice varieties, *Ann. Bot.* 124(3), 447–460.

Mandal, B. B., Tyagi, R. K., Pandey, R., Sharma, N. and Agrawal, A. (2000). In vitro conservation of germplasm of agri-horticultural crops at NBPGR: an overview. In: Razdan, M. K. and Cocking, E. C. (Eds), *Conservation of Plant Genetic Resources In Vitro. Volume 2: Applications and Limitations*, New Hampshire: Science Publishers Inc., pp. 279–307.

Mazur, P. (1990). Equilibrium, quasi-equilibrium, and nonequilibrium freezing of mammalian embryos, *Cell Biophys.* 17(1), 53–92.

McCouch, S. R., McNally, K. L., Wang, W. and Sackville Hamilton, R. (2012). Genomics of gene banks: a case study in rice, *Am. J. Bot.* 99(2), 407–423.

McFerson, J. R. (1998). From in situ to ex situ and back: the importance of characterizing germplasm collections, *HortSci* 33(7), 1134–1135.

Mondoni, A., Probert, R. J., Rossi, G., Vegini, E. and Hay, F. R. (2011). Seeds of alpine plants are short-lived: implications for long-term conservation, *Ann. Bot.* 107(1), 171–179.

Moothoo-Padayachie, A., Varghese, B., Pammenter, N. W., Govender, P. and Sershen (2016). Germination associated ROS production and glutathione redox capacity in two recalcitrant-seeded species differing in seed longevity, *Botany* 94(12), 1103–1114.

Moothoo-Padayachie, A., Varghese, B., Pammenter, N. W., Govender, P. and Sershen (2018). A comparison of partial dehydration and hydrated storage-induced changes in viability, reactive oxygen species production, and glutathione metabolism in two contrasting recalcitrant-seeded species, *Acta Physiol. Plant.* 40(1), 14.

Mroginski, L. A., Sansberro, P. A., Scocchi, A. M., Luna, C. and Rey, H. Y. (2008). A cryopreservation protocol for immature zygotic embryos of *Ilex* (Aquifoliaceae), *Biocell* 32(1), 33–39.

Mycock, D. J. and Berjak, P. (1990). Fungal contaminants associated with several homoiohydrous (recalcitrant) seed species, *Phytophylactica* 22, 413–418.

Mycock, D. J. and Berjak, P. (1993). Cryostorage of coffee somatic embryos. In: Côme, D. and Corbineau, F. (Eds), *Basic and Applied Aspects of Seed Biology*, France: AFSIS, pp. 879–888.

Mycock, D. J. and Berjak, P. (1995). The implication of seed-associated mycoflora during storage. In: Kigel, J. and Galili, G. (Eds), *Seed Development and Germination*, New york: Marcel Dekker, pp. 747–766.

Mycock, D. J., Blakeway, F. C., Watt, M. P. and Bornman, C. H. (2004). General applicability of in vitro storage technology to the conservation and maintenance of plant germplasm, *S. Afr. J. Bot.* 70(1), 31–36.

National Research Council (1991). *The U.S. National Plant Germplasm System*, Washington, DC: The National Academies Press. doi: 10.17226/1583.

Pammenter, N. W. and Berjak, P. (2014). Physiology of desiccation-sensitive (recalcitrant) seeds and the implications for cryopreservation, *Int. J. Plant Sci.* 175(1), 21–28.

Pammenter, N. W., Berjak, P., Farrant, J. M., Smith, M. T. and Ross, G. (1994). Why do stored hydrated recalcitrant seeds die?, *Seed Sci. Res.* 4(2), 187–191.

Panis, B. (1995). Cryopreservation of Banana (*Musa* spp.) Germplasm, Ph.D. Thesis, Catholic University of Leuven, Leuven, Belgium.

Panis, B. and Lambardi, M. (2006). Status of cryopreservation technologies in plants (crops and forest trees). In: Ruane, J. and Sonnino, A. (Eds), *The Role of Biotechnology in Exploring and Protecting Agricultural Genetic Resources*, Rome: Food and Agricultural Organization of the United Nations, pp. 61–78.

Paunescu, A. (2009). Biotechnology for endangered plant conservation: a critical overview, *Rom. Biotech. Lett.* 14(1), 4095–4103.

Pilatti, F. K., Aguiar, T., Simões, T., Benson, E. E. and Viana, A. M. (2011). In vitro and cryogenic preservation of plant biodiversity in Brazil, *In Vitro Cell.Dev. Biol. Plant* 47(1), 82–98.

Pritchard, H. W. and Dickie, J. B. (2003). Predicting Seed Longevity: the use and abuse of seed viability equations. In: Smith, R. D., Dickie, J. D., Linington, S. H., Pritchard, H. W.. and Probert, R. J. (Eds), *Seed Conservation: Turning Science into Practice*, Richmond, UK: Royal Botanic Gardens Kew, pp. 653–721.

Pritchard, H. W., Haye, A. J., Wright, W. J. and Steadman, K. J. (1995). A comparative study of seed viability in *Inga* species: desiccation tolerance in relation to the physical characteristics and chemical composition of the embryo, *Seed Sci. Technol.* 23, 85–100.

Pritchard, H. W. and Nadarajan, J. (2008). Cryopreservation of orthodox (desiccation tolerant) seeds. In: Reed, B. M. (Ed.), *Plant Cryopreservation: A Practical Guide*, New York: Springer, pp. 485–501.

Rao, N. K., Hanson, J., Dulloo, M. E., Ghosh, K., Nowell, D. and Larinde, M. (2006). *Manual of Seed Handling in Genebanks. Handbooks for Genebanks No. 8*, Rome, Italy: Bioversity International.

Ravi, D. and Anand, P. (2012). Production and applications of artificial seeds: a review, *Int. Res. Biol. Sci.* 1, 74–78.

Reed, B. M. (1996). Pretreatment strategies for the cryopreservation of plant tissues. In: Normah, M. N., Narimah, M. K. and Clyde, M. M. (Eds), *In Vitro Conservation of Plant Genetic Resources*, Selangor, Malaysia: Plant Biotechnology Laboratory, Faculty of Life Sciences, Universiti Kebangsaan.

Reed, B. M. (2008). Cryopreservation - Practical considerations. In: Reed, B. M. (Ed), *Plant Cryopreservation: A Practical Guide*, New York: Springer Science and Business Media LLC, pp. 10–13.

Reed, B. M. and Chang, Y. (1997). Medium- and long-term storage of in vitro cultures of temperate fruit and nut crops. In: Razdan, M. K. and Cocking, E. C. (Eds), *Conservation of Plant Genetic Resources In Vitro. Volume 1: General Aspects*, Enfield, NH: Science Publishers Inc., pp. 67–85.

Roberts, E. H. (1973). Predicting the storage life of seeds, *Seed Sci. Technol.* 1, 499–514.

Royal Botanic Gardens Kew (2020) Seed Information Database (SID). Version 7.1, Available at: http://data.kew.org/sid/.

Sakai, A. (2004). Plant cryopreservation. In: Fuller, B., Lane, N. and Benson, E. E. (Eds), *Life in the Frozen State*, London, New York: CRC Press, pp. 329–345.

Sakai, A. and Engelmann, F. (2007). Vitrification, encapsulation-vitrification and droplet-vitrification: a review, *CryoLetters* 28(3), 151–172.

Sakai, A., Hirai, D. and Niino, T. (2008). Development of PVS-based vitrification and encapsulation-vitrification protocols. In: Reed, B. M. (Ed), *Plant Cryopreservation: A Practical Guide*, Berlin, Germany: Springer Verlag, pp. 33–58.

Sakai, A., Niino, T., Yamada, T. and Kohmura, H. (1992). Cryopreservation by vitrification of mulberry meristems, meristematic callus of white clover and multiple bud cultures of asparagus, *Cryobiology* 29, 747–748.

Sakai, A. and Nishiyama, Y. (1978). Cryopreservation of winter vegetative buds of hardy fruit trees in liquid nitrogen, *HortSci* 13, 225–227.

Schnapp, R. Y. and Preece, J. E. (1986). In vitro growth reduction of tomato and carnation microplants, *Plant Cell. Tiss. Org.* 6, 3–8.

Sershen, Pammenter, N. W. and Berjak, P. (2008). Post-harvest behaviour and short- to medium-term storage of recalcitrant seeds and encapsulated embryonic axes of selected amaryllid species, *Seed Sci. Technol.* 36(1), 133–147.

Sershen, Pammenter, N. W., Berjak, P. and Wesley-Smith, J. (2007). Cryopreservation of embryonic axes of selected amaryllid species, *CryoLetters* 28(5), 387–399.

Sershen, Varghese, B., Pammenter, N. W. and Berjak, P. (2012). Cryo-tolerance of zygotic embryos from recalcitrant seeds in relation to oxidative stress – a case study on two amaryllid species, *J. Plant Physiol.* 169(10), 999–1011.

Stanwood, P. C. (1985). Cryopreservation of seed germplasm for genetic conservation. In: Kartha, K. K. (Ed), *Cryopreservation of Plant Cells and Organs*, Boca Raton, FL: CRC Press Inc., pp. 199–226.

Staritsky, G. (1997). Backgrounds and principals of in vitro conservation. In: Razdan, M. K. and Cocking, E. C. (Eds), *Conservation of Plant Genetic Resources In Vitro. Volume 1: General Aspects*, Enfield, NH: Science Publishers Inc., pp. 27–51.

Subbiah, A., Ramdhani, S., Pammenter, N. W., Macdonald, A. H. and Sershen (2019). Towards understanding the incidence and evolutionary history of seed recalcitrance: an analytical review, *Perspect. Plant Ecol.* 37, 11–19.

Sutherland, J. R., Diekmann, M. and Berjak, P. (2002). Forest Tree Seed Health. IPGRI Technical Bulletin 6, Rome: International Plant Genetic Resources Institute.

Taylor, M., Pone, S. and Palupe, A. (1996). Slow growth strategies. In: Normah, M. N., Narimah, M. K. and Clyde, M. M. (Eds), *In Vitro Conservation of Plant Genetic Resources*, Kuala Lumpur: Percetakan Watan, pp. 119–134.

Thakur, S., Tiwari, K. L. and Jadhav, S. K. (2015). In vitro approaches for conservation of *Asparagus racemosus* Willd, *In Vitro Cell.Dev.Biol. Plant* 51(6), 619–625.

Timple, S. E. and Hay, F. R. (2018). High-temperature drying of seeds of wild *Oryza* species intended for long-term storage, *Seed Sci. Technol.* 46(1), 107–112.

Uyoh, E. A., Nkang, A. E. and Eneobong, E. E. (2003). Biotechnology, genetic conservation and sustainable use of bioresources, *Afr. J. Biotechnol.* 2(12), 704–709.

van Treuren, R., de Groot, E. C. and van Hintum, T. J. L. (2013). Preservation of seed viability during 25 years of storage under standard genebank conditions, *Genet. Resour. Crop Evol.* 60(4), 1407–1421.

van Treuren, R. and van Hintum, T. J. L. (2014). Next-generation genebanking: plant genetic resources management and utilization in the sequencing era, *Plant Genet. Resour.* 12(3), 298–307.

Varghese, D., Berjak, P. and Pammenter, N. W. (2009). Cryopreservation of shoot tips of *Trichilia emetica*: a tropical recalcitrant-seeded species, *Cryoletters* 30(4), 280–290.

Vertucci, C. W. (1989). Effects of cooling rate on seeds exposed to liquid nitrogen temperatures, *Plant Physiol.* 90(4), 1478–1485.

Walters, C., Touchell, D. H., Power, P., Wesley-Smith, J. and Antolin, M. F. (2002). A cryopreservation protocol for embryos of the endangered species *Zizania texana*, *CryoLetters* 23(5), 291–298.

Walters, C., Wesley-Smith, J., Crane, J., Hill, L. M., Chiemlarz, P., Pammenter, N. W. and Berjak, P. (2008). Cryopreservation of recalcitrant (i.e. desiccation-sensitive) seeds. In: Reed, B. M. (Ed.), *Plant Cryopreservation: A Practical Guide*, New York: Springer, pp. 465–482.

Walters, C., Wheeler, L. M. and Grotenhuis, J. M. (2005). Longevity of seeds stored in a genebank: species characteristics, *Seed Sci. Res.* 15(1), 1–20.

Wang, M. R., Lambardi, M., Engelmann, F., Pathirana, R., Panis, B., Volk, G. M. and Wang, Q. C. (2020). Advances in cryopreservation of in vitro-derived propagules: technologies and explant sources, *Plant Cell. Tiss. Org.*, doi: 10.1007/s11240-020-01770-0.

Wesley-Smith, J., Vertucci, C. W., Berjak, P., Pammenter, N. W. and Crane, J. (1992). Cryopreservation of desiccation-sensitive axes of *Camellia sinensis* in relation to dehydration, freezing rate and thermal properties of tissue water, *J. Plant Physiol.* 140(5), 596–604.

Wesley-Smith, J., Walters, C., Berjak, P. and Pammenter, N. W. (2004). Non-equilibrium cooling of *Poncirus trifoliata* (L.) embryonic axes at various water contents, *CryoLetters* 25(2), 121–128.

Whitehouse, K. J., Hay, F. R. and Ellis, R. H. (2015). Increases in the longevity of desiccation-phase developing rice seeds: response to high temperature drying depends on harvest moisture content, *Ann. Bot.* 116(2), 247–259.

Whitehouse, K. J., Hay, F. R. and Ellis, R. H. (2018a). Improvement in rice seed storage longevity from high-temperature drying is a consistent positive function of harvest moisture content above a critical value, *Seed Sci. Res.* 28(4), 332–339.

Whitehouse, K. J., Hay, F. R. and Lusty, C. (2020). Why seed physiology is important for genebanking, *Plants* 9(5), 584.

Whitehouse, K. J., Olufemi, A., Owoborode, S., Olaniyi, O., Olaniyan, B., Hay, F. R. and Abberton, M. (2018b). Further evidence that the genebank standards for drying orthodox seeds may not be optimal for subsequent seed longevity, *Biopreserv. Biobank* 16, 327-336.

Withers, L. A. and Engelmann, F. (1997). In vitro conservation of plant genetic resources. In: Altman, A. (Ed), *Biotechnology in Agriculture*, New York: Marcel Dekker Inc., pp. 57–88.

Withers, L. A. and King, P. J. (1980). A simple freezing unit and routine cryopreservation method for plant cell cultures, *CryoLetters* 24, 303-314.

Yamamoto, S., Rafique, T., Fukui, K., Sekizawa, K. and Niino, T. (2012). V-cryo-plate procedure as an effective protocol for cryobanks: case study of mint cryopreservation, *Cryoletters* 33(1), 12–23.

Yamamoto, S., Rafique, T., Priyantha, W. S., Fukui, K., Matsumoto, T. and Niino, T. (2011). Development of a cryopreservation procedure using aluminium cryo-plates, *CryoLetters* 32(3), 256–265.

Yamasaki, F., Domon, E., Tomooka, N., Baba-Kasai, A., Nemoto, H. and Ebana, K. (2020). Thirty-year monitoring and statistical analysis of 50 species' germinability in genebank medium-term storage suggest specific characteristics in seed longevity, *Seed Sci. Technol.* 48(2), 269–287.

Zandvoort, E. A., Hulshof, M. J. H. and Staritsky, G. (1994). In vitro storage of *Xanthosoma* spp. under minimal growth conditions, *Plant Cell. Tiss. Org.* 36(3), 309–316.

Chapter 9

The role of the Svalbard Global Seed Vault in preserving crop genetic diversity

Åsmund Asdal, Nordic Genetic Resource Centre (NordGen), Sweden

1 Introduction

Plant genetic diversity is a precondition for agricultural research and plant breeding needed for securing future food supplies. In particular, plant genetic resources are the raw material needed for development of improved crop plant varieties that farmers and the food industry need for coping with pest and diseases and changing environmental conditions due to climate change (such as drought, floods, salinity) and to increase food production for a growing global population. The loss of these valuable plant genetic resources poses a threat to the world's food security in the long term (FAO, 1998).

In an attempt to safeguard plant genetic resource, an International Board on Plant Genetic Resources (IBPGR) was established in 1974 to coordinate an international plant genetic resources programme for collecting and promoting the conservation and use of plant genetic resources and the creation of genebanks around the world. As a result, a massive campaign for the collection of the disappearing diversity in plant genetics resources around the world was carried out leading to over 7.4 million accessions being conserved in more than 1750 genebanks worldwide (FAO, 2010). These genebanks form part of the global system for the conservation of plant genetic resources.

However, many genebanks are vulnerable to different types of natural disasters, war and conflicts or simply to lack of resources. As the genetic resources they contain are of high value, insurance policies are needed to ensure their protection. One important and simple security measure is to ensure

http://dx.doi.org/10.19103/AS.2020.0085.21

that samples of these valuable genetic materials are conserved in more than one place. In this context, the Svalbard Global Seed Vault offers a free of charge service to store duplicate samples of seeds that are conserved in genebanks worldwide. This chapter provides an overview of this facility in Svalbard and describes its operations as well as its functions and status.

2 Storing seeds in the Arctic: establishing the Svalbard Global Seed Vault

Svalbard is a Norwegian archipelago of islands in the Arctic, halfway between northern Norway and the North Pole. The story of the Svalbard Global Seed Vault started in the Svalbard archipelago in 1984 when the first seeds from the Nordic seed collection conserved at the Nordic Gene Bank were stored on the Spitsbergen Island. The ground in Svalbard provides constant permafrost conditions, providing the ideal conditions for establishing a conservation facility for conserving seeds in sub-zero conditions. The coal mine provided permafrost conditions with a yearly constant temperature of between −3.5°C and −4.0°C which ensured that seeds remain frozen, independent of external energy supply. The success of this endeavour was possible, thanks to good infrastructure in Svalbard and to good collaboration with the local community. At first, a steel container was placed in the coal mine and established as a seed storage unit, and the first seeds were conserved there in November 1984.

The facility gained significant international attention, and the idea of establishing a facility for back-up samples of other national and international genebanks in permafrost was launched for the first time in 1989 (Qvenild, 2005). The idea was not developed further at that time, mainly because the international framework for comprehensive international cooperation on conservation and use of plant genetic resources was not in place. However, the idea was revived when the International Treaty on Plant Genetic Resources for Food and Agriculture (ITPGRFA) came into force in 2004. A report was then commissioned by the Norwegian Ministry of Foreign Affairs (Fowler et al., 2004), to investigate the feasibility of establishing an international seed facility in Svalbard, taking into account political, legal, technical and practical considerations. Based on the recommendation of this report, in 2004 Norway offered to build and manage a global seed vault in Svalbard at the FAO Commission on Genetic Resources for Food and Agriculture meeting in Rome, Italy. With comprehensive international support, the Seed Vault was built and funded by the Norwegian government during 2007 and opened on 26 February 2008. In accordance with Norwegian policy, the Seed Vault was built as a contribution to the international efforts for securing biological and genetic resources. The ultimate goal is that all unique plant genetic material conserved in genebanks is secured and copied in the Svalbard Global Seed Vault. Over the years the Seed Vault has also served a second important goal, namely to contribute to public awareness about the importance of conservation and use of plant genetic resources.

The Nordic Genetic Resource Centre (NordGen) was heavily involved in the planning process of the Seed Vault and has been responsible for its management and operation from the start. Management and operation of the Seed Vault is conducted in accordance with a three-party agreement between the Norwegian Ministry of Agriculture and Food, the Global Crop Diversity Trust and NordGen. Funding for management is provided by the three partners and secured through this three-party agreement. The Svalbard Global Seed Vault is owned by the Norwegian government, which also guarantees the security of the seeds and its long-term management. Thanks to significant joint efforts and financial support from the Global Crop Diversity Trust, many genebanks around the world prepared and shipped seed samples for depositing at the Seed Vault from its opening in 2008. Thanks to these efforts, 320 549 seed accessions from 22 genebanks alone were deposited in the Seed Vault during the first year of operation.

3 The Seed Vault in a global context

From the very beginning of the project, the partners and, in particular, the Norwegian Ministry of Agriculture and Food, have stressed the need for international cooperation as essential to the operation of the Seed Vault. The offer to deposit duplicates of gene bank seed collections is regularly raised in meetings and conferences, for example, in the FAO Commission on Genetic Resources for Food and Agriculture (CGRFA) and in the meetings of the governing body (GB/IT) of the International Treaty on Plant Genetic Resources for Food and Agriculture (ITPGRFA).

The FAO genebank standards (2014) stress that a safety duplicate sample for every original accession should be stored in a geographically distant area, under same or better conditions than those in the original genebank. The standards say that 'Safety duplication' is available at the Svalbard Global Seed Vault on Spitsbergen Island, Norway. Institutions depositing seeds retain ownership and access to samples stored in Svalbard *is granted to the depositor only*.

The GB/IT meeting in Kigali, Rwanda in 2017 highlighted the Seed Vault in its Resolution 12/2017 Cooperation with other international bodies and organizations (ITPGRFA, 2017), Part III: Management of the Svalbard Global Seed Vault. The GB/IT reaffirmed the strong linkages between the Seed Vault and the International Treaty and acknowledged that the Seed Vault is an important element of the Global System for ex situ conservation and use of plant genetic resources for food and agriculture (PGRFA). The resolution further invited the contracting parties, international institutions and other relevant bodies to consider making use of the Seed Vault in their strategy for securing their important seed collections and for the purpose of long-term storage of PGRFA.

The close collaboration between the Seed Vault and the Plant Treaty is further manifested through the fact that the current chair of the GB/IT is invited to be the chairperson of the Svalbard Global Seed Vault International Advisory Panel (IAP). The management of the Svalbard Global Seed Vault is overseen by the International Advisory Panel, which is an important instrument for ensuring the quality and transparency of the Seed Vault operations. The Panel assists the partners in anchoring the Seed Vault in the global plant genetic resource conservation system and in the community of genebanks, national genetic resource programmes and governments involved in conservation and use. Meetings of the IAP are organized on a regular basis with the aim to obtain advice, secure transparency, and being a direct channel for the depositors on issues related to the management and operations of the Svalbard Global Seed Vault.

The opening of the Svalbard Global Seed Vault received great media exposure, which made it to become a great public attraction. It has drawn international attention to the issues on conservation and sustainable utilization of plant genetic resources. Influential politicians and policymakers have regularly paid visits to Svalbard and the Seed Vault and there has been a continuous interest from many of the world's leading media to tell the story of this global undertaking for conservation and utilization of crops and their wild relatives. The Seed Vault is now probably Norway's most well-known building internationally and it has been an iconic symbol of the importance of taking care of plant genetic resources.

Even if the Seed Vault is currently closed to all visitors for security reasons, the interest in visiting the Seed Vault surroundings and to publish stories about the Seed Vault on different media platforms has not decreased. Several thousand visitors visit the Seed Vault entrance every year. NordGen and partners regularly brief the international media and give interviews and lectures on many occasions each year. The Seed Vault mission and the construction itself has gained significant international attention and has significantly increased the overall awareness about the importance of conserving genetic diversity.

4 The Seed Vault facility and its operation

The Seed Vault facility is unlike the initial storage of material in the coal mine, since it is constructed in virgin solid rock without any coal layers. The facility consists of a tunnel of about 100 m, an initial hall and three seed storage chambers, each designed with the capacity to store around 1.5 million seed samples. The Seed Vault is located 130 m above sea level, which is above worst-case climate change scenario for sea level rise. The seed chambers lie between 40 m and 60 m under the surface of the hillside of the Plateau Mountain above Longyearbyen Airport.

Artificial cooling secures that the temperature in the seed storage is maintained at −18°C, which is in accordance with the FAO Genebank Standards' recommendation for long-term conservation of seeds. Permafrost at approximately −3°C in the soils ensures that seeds stay frozen even if the cooling system should fail. During the summer season, the Seed Vault experienced minor water intrusions in the entrance tunnel. In October 2016, Svalbard experienced a heavy rainfall, which was unexpected due to the normally very dry Svalbard climate. This rainfall led to a larger water leak in the tunnel than previous incidents. Due to this incident, and climate change scenarios suggesting that future climate in Svalbard will be warmer and wetter, the Norwegian government decided in 2017 to construct a new watertight entrance tunnel.

The 20 million Euro construction upgrade project was finalized in 2019 and included a new technical building nearby the Seed Vault entrance, new cooling machines and advanced security and surveillance systems in addition to the waterproof entrance tunnel. With this upgrade the Seed Vault partners believe that the Seed Vault is well prepared for securing seeds in a period of climate change. The upgrade also prepared the second storage hall for conserving seeds at −18°C. The first hall was filled up with seed boxes in February 2020 and the first seed samples have now been stored in the second hall.

Conserving seeds in the Seed Vault is a free of charge service. All genebanks making their genetic resources available for research, breeding and education are invited to conserve backup copies of their seed collections in the Seed Vault. The cooperation between the Svalbard Global Seed Vault and the depositors is regulated through a depositor agreement between the Norwegian Ministry of Agriculture and Food and the depositing gene bank. The main conditions for depositing seeds in the Seed Vault imply that the depositing gene bank shall conserve the original and primary seed sample in its own long-term gene bank. Only orthodox seeds of crops and their wild relatives, defined as PGRFA, can be accepted for storage in the Seed Vault. The depositor should also certify that the samples being deposited in the Seed Vault have been safety duplicated in a suitable gene bank at another location, in order to make the Seed Vault the second security back up.

The genetic material deposited in the Seed Vault should be available for breeding and research in accordance with applicable international law. Seeds conserved in the Seed Vault are stored in sealed seed boxes and remain as the property of the depositor. Only the owner of the genebank can have access to the seeds. The seeds can be returned to the depositor on request, in principle at any time when the seeds should be needed by the owner genebank. However, withdrawals of seeds are particularly relevant if the accessions are lost or inaccessible from the genebanks' own or from cooperating genebanks' repositories. Only the depositing institution can obtain access to the seeds it has deposited in the Vault.

Genebanks are invited to ship seeds to Svalbard on three regular pre-announced Seed Vault opening occasions every year. Due to the need for some flexibility, seeds are also brought into the Seed Vault on some more ad hoc occasions, in total normally 4–6 times during the year. The management and operation of seed deposits starts with information to genebank contacts once or twice a year about conditions and plans for receiving seeds at the Seed Vault openings. Current and potential depositor genebanks are informed about Seed Vault opening dates. Potentially new depositors are informed about the process towards signing the depositor agreement. Some basic information about the genebank/institute in question is gathered before the depositor agreement is signed. The NordGen director signs the agreement on behalf of the Norwegian Ministry of Agriculture and Food.

NordGen maintains a web portal database where transparent data about all deposited seed samples, depositor institutes, countries of origin and so on is displayed, that is, the Seed Portal at: https://seedvault.nordgen.org/. For the purpose of keeping the Seed Vault database updated, the gene bank provides an accession list completed with data about all new seed accessions that should be deposited before the seeds are shipped at each seed deposit occasion. An accession list template is developed for this purpose. NordGen uploads the accession list to the database. If information in the provided list contains errors or lack mandatory information, communication with the genebanks is undertaken for correction and clarification.

Two–three weeks' time before the Seed Vault opening date, the genebank ships the seeds to Svalbard, in time for arrival before the Seed Vault opening date. The genebanks cover the shipment costs. Genebanks in developing countries can apply to Crop Trust for financial support for the shipment of seeds. The Seed Vault has a c/o address at the Pole Position Logistics company in Longyearbyen that takes care of the seed boxes or packages temporarily until NordGen staff arrives and brings the seeds into the Seed Vault.

An import permission document issued by the Norwegian Food Authority follows the seed shipment. Depositors are recommended to bring a phytosanitary certificate along with the seeds. Handling of seed boxes in Svalbard includes scanning of the boxes in the security system at Longyearbyen Airport, ensuring that no other items than seeds are stored in the boxes. After scanning, the seed boxes are taken into the Seed Vault, labelled and put on the shelves, at positions assigned by the Seed Portal system. After putting the seeds in place in the Seed Vault, depositor contacts are informed about the process and result, and the Seed Portal is updated. So far, all seed shipments have arrived safely and not a single package or box has been lost. The NordGen operations are reported in annual reports submitted on March 31 each year. Reports are published on NordGen webpage: https://www.nordgen.org/en/our-work/svalbard-global-seed-vault/annual-progress-reports/. The Annual report for 2019 is the last published report.

Twelve years after its opening, the establishment of the Seed Vault has proven to be an unconditional success. At the end of 2020, 88 genebanks located in 63 different countries have deposited security seed samples in the Seed Vault. The depositor genebanks are international agricultural research centres, national and regional genebanks and genebank collections held by universities, breeding and research institutes and NGOs. The current holding of the Seed Vault is 1 060 073 seed samples.

As most genebanks conserve seeds from many countries, there are seeds in the Seed Vault originating from almost every country in the world. Every year since 2008, between 12 and 35 institutes have deposited seeds. Many of these have deposited seed several times, as part of comprehensive programmes for securing their seed collections. About two-thirds of seed deposits have been deposited by International Agricultural Research Centres (IARCs), most of them belonging to the Consultative Group for International Agricultural Research (CGIAR). Four of these have deposited more than 100 000 samples each: CIMMYT (International Maize and Wheat Improvement Center) in Mexico, IRRI (International Rice Research Institute) in the Philippines, ICRISAT (International Crop Research Institute for the Semi-Arid Tropics) in India and ICARDA (International Institute for Agricultural Research in Dry Areas), which until recently had its genebank in Syria. The largest depositors among national genebanks are located in the United States, Germany, Canada, Australia, the Netherlands, South Korea and Switzerland. Genebanks in several developing countries have deposited seeds as well; these include Mali, Nigeria, Sudan, Uganda, Zambia, Burundi, North Korea, Mongolia, Myanmar and Pakistan, among others. The Seed Vault now holds samples of about 5 000 different species. Wheat is represented by more than 212 000 seed samples while there are more than 168000 samples of rice. In total, 19 major cereal, vegetable and forage crops are represented by more than 10000 seed samples (Fig. 1).

5 Case study: the ICARDA withdrawal and re-deposit programme

This case study demonstrates the value of the Seed Vault and the importance of having a genebank accession properly duplicated in another location. In 2015, one of the CGIAR research centres, ICARDA (International Centre for Agricultural Research in Dry Areas) took on a comprehensive programme for re-establishing

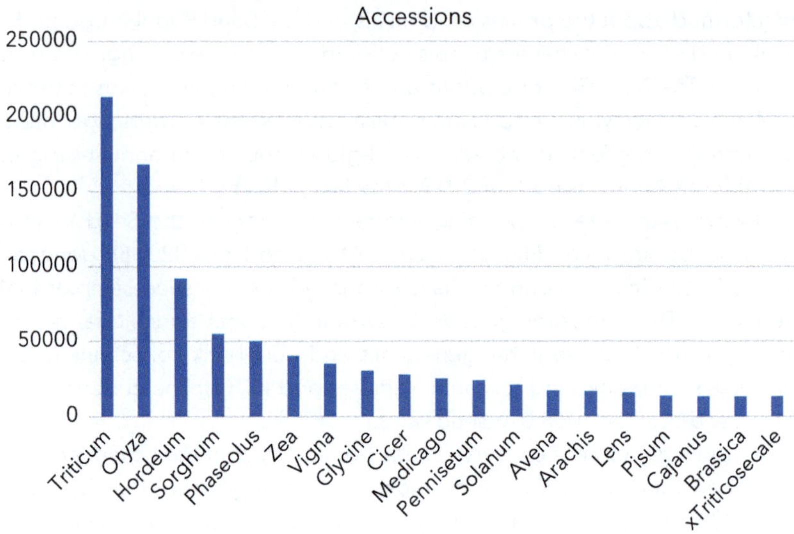

Figure 1 Graph showing the 19 species that have the highest numbers of accessions deposited in the Svalbard Global Seed Vault.

their seed collections previously conserved in Aleppo, Syria. Major parts of the seed collection in Aleppo were deposited in the Seed Vault during 2008–2014 totalling 116 484 seed samples. The conflict in Syria escalated in 2012, and in 2015 it was clear that the genebank in Aleppo was unable to operate and was threatened. Since then the institute has made three withdrawals of seeds from the Seed Vault: the first one in September 2015, the second in 2017 and the last one in August 2019. Through these withdrawals all materials deposited from Aleppo have been sent back and have formed the basis for re-establishing new genebank facilities in ICARDA units in Terbol, Lebanon and in Rabat, Morocco. ICARDA has, with substantial support from Crop Trust, implemented an ambitious programme for multiplication of all seed accessions that originally were conserved in the Aleppo genebank and duplicated in Svalbard. A large part of the withdrawn accessions has already been re-deposited in the Seed Vault since 2017, a majority of these produced by using seed material returned from Svalbard. ICARDA is so far the only genebank that needed to withdraw seed samples deposited in the Svalbard Global Seed Vault. Looking away from the tragedy in Syria and Aleppo, the ICARDA case illustrates the importance of having a global seed conservation facility such as the Svalbard Global Seed Vault.

6 Where to look for further information

More information about the Svalbard Global Seed Vault and it operations can be found on different websites:

- The official website managed by the Norwegian Ministry of Agriculture and Food: www.seedvault.no.
- NordGen Seed Vault website: https://www.nordgen.org/en/our-work/svalbard-global-seed-vault/.
- About the Seed Vault at Crop Trust website: https://www.croptrust.org/what-we-do/svalbard-global-seed-vault/.
- NordGen Seed Portal: https://seedvault.nordgen.org/.

7 References

FAO (1998). *The State of World's Plant Genetic Resources for Food and Agriculture*. Food and Agriculture Organization of the United Nations, Rome, Italy, 510pp.

FAO (2010). *The Second Report on the State of World's Plant Genetic Resources for Food and Agriculture*. Commission on Genetic Resources for Food and Agriculture. Food and Agriculture Organization of the United Nations, Rome, Italy, 510pp.

FAO (2014). *GeneBank Standards for Plant Genetic Resources for Food and Agriculture*. Commission on Genetic Resources for Food and Agriculture. Food and Agriculture Organization of the United Nations, Rome, Italy. Available at http://www.fao.org/3/a-i3704e.pdf.

Fowler, C., George, W., Shands, H. and Skovmand, B. (2004). *Study to Assess the Feasibility of Establishing a Svalbard Arctic Seed Depository for the International Community*. Prepared for the Ministry of Foreign Affairs, Center for International Environment and Development Studies (Noragric) at the Agricultural University of Norway and Nordic Gene Bank, 14.09.04.

Qvenild, M. (2005). Sowing seeds in permafrost: An idea whose time has come. Master of Science Thesis, Department of International Environment and Development Studies, Norwegian University of Life Sciences. Available at: http://www.umb.no/statisk/noragric/publications/master/2006_marte_qvenild.pdf.

The International Treaty on Plant Genetic Resources for Food and Agriculture (2017). Resolution 12/2017. Cooperation with other international bodies and organizations. IT/GB-7/17/Res12.

Part 3

Enhancing conservation and use of plant genetic diversity

Chapter 10

Community-based conservation of crop genetic resources

Stef de Haan, International Potato Center (CIP), Peru

1 Introduction

Community-based conservation of crop genetic resources is both a historical and contemporary phenomenon. Ever since the domestication of crop species initiated approximately 11 000 years ago, and the diversification of intraspecific diversity followed the founder effect, farmers and their communities have been key actors driving ongoing conservation (Brush, 2004). Outcomes include key ecosystem services and benefits to different users ranging from farmers, consumers, entrepreneurs and chefs to breeders (Bellon et al., 2016; Jones, 2017; Moreau and Speight, 2019). There is increased global recognition that the past, present and future contribution of farmers in centers of origin and diversity is key to the ongoing evolution of crop diversity. This acknowledgement has been the basis for including Farmers' Rights in the International Treaty on Plant Genetic Resources for Food and Agriculture, Aichi Target 13 and 17, and the Nagoya Protocol (Halewood, 2016; Visser et al., 2019).

The nature of continued community-based conservation is non-static. While the term 'conservation' is still widely applied for on-farm or in situ management of cultivated species and landraces, the reality is that both the state of conservation (what disappears or is maintained), the drivers and processes underlying conservation, and socioecological interactions are highly dynamic (Almekinders et al., 2019). For communities and farmers, conservation is rarely an objective in itself. Rather, it is the outcome of the active use and utility of genetic resources for food consumption, income generation or

http://dx.doi.org/10.19103/AS.2020.0085.09

cultural purposes. Furthermore, beyond the genetic resources themselves and their utilitarian role, community-based conservation involves the traditional knowledge, culture, spiritual values and traditions that are intrinsically connected (Possey, 1999).

Conservation can be conceptualized to occur at the community level because collectively crop diversity is often–at least partially–managed as a common pool resource. Different communities often uphold their own rules and norms concerning use, exchange and access. Modalities include differential deployment of diversity in communal lands and seed exchange networks (Orlove and Godoy, 1986; Pautasso et al., 2012). However, within communities it is common to encounter considerable heterogeneity when it comes to conservation. At the household level the range and richness of diversity managed can be determined by tradition, social status and well-being. Within households, gender and age can be important factors differentiating practices, uses and knowledge of crop diversity management (Abdelali-Martini et al., 2008; Oakley and Momsen, 2007). In addition, certain key individuals–often referred to as custodians or guardians–can be driven by passion and recognized for managing large stocks of landraces (Gruberg et al., 2013).

How the science community conceptualizes community-based conservation has changed fundamentally during the last few decades. During the 1970s–in the aftermath of the green revolution–the common notion was that landraces would inevitably be replaced by modern varieties (Frankel and Hawkes, 1975). During the 1980s and early 1990s it was mostly anthropologists and agroecologists characterizing the farming systems and livelihoods that continued to actively manage diversity, highlighting the contemporary role of genetic resources on-farm and in smallholder livelihoods (Altieri and Merrick, 1987; Mayer et al., 1992; Riley et al., 1990). Since the early 1990s the volume of research and interventions involving community-based conservation has increased exponentially (Zimmerer and De Haan, 2017). Currently, as we have entered the decade of family farming, attention to planetary boundaries, human and ecosystems health, green infrastructure, and the social inclusiveness of food systems often implicitly involve community-based conservation (Bioversity International, 2017; Herforth et al., 2019).

Also, how the private sector, consumers and civil society act-upon and link to community-based conservation of crop genetic has evolved. In different parts of the world the private and public sectors–including agribusiness, food services and social protection programs–have been spearheading options to promote conservation through direct sourcing of crop diversity for differential markets, sustainable and novel cuisine, public procurement linked to social protection, or through direct private benefit sharing (Beltrame et al., 2019; UNCTAD, 2017). Consumer awareness and urban food literacy around diversity has also shifted and there are multiple examples of the so-called 'neglected and underutilized species' becoming new superfoods (Knorr and WWF, 2019). Civil society concern about crop genetic resources has shifted away from being purely rural development orientated to more of a food governance movement. Examples of the latter include the Food Gastronomic Integration Movement in Bolivia, Liberate Diversity in Europe and the Slow Food movement globally. Despite these changes, little is known about the scale, trade-offs and effects on intraspecific diversity at the community level.

In this chapter we will explore the current state of research, knowledge and practice underlying community-based conservation in the context of continued farmer-driven processes and the emergence of diverse external interventions. The role of science and innovation to monitor the community-based management of genetic resources will be explored. Three questions will guide our discussion following the main topics outlined above: (i) what motivates farmers and communities to conserve crop genetic resources? (ii) what options exist to support community-based conservation and how effective are these? (iii) how can monitoring help to 'take the pulse' of the conservation status on diversity on-farm? The chapter concludes by highlighting selected key challenges that will require new research.

2 Autonomous community-based conservation

First and foremost, the conservation of crop species and varietal diversity should be of value to farmers and communities. Direct benefits to farmers essentially drive continued autonomous community-based conservation. These boil down to ecosystem services that underlay the foundations of resilient, adaptable and sustainable farming and include a range of cultural, provisioning and regulating services (Table 1; Delaquis et al., 2018; Duru et al., 2015; Ficiciyan et al., 2018). In turn, the continued community-based management of diverse crop species, landraces and associated diversity will directly and indirectly result in direct benefits and services for a wide range of other actors and stakeholders in society. These 'third-party' benefits–whether for breeders, chefs, entrepreneurs or consumers–in turn can potentially provide an enabling environment for community-based conservation. Such feedback loops can involve increased market demand and higher farmgate prices.

2.1 Cultural factors

Cultural factors are perhaps among the strongest of the autonomous drivers triggering continued community-based conservation. This is evident from the diversity of links between cultural and crop diversity (e.g. Anderson, 1993; Hamilton, 2003; Johnsson, 1986). Food and cuisine are among the strongest cultural icons fueling demand for specific species and cultivars. Typically, food cultures and preference traits surrounding crop diversity are particularly

Table 1 Ecosystems services types, specific functions and selected examples

Type	Specific function	Examples
Cultural	Identity	Maize culture (Leclerc and D'Eeckenbrugge, 2012; Sandstrom, 1992)
	Prestige	Seed provision (Labeyrie et al., 2013; McGuire, 2008)
	Cuisine	Iconic festive dishes (Avieli, 2005)
	Traditions	Freeze-drying bitter potato landraces (De Haan et al., 2010)
	Rituals	Crops used in offerings (Quiroz and Van Andel, 2018)
	Spiritual value	Rice and Maroon spiritual life (Van Andel et al., 2019)
Provisioning	Food	Seasonality and year-round availability of potato in Andean food systems (De Haan et al., 2018)
	Feed	Use of rice stems, bran and chaff as animal feed (Chunta et al., 2014)
	Human nutrition	Women and child dietary diversity (Jones et al., 2018; Luna-González and Sørensen, 2018)
	Income from sales	Premium prices for landraces (Devaux et al., 2011; Krishna et al., 2010)
	Medicine	Medicinal use of rice (Cabanting and Perez, 2016; Gerards et al., 2015)
Regulating	Nutrient cycling	Nitrogen fixing and phosphorus uptake in maize (Sangabriel-Conde et al., 2013; Van Deynze et al., 2018)
	Pest control	Reduced pest damage deploying varietal diversity (Tooker and Frank, 2012; Mulumba et al., 2012)
	Disease control	Disease reduction in potato (Yang et al., 2019)
	Risk management	Field scattering with different species and varietal portfolios to manage risk (Goland, 1993)
	Yield stability	Variety richness reducing yield variability (Di Falco et al., 2007; Widawsky and Rozelle, 1998)
	Pollination	Production increases with pollination and negative effects of intensification on insect populations (Klein et al., 2007)

strong in centers of crop origin and diversity. Well-known examples include the culinary uses of chillies in Mexico, rice in Laos, spices in India, palms in the Amazon and potatoes in the Andes. Traditional food processing such as fermentation and freeze-drying also commonly require specific varieties.

Food and cuisine–and to a large extent the varietal diversity employed–is directly linked to identity, gender, prestige, power and traditions. Expressions are highly diverse and range from the common use of specific varieties for iconic foods to gendered knowledge for the preparation of festive dishes. Ritual and spiritual practices for growing and using specific crop varieties are common in different contexts. These include the planting of special varieties to the protect other cultivars and offerings (CIP, 2006; Senakun et al., 2018). Custodian farmers can be sought after seed sources of unique varietal diversity (Arce et al., 2018; Subedi et al., 2002). Prestige and power are intrinsically associated with varietal richness at the individual, household or community levels.

2.2 Provisioning functions

Provisioning functions are those that provide communities and households with essential products: food, feed, fuel, medicine and cash. These functions seem straightforward, but frequently involve nuanced considerations involving crop species and varietal choices, including multipurpose functions (e.g. food, feed and straw for roofing), temporal dimensions (e.g. food availability and dietary throughout the year) and spatial considerations (e.g. output on rain-fed vs. irrigated or marginal vs. fertile plots). Crop species and varietal diversity, in combination with locally available resources and management options, allow farmers to plan their production.

Food security is a key aspired cornerstone outcome of community-based conservation. On-farm diversity commonly contributes to the direct availability, access, utilization and stability of food. This in turn, under the right conditions, can translate into dietary diversity and nutrition security at the household level (De Haan et al., 2019; Jones et al., 2018). Conditional factors include land ownership, farm size and carrying capacity, household size, market participation, off-farm employment and social stability (Arce et al., 2016; Steinberg and Taylor, 2002). Self-consumption and income generation can be synergistic. For example, a recent case study in a marginal area in Ghana shows that crop diversity is positively associated with both self-consumption of food crops and cash income from crops sold (Bellon et al., 2019).

2.3 Regulating functions

Regulating functions involve both the conscious and unconscious use of crop genetic resources towards maximizing self-regulation at the plot, farm and agroecosystem levels and minimizing the risks of productivity, food security and economic shortfalls. In their effort to minimize risk, maximize yield stability and balance production output for both home consumption and sales, smallholder farmers commonly employ a suit of strategies. Crop diversity can help the farmer reach their goals even when confronting different disturbances (Kozicka et al., 2019). Environmental adaptability and yield stability are important attributes for farmers' autonomous choice of crop varieties (Asrat et al., 2010).

Likewise, for coping with extreme and out-of-season events in highly exposed environments (Condori et al., 2014; Di Falco and Chavas, 2008). Vanek (2019) shows that trade-offs and synergies between cascades of functional traits impact ecosystem functions and resilience. Farmers therefore navigate a complex web of options, choices and information to guide their decision-making about the genetic resources to be deployed on-farm.

Autonomous community-based conservation continues to be–and has historically been–the *status quo* for the bulk of communities managing genetic resources on-farm. Most communities are not exposed to the exogenous incentives described in the next section. What motivates these communities and farmers to deploy varietal diversity is context specific and involves multiple cultural, provisioning and regulating functions rather than a single driver. For society a key, yet underrated, ecosystem service outcome of farmer-driven conservation involves the ongoing evolution of diversity in a dynamic context. A key characteristic of this type of conservation in the global south relates to its informal nature and disconnect to formal systems. This disconnect is reflected at multiple levels, ranging all the way from variety registration, seed systems to markets. This informal nature is arguably a key strength of community-driven conservation as it is agile, self-regulating, non-static and culturally embedded.

3 Exogenous incentives for community-based conservation

Since the early 1990s, a variety of different approaches have emerged with the intention to support community-based conservation (Jarvis et al., 2011; Scott, 2011). These interventions range from single isolated activities to more integrated approaches offering a range of options to promote the continued cultivation of landraces in farmer communities. They also vary in terms of their philosophical and ideological framework. Interventions range from pure 'market-based approaches' aiming to develop value chains for crop diversity to the so-called 'cultural reaffirmation approaches' targeting to conserve the cosmovision and traditional lifestyles of traditional farmers. A range of different exogenous incentives and interventions are briefly detailed below.

3.1 Biodiversity seed fairs

Biodiversity seed fairs are a practical and relatively inexpensive way to promote seed exchange, incentivize custodian farmers and communities, and monitor diversity (Lemons et al., 2003; Rijal et al., 2000). The fairs often include an element of competition and knowledge exchange. They are also frequently combined with cultural elements, including folk dances and storytelling and exhibitions of traditional dishes. In the Andean region, the number of annual seed fairs has grown exponentially since they were first promoted by projects in the early 1990s (Tapia, 2000; Tapia and Rosas, 1993). In Peru, a combined effort of municipalities, regional governments and NGOs has resulted in hundreds of annual seed fairs throughout the country. They offer the possibility to promote seed exchange, revalue crop diversity and annually monitor intraspecific diversity.

3.2 Repatriation of landraces

The repatriation of landraces is a way to integrate ex situ and in situ approaches (López Noriega et al., 2019). Since 1997 the genebank of the International Potato Center (CIP) has been repatriating landraces to Andean communities. Repatriation commonly starts with a request from a farmer community that has lost local landraces. To date, 89 communities have received over 6000 samples of cultivated potato from CIP's genebank, comprising >1250 accessions (30% of CIP's landrace collection). Repatriation is a potentially impactful approach to restore diversity following varietal loss (Gauchan et al., 2017). Since the CIP genebank started repatriation, the approach has been adopted by numerous other CGIAR and national gene banks (CGIAR, 2017; ICAR, 2017).

3.3 Community seed banks

The establishment of community seed banks (CSB) has been widely promoted in Asia, Africa, Europe and Latin America. CSBs come in different organizational set-ups, but are generally locally governed and designed to conserve and distribute landrace seed for local use. Contrary to other interventions, CSBs have been reasonably well documented (Lewis and Mulvany, 1997; Shrestha et al., 2013; Vernooy et al., 2015). Since their origin, more than 35 years ago, the way they are embedded in the development discourse has ranged from on-farm conservation, seed sovereignty to climate adaptation (Huamán, 2001; Vernooy et al., 2014, 2017). The way information, finances and actual genetic resources are managed by CSBs, and the quality of and timely access to seed of desirable landraces, can be factors determining their success or failure (Bezabih, 2007; Engels et al., 2008; Polreich et al., 2005).

3.4 Market-based approaches

Market-based approaches aiming to link landrace diversity to markets and thereby provide profit-based incentives for farmers to continue to conserve genetic resources have emerged in many different contexts. Interventions range from inclusive retailing of landraces, international superfood value chains, public procurement programs, farmers markets, direct linkages with chefs and restaurant, to agroecotourism (e.g. Meinzen-Dick et al., 2009; Resque et al., 2019). Some studies have suggested that value chains themselves will likely only have a marginal or even negative effect on conservation (Hellin and Higman, 2005; Tobin et al., 2016; Van Dusen and Taylor, 2005). Clearly, the market systems should deviate from the mainstream and offer conditions to deal with mixtures, seasonality, small volumes, heterogeneity and other challenges that typify intraspecific diversity. Jäger et al. (2019) have suggested that enhancing both genetic resources conservation and income generation through markets requires niche market development and signaling true production costs. Collective brands, short circuits, labeling and certification can potentially support such efforts (Garcia-Yi, 2014).

A singular approach to community-based conservation involves the so-called 'cultural affirmation' (Apffel-Marglin, 1998a,b). It developed in the 1980s in the Andean region and was originally inspired by the early works of priests from the liberation theology with a keen interest to document and revalue Quechua and Aymara indigenous cosmovision (Van Den Berg, 1989; Van Kessel and Enriquez, 1989, 2002; Van Kessel and Condori Cruz, 1992). In Peru the PRATEC movement emerged in the 1980s and developed the cultural affirmation approach to document indigenous technology (e.g. Gómez, 1993; Grillo et al., 1994; Rengifo, 1988). With the support of a network of NGOs, PRATEC set about reviving or affirming traditional and ancestral farming practices. When the on-farm conservation of crops and associated traditional knowledge became an international priority, the PRATEC movement applied the principles of cultural reaffirmation to address it.

The principles of cultural affirmation involve accompanying, revigorating and documenting community conservation and associated knowledge. It represents a 'passive' approach targeted at revaluing what exists (or has become lost) and resists promoting new technologies (Shepherd, 2017). It is well recognized that characterizing and revaluing traditional family farming is critical for understanding its rationales and dynamics. Yet, cultural affirmation commonly represents an idealized, static, romanticized, neo-indigenous and often intellectual representation of 'traditional' agriculture. While the approach originated in the Andes, cross-fertilization and adaptation have triggered a growing global interest in revaluing ancestral, traditional and biocultural practices that underlay community-based conservation (e.g. Barthel et al., 2013; Haverkort and Hiemstra, 1999; Haverkort et al., 2003).

3.5 Payment for environmental services

Payment for environmental services (PES) consist of incentives offered to farmers in exchange for conserving crop diversity on-farm and as a compensation or recognition of ecological services provided (Krishna et al., 2013). Such schemes have been applied in Bolivia, India and Peru. Farmer groups are given seeds of lost or threatened landrace and provided with material compensation for conserving these resources. Another model involves Corporate Social Responsibility (CSR) linked to benefit sharing to promote the self-organization of custodian farmers. This is the case of the Association of Potato Biodiversity Guardians (AGUAPAN): a farmer-led organization in Peru. The initiative is funded by Dutch potato seed and breeding companies who directly support the farmers. The association supports its members through annual monetary imbursements to each custodian households to be invested in farm inputs, medical care or child education. In addition, training is provided to custodian leaders, women and youth in topics such as farmers' rights and entrepreneurship.

3.6 Park and protected zone systems

Park and protected zone systems can implicitly or explicitly include or promote community-based conservation of crop genetic resources. In the IUCN's protected area classification this is implicit in the categories of 'protected landscape and seascapes' and 'protected area with sustainable use of natural resources' (Amend et al., 2008; Dudley, 2013). Other international systems also provide the option, including the UNESCO Global Geoparks, Biospheres and World Heritage Sites. Examples of reserves with an implicit on-farm conservation component include the Nor-Yauyos Cochas Landscape Reserve in Peru (IUCN category V), the Rice Terraces of the Cordilleras in the Philippines (UNESCO World Heritage Site), and the Dong Van Karst Plateau Geopark in Peru (UNESCO Geopark). The first protected area explicitly established for the conservation of crop genetic resources was the Biosphere Reserve of Sierra de Manantlán in Mexico in 1986 (Santilli, 2012).

An international landscape-level initiative that also explicitly values the community-based conservation of genetic resources concerns the FAO's Globally Important Agricultural Heritage Systems or GIAHS (FAO, 2018). Some countries like Chile have adopted this global framework into a network of Nationally Important National Agricultural Heritage Systems (NIAHS). Additionally, several regional and national governments have adopted special frameworks that explicitly embrace community-based conservation of crop genetic resources (IDMA, 2018; Ruiz Muller, 2009). This includes the framework of 'agrobiodiversity zones' that was adopted in 2016 by the Ministry of Agriculture and Irrigation of Peru. The main objective of the 'agrobiodiversity zones' is to contribute to improve the living conditions of indigenous peoples, strengthening and consolidating the conservation, sustainable use and local management of the native agrobiodiversity. Park and protected zone systems typically combine several interventions to incentivize community-based conservation, including repatriation, biodiversity seed fairs and agroecotourism.

Peru's potato park is one of the better-known examples of a reserve which explicitly promotes community-based conservation (Argumedo, 2008). The park initiated with the repatriation of hundreds of landraces from the International Potato Center's genebank to catalyze in situ regeneration of lost landraces, thereby creating a dense hotspot of intraspecific diversity. The governance of the park has been strongly guided by NGOs, providing a springboard for political ecological discourse (e.g. Swiderska, 2006). The park is guided by an indigenous biocultural heritage framework (Angé, 2018; Grady, 2012) and integrates market and technological strategies such as links to tourism for income generation and formal seed system to promote virus-free planting materials. From a conservation perspective, the potato park uses a combination of strategies: repatriation, value chain development and integrationist principles combining traditional and modern technologies (Shepherd, 2017).

3.7 Other interventions

In addition to the different exogenous incentives described above, there are numerous other interventions. Participatory and evolutionary breeding using both local and introduced diversity has been promoted in different contexts and at multiple geographical scales to use community-managed genetic resources to adapt to new contexts (e.g. Van Etten et al., 2019; Westengen and Winge, 2019). Another emerging approach involves the linkage of community-based conservation to curriculum development and education at primary and secondary schools. Youth engagement and incorporating local crop diversity and associated traditional knowledge in intercultural education offers a means to stimulate interest among the future generation of farmers.

4 Monitoring community-based conservation

The management of crop genetic resources from one farmer generation to the next results in a range of plant-human adaptive changes that can be used as markers of co-evolving relationships (Zeder, 2015). These involve genotypic, phenotypic and plastic selection of crop genepools that are intertwined with human adaptation, societal advancement and inherent demand for crop diversity. To measure changes in space and time it is indispensable to initially conduct baseline assessments documenting the current state of conservation of crops and landraces. Apart from documenting the key metrics concerning the status of genetic resources, such an assessment would ideally also characterize the main socioeconomic and biophysical drivers affecting the conservation dynamics.

There have been numerous reports highlighting the declining levels of species and intraspecific diversity of crop genetic resources (e.g. IPBES, 2019; Pilling et al., 2020). However, the assessments are rarely supported by hard data. Especially for centers of crop origin in the global south, where community-based conservation is the outcome of continued local demand for diversity, it is rare to count on robust evidence sustaining transitions in the conservation status of crop diversity. Evidence of loss of alleles, landraces or shifts in their frequencies is commonly lacking as a consequence of the non-existence of baseline or timeseries data, fixed monitoring sites, standardized methods or solid partnerships, among other prerequisites needed for systematic monitoring of the conservation status of crop genetic diversity (De Haan et al., 2016).

An urgent challenge to be addressed by the crop conservation community concerns an effective framework for monitoring crop species and landrace diversity. Such a framework needs to consider the relationships between sampling resolution and spatial scale (Mercer et al., 2019). Monitoring networks at the hotspot-level in centers of crop origin can link to national or regional decision-support systems. A range of different methods and metrics have also been proposed (e.g. Aravanopoulos et al., 2015; Padulosi and Dulloo, 2012). For example, implementation of a framework at the hotspot level has been initiated for potato in the Andes (De Haan et al., 2016). It involves complementary methods and metrics at different scales, including hotspot identification, baseline research to determine the total, relative and spatial diversity of landraces, and the documentation of associated collective knowledge (Polreich et al., 2018).

In the context of autonomous community-based management a network approach can help to 'take the pulse' of the overall conservation status of crop genetic resources under conditions that are representative of regional smallholder farming systems. An important trend and step forward in the Andean region has been the gradual time-tagged and spatially explicit baseline documentation of diversity in an ever growing body of in situ catalogues (e.g., Cadima et al., 2004; CIP, 2015, 2017; Scurrah et al., 2013; Moscoe and Blas, 2016a,b; Villagomez and Blanco, 2006). This will at least allow for more robust timeline comparisons in the future. In the context of exogenous incentives for community-based conservation, monitoring tools and methods can provide insights into the effectiveness of different interventions. Indeed, impact assessments of exogenous incentives are much needed as these are virtually non-existent. In all cases, monitoring approaches should not only consider genetic resources but explicitly take on board participatory approaches that capture rationales underlying community-based conservation.

5 Summary and future research outlook

Autonomous community-based conservation has been a constant since the domestication and diversification of crop genetic resources. It is an ongoing phenomenon powered by diverse and changing farmer rationales. From a research perspective it is highly relevant to not solely focus on the conservation of genetic diversity, but especially on the conservation dynamics under on-farm management. This automatically applies a focus on the people who grow diversity and their livelihood portfolios. Understanding how dynamic farmer-driven management of genetic resources, including the loss and incorporation of landraces or shifts in their abundance, helps smallholders and their communities to adapt to changing markets, food systems and climate extremes is essential. A better understanding of the conditioners of dynamic community-based conservation–both between different farmer generations and socioecological environments–will arguably allow for a more intelligent design of policies, incentives and interventions that can support smallholders.

Different trends and shocks have been assumed to cause irreversible loss of crop diversity. Typically, these include the green revolution and push of modern varieties, demographic collapses and wars, and global market liberalization. More recently alleged drivers have come to include food system transitions, climate change and cultural homogenization. One major challenge for researchers involves the resolution and scale of evidence. For example, while in absolute terms global agricultural landscapes and food supply chains have without doubt become more uniform, we often lack data and evidence of how much crop diversity is maintained in agrobiodiversity refugia. The informal nature of community-based conservation means that processes occur under the radar screen and locally rather than globally. How do global tendencies play out in terms of the absolute and relative diversity managed outside the mainstream? How can we bridge spatial scales in research and in effective agrobiodiversity monitoring? What are tipping points, essential farmer-driven evolutionary processes and the minimal population sizes needed for maintaining viable landrace populations? How much is intraspecific diversity on-farm threatened by climate change and how much will it allow farmers to adapt? How will seed exchange networks evolve? These are some of the key research questions to be addressed.

After decades of promoting a portfolio of different exogenous incentives for community-based conservation, there is a growing need to enhance our understanding of the effectiveness and impact of individual and combined interventions. Impact studies, including participatory approaches that capture farmer perspectives, could provide insights into the pros and cons of different methods, their acceptability, replicability and scalability. An important caveat involves site selection. While external interventions to promote community-based conservation would ideally take place in those sites where there is evidence of loss and local demand to restore diversity, it is all too common for interventions to be targeted in diversity hotspots where farmer-driven conservation is intact. Furthermore, it is not uncommon for interventions to be promoted without a proper understanding of local processes or baseline of existing diversity. What makes external intervention succeed or fail? How are they targeted? How are communities consulted and empowered? And, at a deeper level, how can the design of different incentives be tailored towards local demands and realities?

There is an urgent need to mainstream a global-to-local framework for the systematic monitoring of community-based conservation in centers of crop origin. It comes back to the basic principle that one can only manage what one can measure. While monitoring of autonomous farmer-driven management can provide intelligence about the conservation status of landraces under typical conditions of most farmer communities, a monitoring system in a context of external interventions can provide insights into the effectiveness of such incentives. There is enormous scope for innovation, including novel approaches for crowd sourcing and citizen sciences, digital platforms to connect intelligence from the community hotspot to the national and global level, and the development of incentive systems that are not merely extractive but also support community decision-making. Last, but not least, there is an important opportunity to connect community-based and ex situ conservation of crop genetic resources in genebanks more effectively and creatively.

6 References

Abdelali-Martini, M., Amri, A., Ajlouni, M., Assi, R., Sbieh, Y. and Khnifes, A. (2008). Gender dimension in the conservation and sustainable use of agro-biodiversity in West Asia. *The Journal of Socio-Economics* 37(1):365–383.

Almekinders, C. J. M., Stone, G. D., Baranski, M., Carney, J. A., Hanspach, J., Krishna, V. V., Ramirez-Villegas, J., Van Etten, J. and Zimmerer, K. S. (2019). Socioecological interactions amid global change. pp. 116–143. In: Zimmerer, K. S. and De Haan, S. (Eds), *Agobiodiversity: Integrating Knowledge for a Sustainable Future*. Ernst Strungmann Forum, MIT Press, Cambridge, MA.

Altieri, M. A. and Merrick, L. (1987). In situ conservation of crop genetic resources through maintenance of traditional farming systems. *Economic Botany* 41(1):86–96.

Amend, T., Brown, J., Kothari, A., Phillips, A. and Stolton, S. (2008). *Protected Landscapes and Agrobiodiversity Values*. IUCN, GTZ. Kasparek Verlag, Heidelberg.

Anderson, E. F. (1993). *Plants and People of the Golden Triangle: Ethnobotany of the Hill Tribes of Northern Thailand*. Dioscorides Press, Portland, OR.

Angé, O. (2018). Interspecies respect and potato conservation in the peruvian cradle of domestication. *Conservation and Society* 16:30–40.

Apffel-Marglin, F. (1998a). *The Spirit of Regeneration: Andean Culture Confronts Western Notions of Development*. Zed Books, London.

Apffel-Marglin, F. (1998b). From fieldwork to mutual learning: working with PRATEC. *Environmental Values* 11(3):345–367.

Aravanopoulos, F., Tollefsrud, M. M., Graudal, L., Koskela, J., Kätzel, R., Soto, A., Nagy, L., Pilipovic, A., Zhelev, P., Božic, G. and Bozzano, M. (2015). *Genetic Monitoring Methods for Genetic Conservation Units of Forest Trees in Europe. European Forest Genetic Resources Programme (EUFORGEN).* Bioversity International, Rome.

Arce, A., Creed-Kanashiro, H. M., Scurrah, M., Ccanto, R., Olivera, E., Burra, D. and de Haan, S. (2016). The challenge of achieving basal energy, iron and zinc provision for home consumption through family farming in the Andes: a comparison of coverage through contemporary production systems and selected agricultural interventions. *Agriculture and Food Security* 5(1):23.

Arce, A., De Haan, S., Burra, D. D. and Ccanto, R. (2018). Unearthing unevenness of potato seed networks in the high Andes: a comparison of distinct cultivar groups and farmer types following seasons with and without acute stress. *Frontiers in Sustainable Food Systems* 2:43. doi: 10.3389/fsufs.2018.00043.

Argumedo, A. (2008). The Potato Park, Peru: conserving agrobiodiversity in an Andean indigenous biocultural heritage area. pp. 45–58. In: Amend, T., Brown, J., Kothari, A., Phillips, A. and Stolton, S. (Eds), *Protected Landscapes and Agrobiodiversity Values.* IUCN, GTZ, Kasparek Verlag, Heidelberg.

Asrat, S., Yesuf, M. M., Carlsson, F. and Wale, E. (2010). Farmers' preferences for crop variety traits: lessons for on-farm conservation and technology adoption. *Ecological Economics* 69(12):2394–2401.

Avieli, N. (2005). Vietnamese New Year rice cakes: iconic festive dishes and contested national identity. *Ethnology* 44(2):167–187.

Barthel, S., Crumley, C. and Svedin, U. (2013). Bio-cultural refugia–safeguarding diversity of practices for food security and biodiversity. *Global Environmental Change* 23(5):1142–1152.

Bellon, M. R., Ntandou-Bouzitou, G. D. and Caracciolo, F. (2016). On-Farm diversity and market participation are positively associated with dietary diversity of rural mothers in southern Benin, West Africa. *PLoS ONE* 11(9):e0162535.

Bellon, M. R., Kotu, B. H., Azzarri, C. and Caracciolo, F. (2019). To diversify or not to diversify, that is the question. Pursuing agricultural development for smallholder farmers in marginal areas of Ghana. *World Development* 125:104682.

Beltrame, D., Gee, E., Güner, B., O. Lauridsen, N., Samarasinghe, W. L. G., W. Wasike, V., Hunter, D. and Borelli, T. (2019). Mainstreaming biodiversity for food and nutrition into policies and practices: methodologies and lessons learned from four countries. *Anadolu Ege Tarımsal Araştırma Enstitüsü Dergisi* 29(1):25–38.

Bezabih, M. (2007). Agrobiodiversity conservation under an imperfect seed system: the role of Community Seed Banking schemes. *Agricultural Economics* 38(1): 77–87.

Bioversity International (2017). *Mainstreaming Agrobiodiversity in Sustainable Food Systems: Scientific Foundations for an Agrobiodiversity Index.* Bioversity International, Rome.

Brush, S. B. (2004). *Farmers' Bounty: Locating Crop Diversity in the Contemporary World.* Yale University Press, New Haven, CT.

Cabanting, R. M. and Perez, L. M. (2016). An ethnobotanical study of traditional rice landraces (Oryza sativa L.) used for medical treatment in selected local communities of the Philippines. *Journal of Ethnopharmacology* 194:767–773.

Cadima, X., Gonzales, R., Almanza, J., Garcia, W. and Terrazas, F. (2004). *Catálogo de Variedades Locales de Papa y Oca de la Zona de la Candelaria*. PROINPA, CIP, COSUDE, Cochabamba.

CGIAR (2017). *Annual Report 2016 - Genebanks CRP*. CGIAR, Crop Trust, Bonn.

Chunta, S., Prathepha, P. and Jongdee, B. (2014). Nuances of traditional knowledge in utilization of rice landraces by farming communities in North-Eastern Thailand. *Indian Journal of Traditional Knowledge* 13(3):473–483.

Condori, B., Hijmans, R. J., Ledent, J. F. and Quiroz, R. (2014). Managing potato biodiversity to cope with frost risk in the high Andes: a modeling perspective. *PLoS ONE* 9(1):e81510. https://doi.org/10.1371/journal.pone.0081510.

CIP (2006). *Catalogo de Variedades de Papa Nativa de Huancavelica, Perú*. Centro International de la Papa (CIP), Lima.

CIP (2015). *Catalog of Ancestral Potato Varieties from Chugay, La Libertad - Peru*. Centro International de la Papa (CIP), Asociación Pataz, Instituto Nacional de Innovación Agraria (INIA), Lima.

CIP (2017). *Catálogo de Variedades de Papa Nativa del Sureste del Departamento de Junín - Perú*. Centro International de la Papa (CIP), Grupo Yanapai, Instituto Nacional de Innovación Agraria (INIA), Lima.

De Haan, S., Burgos, G., Arcos, J., Ccanto, R., Scurrah, M., Salas, E. and Bonierbale, M. (2010). Traditional processing of black and white *chuño* in the Peruvian Andes: regional variants and effect on the mineral content of native potato cultivars. *Economic Botany* 64(3):217–234.

De Haan, S., Polreich, S., Rodriguez, F., Juarez, H., Plasencia, F., Ccanto, R., Alvarez, C., Otondo, A., Sainz, H., Venegas, C. and Kalazich, J. (2016). A long-term systematic monitoring framework for on-farm conserved potato landrace diversity. pp. 289–297. In: Maxted, N., Dulloo, M. E. and Ford-Lloyd, B. V. (Eds), *Enhancing Crop Genepool Use: Capturing Wild Relative and Landrace Diversity for Crop Improvement*. CABI Publishers, Wallingford.

De Haan, S., Burgos, G., Liria, R., Rodriguez, F., Creed-Kanashiro, H. M. and Bonierbale, M. (2018). The nutritional contribution of potato varietal diversity in Andean food systems: a case study. *American Journal of Potato Research* 96(2):151–163.

Delaquis, E., De Haan, S. and Wyckhuys, K. A. G. (2018). On-farm diversity offsets environmental pressures in tropical agro-ecosystems: a synthetic review for cassava-based systems. *Agriculture, Ecosystems and Environment* 251:226–235.

Devaux, A., Ordinola, M. and Horton, D. (2011). *Innovation for Development: The Papa Andina Experience*. International Potato Center (CIP), Lima.

Di Falco, S., Chavas, J. P. and Smale, M. (2007). Farmer management of production risk on degraded lands: the role of wheat variety diversity in the Tigray region, Ethiopia. *Agricultural Economics* 36(2):147–156.

Di Falco, S. and Chavas, J.-P. (2008). Rainfall shocks, resilience, and the effects of crop biodiversity on agroecosystem productivity. *Land Economics* 84(1):83–96.

Dudley, N. (2013). *Guidelines for Applying Protected Area Management Categories*. International Union for Conservation of Nature and Natural Resources (IUCN), Gland.

Duru, M., Therond, O., Martin, G., Martin-Clouaire, R., Magne, M. A., Justes, E., Journet, E. P., Aubertot, J. N., Savary, S., Bergez, J. E. and Sarthou, J. P. (2015). How to implement biodiversity-based agriculture to enhance ecosystem services: a review. *Agronomy for Sustainable Development* 35(4):1259–1281.

Engels, J. M. M., Polreich, S. and Dulloo, M. E. (2008). Role of community gene / seedbanks in the conservation and use of crop genetic resources in Ethiopia. pp. 263–270. In: Thijssen, M. H., Bishaw, Z., Beshi, A. and de Boef, W. S. (Eds), *Farmers' Seeds and Varieties: Supporting Informal Seed Supply in Ethiopia*. Wageningen International, Wageningen.

FAO (2018). *Globally Important Agricultural Heritage Systems: Combining Agricultural Biodiversity, Resilient Ecosystems, Traditional Farming Practices and Cultural Identity*. Food and Agriculture Organization of the United Nations (FAO), Rome.

Ficiciyan, A., Loos, J., Sievers-Glotzbach, S. and Tscharntke, T. (2018). More than yield: ecosystem services of traditional versus modern crop varieties revisited. *Sustainability* 10(8):2834. doi:10.3390/su10082834.

Frankel, O. H. and Hawkes, J. G. (1975). *Crop Genetic Resources for Today and Tomorrow*. Cambridge University Press, Cambridge.

Garcia-Yi, J. (2014). Market participation and agro-biodiversity loss: the case of native chili varieties in the Amazon rainforest of Peru. *Sustainability* 6(2):615–630.

Gauchan, D., Joshi, B. K., Ghimire, K. H., Poudyal, K., Sapkota, S., Sharma, S., Singh Dangol, D. M., Sthapit, S., Gautam, S., Khatiwada, S. and Jarvis, D. I. (2017). Process of rebuilding local seed system after 2015 Nepal earthquake: rescue collection, conservation and repatriation. pp. 32–40. In: Joshi, B. K. and Gauchan, D. (Eds), *Rebuilding Local Seed System of Native Crops in Earthquake Affected Areas of Nepal*. Proceedings of the Sharingshop, NAGRC, BI and Crop Trust, Kathmandu.

Gerards, M. C., Terloub, R. J., Yuc, H., Koksc, C. H. W. and Gerdes, V. E. A. (2015). Traditional Chinese lipid-lowering agent red yeast rice results in significant LDL reduction but safety is uncertain - a systematic review and meta-analysis. *Atherosclerosis* 240(2):415–423.

Goland, C. (1993). Field scattering as agricultural risk management: a case study from Cuyo Cuyo, department of Puno, Peru. *Mountain Research and Development* 13(4):317–338.

Gómez, C. (1993). *Afirmación Cultural Andina*. Proyecto Andino de Tecnologías Campesinas (PRATEC), Lima.

Grady, T. G. (2012). Regarding biocultural heritage: in situ political ecology of agricultural biodiversity in the Peruvian Andes. *Agriculture and Human Values* 30(4):587–604.

Grillo, E., Quiso, V., Rengifo, G. and Valladolid, J. (1994). *Crianza Andina de la Chacra*. Proyecto Andino de Tecnologías Campesinas (PRATEC), Lima.

Gruberg, H., Meldrum, G., Padulosi, S., Rojas, W., Pinto, M. and Crane, T. (2013). *Towards a Better Understanding of Custodian Farmers and Their Roles: Insights from a Case Study in Cachilaya*. Bioversity International, Rome.

Halewood, M. (2016). *Farmer's Crop Varieties and Framer's Rights: Challenges in Taxonomy and Law*. Routledge, Earthscan, New York.

Hamilton, R. W. (2003). *The Art of Rice: Spirit and Sustenance in Asia*. UCLA Fowler Museum of Cultural History, Los Angeles.

Haverkort, B. and Hiemstra, W. (1999). *Food for Thought: Ancient Visions and New Experiments of Rural People*. Zed Books, London.

Haverkort, B., Van't Hooft, K. and Hiemstra, W. (2003). *Ancient Roots, New Shoots: Endogenous Development in Practice*. Zed Books, London.

Hellin, J. and Higman, S. (2005). Crop diversity and livelihood security in the Andes. *Development in Practice* 15(2):165–174.

Herforth, A., Johns, T., Creed-Kanashiro, H. M., Jones, A. D., Khoury, C. K., Lang, T., Maundu, P., Powell, B. and Reyes-García, V. (2019). Agrobiodiversity and Feeding the

World: more of the same will result in more of the same. pp. 185–211. In: Zimmerer, K. S. and De Haan, S. (Eds), *Agobiodiversity: Integrating Knowledge for a Sustainable Future*. Ernst Strungmann Forum, MIT Press, Cambridge, MA.

Huamán, Z. (2001). Semilleros comunales de papas nativas del Perú. *Revista Biodiversidad* 30:20–22.

ICAR (2017). *ICAR-CGIAR Agricultural Cooperation*. Indian Council of Agricultural Research (ICAR), New Delhi.

IDMA (2018). *Agrobiodiversidad para Alimentar al Peru y al Mundo: experiencias exitosas de la conservación de la ABD en comunidades andinas de Huanuco y Lima*. Instituto de Desarrollo y Medio Ambiente (IDMA), Lima.

IPBES (2019). *Global Assessment Report on Biodiversity and Ecosystem Services*. Intergovernmental Science-Policy Platform on Biodiversity and Ecosystem Services (IPBES), Bonn.

Jäger, M., Van Loosen, I. and Giuliani, A. (2019). How have markets affected the governance of agrobiodiversity? pp. 307–323. In: Zimmerer, K. S. and De Haan, S. (Eds), *Agobiodiversity: Integrating Knowledge for a Sustainable Future*. Ernst Strungmann Forum, MIT Press, Cambridge, MA.

Klein, A. M., Vaissière, B. E., Cane, J. H., Steffan-Dewenter, I., Cunningham, S. A., Kremen, C. and Tscharntke, T. (2007). Importance of pollinators in changing landscapes for world crops. *Proceedings Biological Sciences* 274(1608):303–313.

Knorr and WWF (2019). *Future 50 Foods: 50 Food for Healthier People and a Healthier Planet*. Knorr, World Wildlife Fund (WWF), United Kingdom.

Kozicka, M., Groot, J. C. J. and Gotor, E. (2019). Can crop diversity strengthen small-scale farmers' resilience? Modelling future global biophysical and economic trends to understand individual farmers' resilience options. pp. 135–144. In: Bioversity International (Ed.), *Agrobiodiversity Index Report 2019: Risk and Resilience*. Bioversity International, Rome.

Krishna, V. V., Pascual, U. and Zilberman, D. (2010). Assessing the potential of labeling schemes for *in situ* landrace conservation: an example from India. *Environment and Development Economics* 15(2):127–151.

Krishna, V. V., Drucker, A. G., Pascual, U., Raghu, P. T. and King, E. D. I. O. (2013). Estimating compensation payments for on-farm conservation of agricultural biodiversity in developing countries. *Ecological Economics* 87:110–123.

Jarvis, D. I., Hodgkin, T., Sthapit, B. R., Fadda, C. and Lopez-Noriega, I. (2011). An heuristic framework for identifying multiple ways of supporting the conservation and use of traditional crop varieties within the agricultural production system. *Critical Reviews in Plant Sciences* 30(1-2):125–176.

Johnsson, M. (1986). *Food and Culture among the Aymara: Symbolic Expressions of Social Relations. Uppsala Studies in Cultural Anthropology*. Almqwist & Wiksell International, Stockholm.

Jones, A. D. (2017). Critical review of the emerging research evidence on agricultural biodiversity, diet diversity, and nutritional status in low- and middle-income countries. *Nutrition Reviews* 75(10):769–782.

Jones, A. D., Creed-Kanashiro, H., Zimmerer, K. S., De Haan, S., Carrasco, M., Meza, K., Cruz-Garcia, G. S., Tello, M., Plasencia Amaya, F., Marin, R. M.. and Ganoza, L. (2018). Farm-level agricultural biodiversity in the Peruvian Andes is associated with greater odds of women achieving a minimally diverse and micronutrient adequate diet. *The Journal of Nutrition* 148(10):1625–1637. https://doi.org/10.1093/jn/nxy166.

Labeyrie, V., Rono, B. and Leclerc, C. (2013). How social organization shapes crop diversity: an ecological anthropology approach among Tharaka farmers of Mount Kenya. *Agriculture and Human Values* 31(1):97–107.

Leclerc, C. and D'Eeckenbrugge, G. C. (2012). Social organization of crop genetic diversity. The G × E × S interaction model. *Diversity* 4(1):1–32.

Lemons, J., Victor, R. and Schaffer, D. (2003). *Conserving Biodiversity in Arid Regions: Best Practices in Developing Nations.* Kluwer Academic Publishers, Boston, MA.

Lewis, V. and Mulvany, P. M. (1997). *A Typology of Community Seed Banks.* National Resources Institute (NRI), Intermediate Technology Development Group (ITDG), Chatham.

López Noriega, I. L., Halewood, M., Abberton, M., Amri, A., Angarawai, I. I., Anglin, N., Blümmel, M., Bouman, B., Campos, H., Costich, D., Ellis, D., Gaur, P. M., Guarino, L., Hanson, J., Kommerell, V., Kumar, L., Lusty, C., Ndjiondjop, M. N., Payne, T., Peters, M., Popova, E., Prakash, G., Sackville-Hamilton, R., Tabo, R., Upadhyaya, H., Yazbek, M. and Wenzl, P. (2019). CGIAR operations under the Plant Treaty framework. *Crop Science* 59(3):819–832.

Luna-González, D. V. and Sørensen, M. (2018). Higher agrobiodiversity is associated with improved dietary diversity, but not child anthropometric status, of Mayan Achí people of Guatemala. *Public Health Nutrition* 21(11):2128–2141. doi:10.1017/S1368980018000617.

Mayer, E., Glave, M., Brush, S. B. and Taylor, J. E. (1992). *La Chacra de la Papa: economía y ecología.* Centro Peruano de Estudios Sociales (CEPES), Lima.

McGuire, S. J. (2008). Securing access to seed: social relations and sorghum seed exchange in eastern Ethiopia. *Human Ecology* 36(2):217–229.

Meinzen-Dick, R. S., Devaux, A. and Antesana, I. (2009). Underground assets: potato biodiversity to improve the livelihoods of the poor. *International Journal of Agricultural Sustainability* 7(4):235–248.

Mercer, K. L., Vigouroux, Y., Castañeda-Alvarez, N. P., De Haan, S., Hijmans, R. J., Leclerc, C., McKey, D. and Vanek, S. J. (2019). Crop evolutionary agroecology: genetics and functional dimensions of agrobiodiversity and associated knowledge. pp. 21–62. In: Zimmerer, K. S. and De Haan, S. (Eds), *Agobiodiversity: Integrating Knowledge for a Sustainable Future.* Ernst Strungmann Forum, MIT Press, Cambridge, MA.

Moreau, T. and Speight, D. (2019). Cooking up diverse diets: advancing biodiversity in food and agriculture through collaborations with chefs. *Crop Science* 59(6):2381–2386.

Moscoe, L. and Blas, R. (2016a). *Catálogo de la Diversidad de la Oca de Amaru.* University of Wisconsin-Madison, Universidad Agraria La Molina, Lima.

Moscoe, L. and Blas, R. (2016b). *Catálogo de la Diversidad de la Oca de Viacha.* University of Wisconsin-Madison, Universidad Agraria La Molina, Lima.

Mulumba, J. W., Nankya, R., Adokorach, J., Kiwuka, C., Fadda, C., De Santis, P. and Jarvis, D. I. (2012). A risk-minimizing argument for traditional crop varietal diversity use to reduce pest and disease damage in agricultural ecosystems of Uganda. *Agriculture, Ecosystems and Environment* 157:70–86.

Oakley, E. and Momsen, J. H. (2007). Women and seed management: a study of two villages in Bangladesh. *Singapore Journal of Tropical Geography* 28(1):90–106.

Orlove, B. S. and Godoy, R. (1986). Sectoral fallowing in the central Andes. *Journal of Ethnobiology* 6(1):169–204.

Padulosi, S. and Dulloo, E. (2012). Towards a viable model for monitoring agrobiodiversity on-farm: a proposed new approach for red listing of cultivated plant species. pp.

171–200. In: Padulosi, S., Bergamimi, N. and Lawrence, T. (Eds), *On Farm Conservation of Neglected and Underutilized Species: Trends and Novel Approaches to Cope with Climate Change*. Bioversity International, Rome.

Pautasso, M., Aistara, G., Barnaud, A., Caillon, S., Clouvel, P., Coomes, O. T., Delêtre, M., Demeulenaere, E., De Santis, P., Döring, T., Eloy, L., Emperaire, L., Garine, E., Goldringer, I., Jarvis, D., Joly, H. I., Leclerc, C., Louafi, S., Martin, P., Massol, F., McGuire, S., McKey, D., Padoch, C., Soler, C., Thomas, M. and Tramontini, S. (2012). Seed exchange networks for agrobiodiversity conservation: a review. *Agronomy for Sustainable Development* 33(1):151–175. doi: 10.1007/s13593-012-0089-6.

Pilling, D., Bélanger, J. and Hoffmann, I. (2020). Declining biodiversity for food and agriculture needs urgent global action. *Nature Food* 1(3):144–147.

Polreich, S., Yemane, T., Becker, H. C. and Maass, B. L. (2005). Assessing the effectiveness of the community-based seed supply system for in situ conservation of local wheat varieties. Proceedings of the Tropentag Conference on International Agricultural Research for Development, Stuttgart-Hohenheim.

Polreich, S., Rodriguez, F., Nuñez, J., Roca, L., Juarez, H., Arce, A., Plasencia, F., Luziatelli, G. and de Haan, S. (2018). *Protocolo para Establecer una Línea Base de Diversidad de Papa Cultivada en sus Microcentros de Alta Diversidad*. International Potato Center, Lima, Peru.

Possey, D. A. (1999). *Cultural and Spiritual Values and Biodiversity*. United Nations Environmental Program (UNEP), Nairobi.

Quiroz, D. and Van Andel, T. (2018). The cultural importance of plants in western African religions. *Economic Botany* 72(3):251–262.

Rengifo, G. (1988). *Recursos Filogenéticos Andinos.* Conversaciones en el Marco del VI Congreso Internacional sobre Cultivos Andinos. Proyecto Andino de Tecnologías Andinas (PRATEC), Lima.

Resque, A. G. L., Coudel, E., Piketty, M. G., Cialdella, N., Sá, T., Piraux, M., Assis, W. and Le Page, C. (2019). Agrobiodiversity and public food procurement programs in Brazil: influence of local stakeholders in configuring green mediated markets. *Sustainability* 11(5):1425.

Rijal, D., Rana, R., Subedi, A. and Sthapit, B. (2000). Adding value to landraces: community-based approaches for in-situ conservation of plant genetic resources in Nepal. pp. 166–172. In: Friis-Hansen, E. and Sthapit, B. (Eds), *Sthapit, Participatory Approaches to the Conservation and Use of Plant Genetic Resources*. International Plant Genetic Resources Institute (IPGRI), Rome.

Riley, K. W., Mateo, N., Hawtin, G. C. and Yadav, R. (1990). *Mountain Agriculture and Crop Genetic Resources*. International Research Development Centre (IDRC), London.

Ruiz Muller, M. (2009). *Las Zonas de Agrobiodiversidad y los Registros de Cultivos Nativos. Sociedad Peruana de Derecho Ambiental (SPDA)*. Bioversity International, Lima.

Sandstrom, A. R. (1992). *Corn Is Our Blood: Culture and Ethnic Identity in a Contemporary Aztec Indian Village*. University of Oklahoma Press, Norman, OK.

Sangabriel-Conde, W., Negrete-Yankelevich, S., Maldonado-Mendoza, I. E. and Trejo-Aguilar, D. (2013). Native maize landraces from los Tuxtlas, Mexico show varying mycorrhizal dependency for P uptake. *Biology and Fertility of Soils* 50(2):405–414.

Santilli, J. (2012). *Agrobiodiversity and the Law: Regulating Genetic Resources, Food Security and Cultural Diversity*. Earthscan Publishers, Oxon.

Scott, G. J. (2011). Plants, people, and the conservation of biodiversity of potatoes in Peru. *Natureza and Conservação* 9(1):21–38.

Scurrah, M., De Haan, S. and Winge, T. (2013). Cataloguing potato varieties and traditional knowledge from the Andean Highlands of Huancavelica, Peru. pp. 65-79. In: Anderson, R. and Winge, T. (Eds), *Realising Farmers' Rights to Crop Genetic Resources: Success Stories and Best Practices*. Earthscan Publishers, Oxon.

Senakun, C., Chunta, S., Somboonwattanakul, I., Yodsiri, S., Kurukodt, J. and Senakun, A. (2018). Diversity, utilization and cultural significance of purple rice in northeastern Thailand. *International Journal of Agricultural Technology* 14(7):1893-1904.

Shepherd, C. (2017). Andean Cultural Affirmation and Cultural Integration in Context: reflections on indigenous knowledge for the in situ conservation of agrobiodiversity. pp. 130-146. In: Sillitoe, P. (Ed.), *Indigenous Knowledge: Enhancing Its Contribution to Natural Resources Management*. CABI Publishers, Surrey.

Shrestha, P., Vernooy, R. and Chaudhary, P. (2013). Community seed Banks in Nepal: past, present and future. Proceedings of the a National Workshop. LI-BIRD, the development fund. IFAD, Bioversity International, Pokhara.

Steinberg, M. K. and Taylor, M. (2002). The impact of political turmoil on maize culture and diversity in Highland Guatemala. *Mountain Research and Development* 22(4):344-351.

Subedi, A., Chaudhary, P., Baniya, B. K., Rana, R. B., Tiwari, R. K., Rijal, D. K., Sthapit, B. R. and Jarvis, D. I. (2002). Who maintains crop genetic diversity and how?: implications for on-farm conservation and utilization. *Culture and Agriculture* 25(2):41-50.

Swiderska, K. (2006). Protecting Traditional Knowledge: A framework based on customary laws and bio-cultural heritage sustainable agriculture, biodiversity and livelihoods programme. IIED Paper for the International Conference on Endogenous Development and Bio-Cultural Diversity, 3-5 October 2006, Geneva.

Tapia, M. E. (2000). Mountain agrobiodiversity in Peru: seed fairs, seed banks, and mountain-to-mountain exchange. *Mountain Research and Development* 20(3):220-225.

Tapia, M. E. and Rosas, A. (1993). Seed Fairs in the Andes: a strategy for local conservation of plant genetic resources. pp. 111-118. In: de Boef, W., Amanor, K., Wellard, K. and Bebbington, A. (Eds), *Cultivating Knowledge: Genetic Diversity, Farmer Experimentation and Crop Research*. Intermediate Technology Publications, London.

Tobin, D., Bates, R., Brennan, M. and Gill, T. (2016). Peru potato potential: biodiversity conservation and value chain development. *Renewable Agriculture and Food Systems* 33(1):19-32.

Tooker, J. F. and Frank, S. D. (2012). Genotypically diverse cultivar mixtures for insect pest management and increased crop yields. *Journal of Applied Ecology* 49(5):974-985.

UNCTAD (2017). *BioTrade and Access and Benefit Sharing: From Concept to Practice, a Handbook for Policymakers and Regulators*. United Nations Conference on Trade and Development (UNCTAD), New York.

Van Andel, T., Veltman, M. A., Bertin, A., Maat, H., Polime, T., Hille Ris Lambers, D., Tjoe Awie, J., De Boer, H. and Manzanilla, V. (2019). Hidden rice diversity in the Guianas. *Frontiers in Plant Science* 10:1161. https://doi.org/10.3389/fpls.2019.01161.

Van Den Berg, H. (1989). *La Tierra no da así Nomás: los ritos agrícolas en la religión de los aymara-cristianos*. Centrum voor Studie en Documentatie van Latijns Amerika (CEDLA), Amsterdam.

Van Deynze, A., Zamora, P., Delaux, P. M., Heitmann, C., Jayaraman, D., Rajasekar, S., Graham, D., Maeda, J., Gibson, D., Schwartz, K. D., Berry, A. M., Bhatnagar, S., Jospin, G., Darling, A., Jeannotte, R., Lopez, J., Weimer, B. C., Eisen, J. A., Shapiro, H. Y., Ané,

J. M. and Bennett, A. B. (2018). Nitrogen fixation in a landrace of maize is supported by a mucilage-associated diazotrophic microbiota. *PLoS Biology* 16(8):e2006352. Crossref.

Van Dusen, M. E. and Taylor, J. E. (2005). Missing markets and crop diversity: evidence from Mexico. *Environment and Development Economics* 10(4):513–531.

Van Etten, J., De Sousa, K., Aguilar, A., Barrios, M., Coto, A., Dell'Acqua, M., Fadda, C., Gebrehawaryat, Y., Van de Gevel, J., Gupta, A., Kiros, A.Y., Madriz, B., Mathur, P., Mengistu, D.K., Mercado, L., Mohammed, J.N., Paliwal, A., Pè, M.E., Quirós, C.F., Rosas, J.C., Sharma, N., Singh, S.S., Solanki, I.S. and Steinke, J. (2019). Crop variety management for climate adaptation supported by citizen science. *Proceedings of the National Academy of Sciences* 116(10):4194–4199.

Van Kessel, J. and Enriquez, P. (1989). La expansión de la frontera agrícola andina hacia arriba: la lucha contra heladas y granizadas. pp. 345–379. In: MUSIRO (Ed.), *En La visión india: tierra, cultura, lengua, derechos humanos*. Musiro Fundación para la Cultura Indígena, Leiden.

Van Kessel, J. and Condori Cruz, D. (1992). *Criar la Vida: trabajo y tecnología en el mundo andino*. Vivarium, Santiago de Chile.

Van Kessel, J. and Enriquez, P. (2002). *Señas y Señaleros de la Santa Tierra: Agronomía Andina*. Abya Yala - IECTA, Iquique.

Vanek, S. (2019). Crop and varietal diversity impacts on agroecosystem functions and resilience. pp. 79–102. In: Zimmerer, K. S. and De Haan, S. (Eds), *Agobiodiversity: Integrating Knowledge for a Sustainable Future*. Ernst Strungmann Forum, MIT Press, Cambridge, MA.

Vernooy, R., Sthapit, B., Galluzzi, G. and Shrestha, P. (2014). The multiple functions and services of community seedbanks. *Resources* 3(4):636–656.

Vernooy, R., Shrestha, P. and Sthapit, B. (2015). *Community Seed Banks: Origins, Evolution and Prospects*. Earthscan Publications, Bioversity International, Oxon.

Vernooy, R., Sthapit, B., Otieno, G., Shrestha, P. and Gupta, A. (2017). The roles of community seed banks in climate change adaption. *Development in Practice* 27(3):316–327.

Villagomez, J. and Blanco, E. (2006). *Catálogo de Ají de Ecotipos Conservados en Campos de Agricultores*. PROINPA, PNUD, Cochabamba.

Visser, B., Brush, S. B., Aistara, A., Andersen, R., Jäger, M., Nemogá, G., Padmanabhan, M. and Sherwood, S. G. (2019). The governance of agrobiodiversity. pp. 283–306. In: Zimmerer, K. S. and De Haan, S. (Eds), *Agobiodiversity: Integrating Knowledge for a Sustainable Future*. Ernst Strungmann Forum, MIT Press, Cambridge, MA.

Westengen, O. T. and Winge, T. (2019). *Farmers and Plant Breeding: Current Approaches and Perspectives*. Routledge, Abingdon, UK.

Widawsky, D. and Rozelle, S. (1998). Varietal diversity and yield variability in Chinese rice production. pp. 159–172. In: Smale, M. (Ed). *Farmers, Gene Banks and Crop Breeding*. Kluwer Academic Publishers, Boston, MA.

Yang, L. N., Pan, Z. C., Zhu, W., Wu, E. J., He, D. C.,Yuan, X., Qin, Y. Y., Wang, Y., Chen, R. S., Thrall, P. H., Burdon, J. J., Shang, L. P., Sui, Q. J. and Zhan, J. (2019). Enhanced agricultural sustainability through within-species diversification. *Nature Sustainability* 2(1):46–52.

Zeder, M. A. (2015). Core questions in domestication research. *Proceedings of the National Academy of Sciences of the United States of America* 112(11):3191–3198.

Zimmerer, K. S. and De Haan, S. (2017). Agrobiodiversity and a sustainable food future. *Nature Plants* 3(4):17047.

Chapter 11

Participatory plant breeding programs to optimize use of crop genetic resources[1]

Margaret Smith, Cornell University, USA; and J. C. Dawson, University of Wisconsin-Madison, USA

1 Introduction

Throughout the history of settled agriculture, farmers have exploited crop genetic diversity to tailor varieties to their local agroecologies, cropping systems, and crop uses. The earliest farmers were choosing which seed to save from what were probably highly genetically diverse populations of plants, little different from wild populations. Those early selection efforts gave rise, gradually and over many years, to crops that resemble what we consider our cultivated agricultural plants today. Throughout that process, crop varieties have been continually evolving entities, exposed annually to both natural and human selection. While a variety may retain its name as an identifier, if it is a farmer-maintained variety, it is not a genetically static entity. Each generation of seed saving is accompanied by natural and sometimes intentional human selection pressures and the consequent genetic changes. Those annual changes add up to gradual shifts in the genetic constitution of a variety over time, allowing crop varieties to adapt to changes in climate, evolution of pest populations, alterations in crop management practices, and changing crop uses and culinary preferences.

1 Topic Statement: Involving local farmers in breeding programs enhances agricultural production and local crop genetic resources.

http://dx.doi.org/10.19103/AS.2020.0085.10

The ex-situ conservation of crop genetic diversity essentially aims to preserve varieties that have been historically important, that come from unique environments, and/or that possess unusual characteristics. This is, indeed, an important aspect of conservation; however, it may be likened to the 'museum' approach rather than the 'living laboratory' approach. While varieties have never been static entities, those conserved in gene banks are essentially that – a snapshot in time of the genetic profile of a variety that, if it had continued to be cultivated, would have been subject to on-going genetic change through natural and human selection. This genetic change would have ensured the variety's adaptation to current conditions and needs, while also ensuring the conservation of useful alleles and allele assemblages via active in situ conservation. While a given variety's genetic constitution at some moment in time would not necessarily be conserved, the genetic variation of relevance and use would continue to be conserved and the forces of selection ensuring its relevance and use would be allowed to continue. Conservation of useful traits (and thus useful genetic variation) is a natural outcome of this on-going selection process.

In addition, gene bank accessions are not necessarily accurate representations of the diversity that was present in farmers' fields at the moment of collection. Gene banks make a heroic effort to regenerate accessions according to best practices; however, they are chronically underfunded and may not be able to regenerate all accessions as frequently as they need to. This can mean that lower germination rates lead to smaller population sizes during regeneration, or simply a lack of adequate funding may mean smaller plots and fewer plants regenerated, which can lead to genetic drift and loss of potentially useful alleles. The resulting regenerated accessions may have lower levels of diversity than those cultivated in farmers' fields, and some may be essentially pure lines depending on the species (Thomas et al. 2012, 2015, Khan et al. 2020).

Formal plant breeding efforts have built on the useful genetic variation found on farms as well as that found in gene banks in several ways. Formal plant breeders engaged in farmer-participatory efforts have introduced novel genetic variation that was lacking in farmers' populations, and then allowed on-farm selection to hone these new genetic mixtures, perpetuating the historical process of farmer innovation and selection. This process has been well documented for maize varieties in Mexico and termed 'creolization' or 'rustication' (e.g. Bellón and Brush 1994). Advantages in terms of broader adaptation resulted from introgression with modern varieties of millet in Rajasthan, India (vom Brocke et al. 2003). However, these authors also noted that where rainfall was more limited, landraces had better productivity. The value of 'creolization' is clearly context dependent.

In formal breeding that targets large-scale commercialization, genetic variation has been tapped for both trait understanding as well as trait introgression. The reservoirs of genetic variation found in farmer varieties have allowed greater understanding of the genetic control of important adaptive traits. While these are often complex in their inheritance, innovations in experimental design and statistics have helped to improve trait heritability and thus the ability to effectively select for even genetically complex traits. Improved data management has facilitated broad multi-location evaluation to assess stability over varying environmental conditions and to assess wide adaptation – a characteristic of considerable interest to the seed industry. These assessments utilize multiple environments to represent both geographic and temporal environmental variation. While farmer-maintained varieties have been included in some of these multi-location evaluations that target wide adaptation, results have not often revealed wide adaptation in locally cultivated and locally conserved varieties, perhaps for obvious reasons. Taken together, these formal breeding outcomes and products represent a form of conserving useful allele assemblages, but primarily to the extent that those allele assemblages benefit commercially oriented variety development.

Interest in farmer-participatory approaches in utilizing genetic variation has been spurred by the growing realization that selection in formal breeding programs, which typically carry out selection under optimal conditions with a focus on broad geographic adaptation to target larger seed markets, may not be the best strategy for addressing the needs of smaller-scale and diverse farmers. It is clear that genetic improvement under optimal conditions does not improve stress tolerance other than for relatively mild levels of stress (Edmeades and Bänziger 1996). In many target areas, locally variable environments, socioeconomic conditions, and crop use needs and preferences dictate a need for locally tailored varieties, which are by definition not broadly adapted. Finally, locally adaptive traits may be lost due to repulsion phase linkages with genes strongly selected for in formal breeding programs or in situations where these traits are not needed across a broad-enough target area to merit priority by a seed company. These situations are likely to compromise traits essential to local field performance and farmer acceptability, making the resulting varieties unlikely to be adopted or used.

This chapter summarizes a sample of variety evaluation, experimental design, and breeding method innovations that have served as solid approaches for participatory plant breeding (PPB) efforts. With success in PPB comes success in conservation at a local level of useful alleles and allele assemblages in the form of on-farm crop genetic resources. PPB programs of this sort have the potential to add value to local or traditional varieties that might otherwise be abandoned, thus promoting their *in situ* conservation (Sthapit et al. 2001). This chapter will briefly touch on methodologies to assess farmers' variety preferences – a commonly used process for eliciting trait needs and aspirations for different groups of farmers or different end uses for crops (Weltzien et al.

2020). Following sections will highlight some experimental designs for on-farm variety evaluation and farmer-participatory breeding methods for combining in-situ conservation with genetic improvement. Finally, some of the challenges that may limit genetic gain from PPB programs are noted - problems that increase the risk of wholesale replacement of on-farm genetic diversity rather than conservation through improvement.

2 Methodologies to assess variety preference

A clear and detailed understanding of farmers' varietal trait needs and preferences is essential to identifying required and desired traits as guides to breeding objectives (Weltzien et al. 2020). If the array of traits that are important to farmers or the nature of change desired in those traits (e.g. improvement vs. maintenance of status quo) is inaccurate or incomplete, the products of breeding efforts (whether participatory or not) are unlikely to be accepted. That outcome can lead to short-term conservation through continued use of existing farmer varieties since introduced varieties do not meet farmers' needs. However, it may not result in longer-term conservation if varietal performance is inadequate to support continued farm viability, leading to the abandonment of both the farm and the varieties being conserved there. Alternatively, that outcome can lead to wholesale replacement of farmers' varieties as their performance lags and ultimately becomes unacceptable, thus losing the standing genetic diversity that they comprised.

To encourage effective genetic improvement that incorporates useful existing diversity from in situ genetic resources, methodologies and designs have been developed for farmer-participatory assessment of trait needs/preferences based on varietal ranking with farmers. It is important to spend time on this stage and to use methods developed to solicit comprehensive responses rather than simply asking farmers what their main trait needs are. This can lead to an emphasis on traits that are currently lacking or less than desired but little discussion on the traits that need to be maintained, which farmers may not mention because they feel they are obvious (eating quality, for example). Other aspects of the system like labor constraints and how the crop under consideration fits into the farmer's current agricultural system are also important. Using existing varieties and comparisons with alternatives often gives farmers a more concrete set of options to discuss and leads to a better understanding of farmers' needs. A few examples are briefly described below.

2.1 Gathering farmer input

Participatory variety selection that involves farmers in identifying the most promising among a set of varieties is a well-established practice (e.g. see Witcombe and Joshi 1996). However, farmer input can be difficult to collect and summarize in a systematic way given the unbalanced and subjective nature of many such evaluations and the complexity of teasing out variation in preferences among different subsets of farmers. Evidence is abundant that trait needs and preferences do vary among categories such as farmer groups, crop management systems, and end uses.

Sperling and Scheidegger (1995) noted that different communities in Rwanda identified different bean entries in a participatory selection exercise. Cassava varieties that are bitter (and consequently toxic in the absence of processing) were preferred by Malawian farmers, despite awareness and availability of non-toxic varieties - an apparently paradoxical result (Chiwona-Karltun et al. 1998). The logic of these preferences became apparent through detailed exploration - bitterness protected farmers' crops from theft, animal damage, and unplanned harvest for consumption. Bean farmers in Rwanda preferred attributes related to how productive and well adapted the variety is when intercropped with maize, such as limited vigor, good adaptation to the intercrop situation, and competitive ability (Isaacs et al. 2016). These are traits unlikely to be emphasized in researcher-led bean breeding efforts that typically target a single crop in isolation. A recent review demonstrated that lack of emphasis on improving consumer-preferred traits accounted for limited adoption of improved varieties of roots, tubers, and bananas (Thiele et al. 2020).

Many studies have noted gendered differences in trait preferences. Ratings differed between female and male respondents evaluating sorghum in Burkina Faso, with females placing more emphasis on grain-quality traits (vom Brocke et al. 2010). Similarly, a study in Nigeria showed that women placed more emphasis than men on cassava cooking and product-quality traits, although the preference patterns varied by geographic region and proximity to processing facilities (Teeken et al. 2018). As concluded in a recent summary of evidence regarding farmer varietal trait preferences (Weltzien et al. 2020), preferences are often different between women and men, particularly when household roles and responsibilities related to crop production and post-harvest activities differ between these groups. Other farmer groupings based on farming systems, crop management options, agroecologies, or crop uses are likely to lead to similar differences in trait preferences among groups. Clearly, farmer rating and ranking data can embody important subtleties that are essential in identifying appropriate breeding objectives.

Since farmer evaluation of variety trials is a commonly used approach to elucidate trait needs and preferences (Weltzien et al. 2020), it is critical to consider mechanisms for simplifying this process from the farmers' and the researchers' perspectives while ensuring its robustness and validity in terms of data collection and interpretation. Marking plants with two different colors of ribbons to identify varieties/plants that should be further tested and those that should be eliminated was used as an efficient way to collect multiple opinions on a larger group of varieties (Sperling and Scheidegger 1995). Notably, these authors observed that farmers' responses did not appear to be especially influenced by the 'votes' of others when making their choices. To simplify and sort trait importance, Bellón (2001) suggested the use of cards depicting different traits that could be laid out on a grid, either to reflect relative importance of traits or to characterize individual varietal strengths and weaknesses (e.g. Fig. 1). Group ranking exercises were carried out with maize and climbing bean farmers in Honduras, leading to charts indicating farmers' trait needs and preferences (Meitzner 1998) (Fig. 2).

Inherent in these techniques are means for cross-classifying trait needs as they intersect with farmer groupings, crop uses, cropping systems, and so on. PPB that is aiming to optimize the use of crop genetic resources must ensure that traits valued by farmers and end users are maintained or improved, and that perceived deficiencies are targeted for improvement to enhance the likelihood of continued in situ conservation of these resources. Consequently, nuanced and complex understanding of needs and wants is essential to identify appropriate starting populations and selection criteria. One approach to this challenge is to employ the concept of 'use categories' in evaluations. Use categories reflect information about agronomic adaptation combined with the various culinary, commercial, and economic objectives farmers have for a given cultivar (Zimmerer 1991). Other authors suggest a diagnostic framework that cross-classifies farmer demographics (gender, market integration, land-holding size - whichever are most germane in a given situation) with aspects such as adaptation, required processing qualities, management options, or

Figure 1 Example of card-based system for classifying varietal traits of importance and evaluating individual varieties (adapted from Bellón 2001).

Figure 2 Maize and climbing bean farmers' group evaluation of variety trait needs and preferences. Photo credit: Laura Meitzner-Yoder.

commercialization traits (Christinck and Weltzien 2005). The specific approach taken can vary widely - what is essential is to identify key farmer/user groups for whom varietal needs differ, and to systematically identify the product characteristics that will meet those various niches.

2.2 Experimental designs for evaluation

While on-farm variety evaluation has a long history of use, researchers have often struggled to ensure that evaluations are not overly demanding of land, labor, and farmers' time and effort, particularly for target environments dominated by smallholder and/or resource-poor farmers. Similar constraints exist for highly diversified farmers in countries with more industrialized and market-based agriculture. While they may have access to good land and more resources, their time is highly constrained during the growing season, so experimental designs must be simple in order to be implemented. Although they are embedded in a more industrialized system, these farmers do not represent a primary seed market and so often their specific trait needs are not addressed by larger seed companies.

It is exactly these sorts of farmers, whether in traditional or industrialized agricultural systems, who are most likely to be growing local varieties that represent unique genetic resources, making the challenge of appropriately scaled on-farm variety evaluation especially pertinent to PPB that aims to enhance in situ conservation efforts. The need for small scale when designing on-farm evaluations suggests limiting numbers of varieties evaluated and minimizing or even eliminating replication. However, reducing these numbers simultaneously reduces precision of measurements (whether quantitative or qualitative) in typical experimental designs. Several innovative designs have been developed and used to address these concerns in the context of PPB and in situ conservation efforts. Two such designs are briefly described below.

2.3 Mother-baby trial design

Mother-baby trial designs were described by Snapp (2002). These designs combine two elements: (1) a researcher-managed replicated trial to evaluate larger numbers of treatments and test hypotheses, and (2) numerous satellite trials comparing small subsets of the treatments on farms in unreplicated, larger plots. This design allows evaluation of more complex questions in the central 'mother' trial, which should be located near the satellite 'baby' trials, and which uses replication to increase precision. The 'baby' trials provide real-world evaluation that samples farmers' actual soils, microenvironments, management practices, and so on; direct comparison to multiple examples of farmers' own practices (management, varieties, etc.); an opportunity to collect farmer rankings and perceptions across multiple overlapping subsets of the treatments; and an opportunity for farmers to gain hands-on experience with the new treatments on their own farms. Farmers frequently participate in evaluating the central 'mother' trial as well as their own 'baby' trial.

This design was initially used to evaluate soil fertility management options (Kamanga et al. 2001) and has subsequently been successfully used for variety evaluation in many crops including sweet potato (Baafi et al. 2020, Odhiambo et al. 2011), maize (Abeba 2005, Baluti et al. 2011), mungbean (Bari and Islam 2009), tomato (Testen et al. 2016), and organic vegetables (Lyon et al. 2020).

Using mother-baby common bean evaluations in Ethiopia, Gurmu (2013) determined that farmers' selection criteria placed relatively higher priority on qualitative traits (e.g. seed color, marketability, seed size, drought tolerance, insect and disease resistance) than they placed on yield. One of the top two preferred varieties reported in this study was the local variety, preferred for its seed color and marketability even though it was one of the lower-yielding varieties evaluated. Various researchers have similarly noted that farmers' definitions of needed traits are more multivariate than what breeders tend to identify (e.g. vom Brocke et al. 2010). Mother-baby approaches to evaluating varieties and gathering feedback on their positive and negative traits allow for the identification of locally relevant preferences based on complex, multivariate trait needs, facilitating conservation of unique varieties via their continued use or conservation of important allele assemblages from such varieties through breeding to add value to local varieties that might otherwise be abandoned (Sthapit et al. 2002).

2.4 Tricot design

Another design that is particularly well suited to gathering farmer input on varieties is called 'triadic comparisons of technologies' or the tricot design (Steinke et al. 2017). Much information gathered from farmers' participatory evaluations is in the form of ranking data. Although this data is subjective and comparative information, that does not preclude rigorous statistical analysis to ensure validity of the data. A thorough description of methods for analyzing ranking and rating data is provided by Coe (2002). Of particular interest in the use of ranking data is the ranking of a larger set of varieties than any one farmer can test on their own, which is the goal of tricot methodologies. By using farms as incomplete blocks where subsets of the total number of varieties are tested, each farm can have a manageable set of varieties while contributing to a much larger comparison. Methods to analyze incomplete block designs with pairwise comparisons among ranked treatments have existed for decades (Bradley and Terry 1952, Pendergrass and Bradley 1960). However, their use in participatory trialing and selection has gained ground in recent years with the development of software to make it possible for very large numbers of farmers to participate in trials and the development of generalizations and analysis tools that make it much easier to design and implement a participatory trial.

The tricot approach is based on providing many farmers with three varieties each out of a larger evaluation set. Each farm is thus an incomplete block in the larger trial, and varieties are assigned to farms to ensure that all pairwise comparisons can be calculated. It is important to have large numbers of farmers for comparisons to be reliable, but the methods of analysis are robust despite missing incomplete blocks (when a farmer drops out for example). Each farmer plants the three varieties using their own management and ranks them (using a simple best - middle - worst scale) for traits of interest. Varieties can be added into the mix or dropped out of the mix on a seasonal basis, allowing for on-going comparison of the most relevant genetic resources. The relative simplicity of this design makes it possible to evaluate a wide array of traits on a large number of farms. Variety rankings highlight farmers' wisdom about what is likely to be most effective and most useful in their particular situation - wisdom that is often elusive from a researcher's perspective.

Particularly useful for variety ranking data is the Plackett Luce model, a generalization of the Bradley-Terry model for pairwise comparisons of ranked treatments in an incomplete block design (Luce 1959, Plackett 1975). This model determines a 'worth' value for each variety, and from there a probability that a given variety outperforms all other varieties evaluated. Importantly, this model has been expanded to provide for Plackett Luce trees, accounting for variation in worth rankings among different subsets such as specific types of farmers, soil types, management systems, or product end uses (Turner et al. 2020). This allows identification of human, agroecological, or end use domains in which a given variety fits best. With enough farmers participating, rankings across many farms can be disaggregated by types of farmers, soil types, management systems, crop end uses, and so on, thus allowing for identification of domains where different varieties are most apt. While individual farmer rankings are low in reliability, study has found that the accuracy of overall results is good (Steinke et al. 2017). Tricot designs have the advantage of not requiring a local, researcher-managed central trial (like the 'mother' trial in a mother-baby design), making this approach more broadly scalable (van Etten et al. 2019).

Outcomes from tricot evaluation are more likely to capture the context-specific advantages of local varieties, which often are not widely adapted in the agronomic sense or ideal for a wide array of end uses, and thus may fare poorly in rankings that are aggregated across these groupings. Teasing out best-fit niches for local varieties can encourage their use in targeted PPB programs, thus also encouraging in situ conservation of the unique and beneficial genetic diversity they embody. Furthermore, having only farmer-managed small trials simplifies the inclusion of many farms that sample a broad range of types of farmers, agroecologies, and management systems. With such a data set, breeders can begin to understand the parameters that determine where and in what circumstances a particular variety is well suited. With local varieties, this may allow their diffusion beyond the immediate area and thus increase the chances that they will be conserved in situ and not lost as a result of a local problem such as crop failure or conflict.

Tricot evaluation combined with Plackett Luce analysis has been effective in identifying valuable varieties as genetic resources, and is particularly effective with diverse sets of genetic materials rather than sets of elite, very similar varieties (van Etten et al. 2019). Given that traditional and local varieties often encompass a greater range of types and traits than elite materials, this makes the tricot design especially useful for these in situ conserved genetic resources.

3 Participatory plant breeding based on traditional/ local varieties

While the variety evaluation methods described above can help to identify which genetic resources are most useful to given sets of farmers, they are also helpful to tease out what traits could be improved to add value to these varieties. This is particularly important for genetic resource conservation, since the improvement of unique varieties and land races helps to ensure that they continue to be of value to farmers and thus are conserved in situ. Unless these varieties provide unique traits and trait value to farmers, they will be replaced and lost. As noted by Sthapit and colleagues (Sthapit et al. 2002), 'Participatory plant breeding . . . is a strategy used to strengthen on-farm conservation by encouraging farmers to continue to search, select, and manage local seed supply systems.'

Key to any breeding effort is careful identification of selection objectives to ensure relevance and utility of the product. This is true also of PPB, which is why the careful, cross-classification-based understanding of crop variety preferences and rankings described in the previous section is so critical. Once relevant breeding objectives are identified, many approaches can be taken for participatory improvement towards these objectives. PPB is most likely to meet with success, however, if the approaches are kept simple and manageable for both farmers and researchers. A few examples of simple strategies emphasized in on-farm PPB are briefly described below.

3.1 Mass selection

Mass selection is one of the simplest possible approaches to breeding, and mirrors what farmers have done in selecting their own varieties for thousands of years. For PPB, mass selection would involve participating farmers growing a segregating population and selecting target plant types from the field. Selection can be done by farmers alone, or by farmers and breeders working together. Either way, the wisdom of farmers in recognizing valuable plants/traits is key to the success of the PPB approach as distinct from on-station breeding using farmers' local varieties as starting populations.

Mass selection in F4 bulks of upland rice in eastern India was conducted on two populations derived from a long-grown variety released in 1983 (not a farmer-developed variety, but certainly could be considered 'local' after years of use) crossed with two different elite lines from the IRRI's rice breeding program (Virk et al. 2003). Mass selection by farmers in one of these populations was effective in identifying a variety that topped the local checks in terms of yield and early maturity, and had a number of other desirable agronomic and quality traits. This PPB effort led to the first upland rice variety that was highly accepted by farmers in this rainfed upland region of eastern India (Witcombe et al. 2005a). Mass selection in farmer varieties of maize was initiated in Honduras (Smith et al. 2001). In this work, pollination control was achieved through joint farmer-breeder plant selection at flowering, with the breeder creating pollen bulks from selected plants that were used to pollinate other selected plants. While results were mixed after 2–4 selection cycles on four farms, the farmer selections appeared to have more advantage over on-station breeder selections when evaluated on-farm at the farm with the lowest inherent yield potential. This suggests that PPB may be most useful in stress conditions that are most divergent from typical experiment station selection conditions.

3.2 Stratified mass selection

Stratification (also called gridding) involves selecting a certain proportion of the plants within each subsection of the entire field. It can be used with mass selection to minimize environmental variation within a field subsection, thereby increasing the chance that selections are based on genetic differences among plants. The more variable the field environment, the more powerful stratification can be for improving gains from selection. Stratified mass selection has proven effective in some cross-pollinated crops, even without the option of pollen control (meaning that selection is exercised only on female gametes). While gains per cycle were small, stratified mass selection in local varieties grown by five maize farmers in Mexico led to gains in yield and number of ears per plant (statistically significant at two of the five farms) over as few as three selection cycles (Smith et al. 2001).

3.3 Selection among vs. within populations

While formal plant breeders tend to focus on creating segregating populations within which to exercise selection, many PPB efforts have focused instead on farmer selection among either fixed breeding lines or segregating populations. As Ceccarelli and colleagues noted regarding their barley PPB efforts in Syria, Morocco, and Tunisia (Ceccarelli et al. 2001), most farmers in these countries grow heterogeneous land races; however, they observed no evidence that farmers perceived benefit to selection within populations. They were clearly enthusiastic about selecting among diverse breeding materials. Farmers' participation tended to maintain greater diversity than was maintained by non-participatory breeders' efforts. Thus even without selection within segregating populations, PPB served a vital role in conserving genetic diversity in this case.

3.4 Selection in later stages of the breeding process

Much of the original PPB work was similarly focused on selection by farmers among populations, traditional varieties, and/or products of breeding efforts (e.g. Sperling and Scheidegger 1995). Some authors clearly distinguish this as 'participatory variety selection' rather than PPB. In the case of farmers selecting only among varieties from a formal breeding program created without farmer input, this distinction may be critical. In the context of participatory approaches to on-farm conservation of genetic resources, there is often far more genetic diversity in the varieties tested, so either approach has the potential to encourage conservation of useful genetic diversity on farm.

In especially, farmer selection in the later stages of the breeding process can be particularly helpful to support the activity of independent breeders or small regional seed companies. Independent plant breeders, regional seed companies and seed savers' networks often have novel genetic diversity despite having smaller breeding programs. They may have multiple varieties they have created but may not have resources to trial their breeding lines across diverse environments, collaborate with farmers, expose varieties to potential seed company partners, or navigate the commercialization process. On-farm trialing and farmer selection among these advanced lines or pre-commercialization varieties may help the independent breeder determine the range of adaptation of their lines, and develop a market for a variety that excels in a particular region. Particularly in countries with industrialized agricultural systems, farmers may not wish to become seed producers, but do want to participate in variety selection and purchase seeds from companies who are focused on their region or their agricultural system.

Experimental designs such as the tricot method may be particularly helpful for smaller-scale programs without access to multi-locational trial sites such as those run by larger breeding programs in the public and private sector. This can help support their continued breeding work, and may maintain a large pool of genetic diversity through the existence of many smaller regionally focused breeders and companies. One of the principal barriers to this type of network approach in expanding trialing is in the logistics and analysis of the resulting data. Software such as Biodiversity International's ClimMob (climmob .net) and Seedlinked (seedlinked.com) are being developed to fill this gap, enabling independent breeders and smaller regional seed companies to build their own testing networks. If successful, this decentralized approach to variety development and testing involving many smaller-scale breeding programs will almost certainly increase the genetic diversity within a crop and region and contribute to in-situ conservation efforts, combining improved performance with increased diversity.

4 Challenges and limitations

PPB has the potential to enhance the value of local varieties by identifying domains in which they are especially well suited and by improving traits that are lacking in them. Both of these avenues increase the likelihood of in situ genetic resources conservation of these varieties, or their unique traits and the allele assemblages underlying those traits. However, PPB approaches have their challenges and limitations, noted in the following paragraphs. Recognizing these limitations and mitigating their impacts where possible will increase the chance of success from PPB programs based on local genetic resources.

Perhaps the most obvious challenge to PPB efforts aimed at genetic resources conservation is that they often target the most resource-poor environments, where so-called modern varieties have not been widely adopted. These tend to be stress environments where environmental variability is high and trait heritability correspondingly low. Low-trait heritability compromises the effectiveness of selection, particularly when exercised on a single plant basis. If progress from selection is erratic or too gradual to be readily observed, farmers may not see the value in engaging with PPB. Limited progress that does not maintain or enhance the value of local varieties to the farmers cultivating them leaves those varieties vulnerable to abandonment or replacement – obvious risks to genetic resources conservation. This can be partially overcome by working to find uniform fields within a stress environment, as heritability can be comparable in stress environments compared to non-stress environments provided that the germplasm evaluated is appropriate and the within-field heterogeneity is minimized (Zystro et al. 2019). If lines selected in non-stress environments are used, they may all do equally poorly in a stress environment, reducing heritability to near zero. Similarly if the field is so heterogeneous that even replication is not sufficient to remove the confounding of micro-environmental conditions and genetics, heritability will be near zero. But with more uniform (if stressful) field conditions and adapted germplasm, progress can be good. Principles of good experimental design take time to learn, even for scientists, and sometimes learning requires a few experiences with unusable data. Both farmers and breeders need to be patient with the process when trying to build participatory selection programs.

Similar risks result from the challenges of year-to-year variation on individual farms. Increasingly erratic weather patterns in many agricultural areas create inconsistent or even diametrically opposing selection pressures from year to year. Clearly successful varieties must be able to perform well over the range of conditions encountered on a farm over time. The approach to breeding for temporal variability of this sort is to retain large population sizes through each selection generation, to ensure that sufficient genetic variability is present in the subsequent generation to provide for a reasonable number of successful plants in whatever conditions present themselves. Population sizes that are manageable in a PPB effort may not be sufficiently large to allow effective selection over time when selection pressures are highly variable. In this case, developing a larger testing network using triadic methods and many farms may be an essential strategy, as varieties in development will be tested over a wider range of environmental conditions each year, perhaps sampling some of the same conditions as are present in year-to-year variation within a single farm. In addition, the network approach allows for larger numbers of entries and population sizes from one year to the next, particularly if selected entries are not just continued on one or a few farms, but are brought back to the breeder for recombination and recirculation to the testing network.

Institutionally, PPB has been challenged to scale-up beyond single projects with very limited numbers of farmers in a few selected locations. Individual researchers can only work with a limited number of farmers. Varieties selected through PPB often remain untested in a broader array of locations, and thus any genetic gains made remain only local. As long as a variety is cultivated only locally, it remains vulnerable to loss from local problems such as crop failure or conflict. The tricot methodology offers potential to overcome the problem of scaling-up in two ways. First, the simplicity of the on-farm design allows a researcher to work with larger numbers of farmers. Second, the ability to tease apart and understand the factors that condition success for a given variety (i.e. the domain where it fits well) allows for expanding the use of PPB-derived varieties into other areas, thus increasing the chances that their unique allele assemblages will be conserved. While we often think of farmer varieties as particularly well adapted in very local niches, there is evidence that PPB products can find utility more broadly. For example, certain varieties of rice developed through PPB efforts in Nepal were found to be preferred also by farmers in Bangladesh for high yield and good grain quality (Witcombe et al. 2005b). The same has been true in our experience in the northern tier of the United States where vegetable varieties developed by independent breeders for organic systems perform very well on similarly managed farms outside of the original region where they were developed. Trialing networks make it possible to identify these varieties, and for farmers to access them even though they were initially targeted to a different region.

5 Summary

PPB clearly has a role to play in the overall portfolio of activities needed for successful conservation of plant genetic resources. While not all PPB efforts address conservation goals, many of them do, whether by intent or coincidentally. If a given PPB effort is based on local varieties or land races, it has the potential to ensure their future utility and possibly enhance their value to farmers, thus increasing the likelihood that these varieties will continue to be cultivated and thus conserved in situ. If the PPB effort is founded on populations that combine local or land race varieties with improved germplasm, it has the potential to conserve allele assemblages that provide the local or land race varieties with unique traits such as microenvironment adaptation, special quality features, biotic and abiotic stress tolerances, and particular end-use traits. Conserving these allele assemblages is an important part of genetic resources conservation, and truly mirrors how genetic resources were conserved (while also evolving through human and natural selection) over evolutionary time.

Critical to success of PPB efforts is profound understanding of farmers' trait needs and desires, not as a broad aggregate but rather disarticulated by different farmer groups, end uses, and production systems. Several powerful approaches in achieving this intersectional understanding of trait needs have been developed and validated. Their success depends on efficient and statistically robust experimental approaches, which also have been developed and validated. PPB itself is best approached with fairly simple methods that minimize labor and resource demands on both farmers and researchers. The complex genetic nature of some traits may make progress difficult with such simple approaches - a risk that must be mitigated if the PPB effort is not to fail and thus compromise the genetic resources' conservation goals it might otherwise have achieved. When carefully oriented towards farmers' trait needs and thoughtfully designed, PPB strategies can contribute strongly to the conservation of in situ genetic resources, either in the form of varieties as a whole or allele assemblages of unique value.

6 Where to look for further information

Project updates, methodological insights, and further information can be found on web sites of a number of programs engaged in various aspects of PPB. The following are worthwhile places to start, but only scratch the surface - many others could be listed.

- Seed to Kitchen Collaborative (University of Wisconsin): https://seedtokitchen.horticulture.wisc.edu.
- Northern Organic Vegetable Improvement Collaborative (NOVIC): https://eorganic.info/index.php?q=novic.
- MasAgro Project (INIFAP and CIMMYT in Mexico): https://www.cimmyt.org/news/preserving-native-maize-and-culture-in-mexico/.
- Examples of PPB programs at ICRISAT: http://exploreit.icrisat.org/profile/crop%20improvement/30.
- Gender Responsive Researchers Equipped for Agricultural Transformation (GREAT Program): https://www.greatagriculture.org.

7 References

Abeba, G. (2005). Participatory selection of drought tolerant maize varieties using mother and baby methodology: A case study in the semi arid zones of the Central Rift Valley of Ethiopia. Paper presented at the African Crop Science Conference, Entebbe, Uganda.

Baafi, E., Akom, M., Agyeman, A., Darko, C. and Carey, T. (2020). Breeding farmer and consumer preferred sweetpotatoes using accelerated breeding scheme and mother-baby trials. *Open Agriculture* 5(1), 548–557.

Baluti, M. O., Ngwira, A. R. and Mwale, C. D. (2011). Farmer's voices on mother-baby trials. Paper presented at the 10th African Crop Ccience Conference, Maputo, Mozambique.

Bari, M. N. and Islam, M. R. (2009). Selection of mungbean variety for rice-wheat-fallow cropping system: A participatory research and development (PR&D) approach. *Journal of Agriculture and Rural Development* 7(1), 33–40.

Bellón, M. R. (2001). *Participatory Research Methods for Technology Evaluation: A Manual for Scientists Working with Farmers*. México, D.F.: CIMMYT.

Bellón, M. R. and Brush, S. B. (1994). Keepers of maize in Chiapas, Mexico. *Economic Botany* 48(2), 196–209.

Bradley, R. A. and Terry, M. E. (1952). Rank analysis of incomplete block designs. I. The method of paired comparisons. *Biometrika* 39, 324–345.

Ceccarelli, S., Grando, S., Bailey, E., Amri, A., El-Felah, M., Nassif, F., Rezgui, S. and Yahyaoui, A. (2001). Farmer participation in barley breeding in Syria, Morocco, and Tunisia. *Euphytica* 122(3), 521–536.

Chiwona-Karltun, L., Mkumbira, J., Saka, J., Bovin, M., Mahungu, N. M. and Rosling, H. (1998). The importance of being bitter: a qualitative study on cassava cultivar preference in Malawi. *Ecology of Food and Nutrition* 37(3), 219–245.

Christinck, A. and Weltzien, E. (2005). Identifying target regions and target groups. In: Christinck, A., Weltzien, E. and Hoffmann, V. (Eds), *Setting Breeding Objectives and Developing Seed Systems with Farmers: A Handbook for Practice Use in Participatory Plant Breeding Projects* (pp. 25–40). Weikersheim, Germany: Margraf Publishers GmBH.

Coe, R. (2002). Analyzing ranking and rating data from participatory on-farm trials. In: Bellon, M. R. and Reeves, J. (Eds), *Quantitative Analysis of Data from Participatory Methods in Plant Breeding* (pp. 44–65). México, DF: CIMMYT.

Edmeades, G. O. and Bänziger, M. (1996). Conclusions: what have we learned and where do we go? Paper presented at the Developing Drought and Low N Tolerant Maize. CIMMYT, El Batán, México.

Gurmu, F. (2013). Assessment of farmers' criteria for common bean variety selection: the case of Umbullo watershed in Sidama zone of the southern region of Ethiopia. *Ethiopian e-Journal for Research and Innovation Foresight* 5(2), 4–13.

Isaacs, K. B., Snapp, S. S., Kelly, J. D. and Chung, K. R. (2016). Farmer knowledge identifies a comptititve bean ideotype for maize-bean intercrop systems in Rwanda. *Agriculture and Food Security* 5(1), 15.

Kamanga, B. C. G., Kanyama-Phiri, G. Y. and Snapp, S. (2001). *Experiences with Farmer Participatory Mother-Baby Trials and Watershed Management to Improve Soil Fertility Options in Malawi*. Retrieved from México, DF.

Khan, A. R., Goldringer, I. and Thomas, M. (2020). Management practices and breeding history of varieties strongly determine the fine genetic structure of crop populations: a case study based on European wheat populations. *Sustainability* 12, 613.

Luce, R. D. (1959). *Individual Choice Behavior: A Theoretical Analysis*. New York, NY: Wiley.

Lyon, A., Tracy, W., Colley, M., Culbert, P., Mazourek, M., Myers, J., Zystro, J. and Silva, E. M. (2020). Adaptability analysis in a participatory variety trial of organic vegetables. *Renewable Agriculture and Food Systems* 35(3), 296–312.

Meitzner, L. S. (1998). Opportunities for plant improvement? A study of Honduran climbing bean seed systems and selection practices. (M.P.S.). Cornell University.

Odhiambo, G. O., Kwach, J. K., Gichuki, S. G. and Dida, M. M. (2011). Participatory selection of sweet potato (*Ipomoea batatis* (L.) Lam.) cultivars using mother and baby trial in western Kenya. *Acta Horticulturae* 911, 487–495.

Pendergrass, R. N. and Bradley, R. A. (1960). Ranking in triple comparisons. In: Olkin, O. et al. (Eds.), *Contributions to Probability and Statistics* (pp. 331–351). Stanford University Press.

Plackett, R. L. (1975). The analysis of permutations. *Journal of the Royal Statistical Society, Applied Statistics, Series C* 24(2), 193–202.

Smith, M. E., Castillo, G., and Gómez, F. (2001). Participatory plant breeding with maize in Mexico and Honduras. *Euphytica* 122, 551–565.

Snapp, S. (2002). Quantifying farmer evaluation of technologies: the mother and baby trial design. In: Bellon, M. R. and Reeves, J. (Eds), *Quantitative Aspects of Data from Participatory Methods in Plant Breeding* (pp. 9–16). México, DF: CIMMYT.

Sperling, L. and Scheidegger, U. (1995). Participatory selection of beans in Rwanda: Results, methods and institutional issues. *IIED Gatekeeper Series*, 51, 18.

Steinke, J., van Etten, J. and Zelan, P. M. (2017). The accuracy of farmer-generated data in an agricultural citizen science methodology. *Agronomy for Sustainable Development* 37(4), 32.

Sthapit, B. R., Joshi, K., Gyawali, S., Subedi, A., Shrestha, K., Chaudhary, P. R., Rana, R., Rijal, D., Upadhaya, M. and Jarvis, D. (2002). Participatory plant breeding: setting breeding goals and choosing parents for on-farm conservation. In: Bellón, M. R. and Reeves, J. (Eds), *Quantitative Analysis of Data from Participatory Methods in Plant Breeding* (pp. 104–112). México, DF: CIMMYT.

Sthapit, B. R., Joshi, K. D., Rana, R. B. and Upadhaya, M. P. (2001). Enhancing biodiversity and production through PPB. Paper presented at the An exchange of experiences from South and South East Asia. Proceedings of the International Symposium on Participatory Plant Breeding and Participatory Plant Genetic Resource Enhancement, Pokhara, Nepal.

Teeken, B., Olaosebikan, O., Haleegoah, J., Oladejo, E., Madu, T., Bello, A., Parkes, E., Egesi, C., Kulakow, P., Kirscht, H. and Tufan, H. A. (2018). Cassava trait preferences of men and women farmers in Nigeria: implication for breeding. *Economic Botany* 72(3), 263–277.

Testen, A. L., Mamiro, D. P., Mtui, H. D., Nahson, J., Mbega, E. R., Francis, D. M. and Miller, S. A. (2016). Introduction and evaluation of tomato germplasm by participatory mother and baby trials in the Morogoro region of Tanzania. *HortScience* 51(12), 1467–1474.

Thiele, G., Dufour, D., Vernier, P., Mwanga, R. O. M., Parker, M. L., Geldermann, E. S., Teeken, B., Wossen, T., Gotor, E., Kikulwe, E., Tufan, H., Sinelle, S., Kouakou, A. M., Friedmann, M., Polar, V. and Hershey, C. (2020). A review of varietal change in roots, tubers and bananas: consumer preferences and other drivers of adoption and implications for breeding. *International Journal of Food Science and Technology*. doi:10.1111/ijfs.14684.

Thomas, M., Thépot, S., Galic, N., Jouanne-Pin, S., Remoué, C. and Goldringer, I. (2015). Diversifying mechanisms in the on-farm evolution of crop mixtures. *Molecular Ecology* 24, 2937–2954.

Thomas, M., Demeulenaere, E., Dawson, J. C., Khan, A. R., Galic, N., Jouanne-Pin, S., Remoue, C., Bonneuil, C. and Goldringer, I. (2012). On-farm dynamic management of genetic diversity: the impact of seed diffusions and seed saving practices on a population-variety of bread wheat. *Evolutionary Applications* 5, 779–795.

Turner, H. L., van Etten, J., Firth, D. and Kosmidis, I. (2020). Modelling rankings in R: the PlackettLuce package. *Computational Statistics* 35(3), 1027–1057.

van Etten, J., de Sousa, K., Aguilar, A., Barrios, M., Coto, A., Dell'Acqua, M., Fadda, C., Gebrehawaryat, Y., van de Gevel, J., Gupta, A., Kiros, A. Y., Madriz, B., Mathur, P., Mengistu, D. K., Mercado, L., Nurhisen Mohammed, J., Paliwal, A., Pè, M. E., Quirós, C. F., Rosas, J. C., Sharma, N., Singh, S. S., Solanki, I. S. and Steinke, J. (2019) Crop variety management for climate adaptation supported by citizen science. *Proceedings of the National Academy of Sciences of the United States of America* 116(10), 4194–4199.

Virk, D. S., Singh, D. N., Prasad, S. C., Gangwar, J. S. and Witcombe, J. R. (2003). Collaborative and consultative participatory plant breeding of rice for the rainfed uplands of eastern India. *Euphytica* 132(1), 95–108.

vom Brocke, K., Trouche, G., Weltzien, E., Barro-Kondombo, C. P., Gozé, E. and Chantereau, J. (2010). Participatory variety development for sorghum in Burkina Faso: farmers' selection and farmers' criteria. *Field Crops Research* 119(1), 183–194.

vom Brocke, K., Weltzien, E., Christinck, A., Presterl, T. and Gieger, H. H. (2003). Effects of farmers' seed management on performance and adatpation of pearl millet in Rajasthan, India. *Euphytica* 130(2), 267–280.

Weltzien, E., Rattunde, F., Christinck, A., Isaacs, K. and Ashby, J. (2020). Gender and farmer preferences for varietal traits: evidence and issues for crop improvement. *Plant Breeding Reviews* 43, 243–278.

Witcombe, J. and Joshi, A. (1996). Farmer participatory approaches for varietal breeding and selection and linkages to the formal seed sector. Paper presented at the Participatory Plant Breeding, Wageningen, The Netherlands.

Witcombe, J. R., Gyawali, S., Sunwari, S., Sthapit, B. R. and Joshi, K. D. (2005a). Participatory plant breeding is better described as highly client-oriented plant breeding. II. Optional farmer collaboration in the segregating generations. *Experimental Agriculture* 42(1), 79–90.

Witcombe, J. R., Joshi, K. D., Gyawali, S., Musa, A. M., Johansen, C., Virk, D. S. and Sthapit, B. R. (2005b). Participatory plant breeding is better described as highly client-oriented plant breeding. I. Four indicators of client-orientation in plant breeding. *Experimental Agriculture* 41(3), 299–319.

Zimmerer, K. S. (1991). Managing diversity in potato and maize fields of the Peruvian Andes. *Journal of Ethnobiology* 11(1), 23–49.

Zystro, J., Colley, M. and Dawson. J. C. 2019. Alternative experimental designs for plant breeding. *Plant Breeding Reviews* 42, 87–117.

Chapter 12

Seed systems and diversity

Niels Louwaars, Plantum and Wageningen University, Law Group, The Netherlands

1 Introduction

1.1 Seed systems

Almost all farmers commonly spend significant attention to the seeds that they use. Variety lists are carefully analysed, the germination rate of home-produced seed is checked, or neighbours or farmers who are well-known as seed experts (Sperling et al., 1993) in the community are approached to obtain good seed. Good seed is defined by farmers as seed that will germinate, that does not contain weeds or, diseased seeds and in more advanced systems, that are well designed or treated to fit mechanical planters and produce vigorous seedlings that germinate uniformly and that withstand pests and diseases. In relation to genetics, good seeds mean seed with the genetic potential to provide the best chance for a good harvest and in some cases to maintain options for the future. Very poor farmers, on the other hand, commonly have to use any seed that is available to them, often from local grain markets (Lipper et al., 2010) or aid organizations (Sperling et al., 2008).

Seed systems are defined as the complex of actions that lead to the seed that farmers use to produce their crops. This includes breeding and selection, multiplication and storage, and exchange and delivery mechanisms. Traditionally, two systems are recognized: farmers' system and formal seed system (Cromwell et al., 1993; Almekinders et al., 1994). Farmers' seed systems

http://dx.doi.org/10.19103/AS.2020.0085.11

consist of the traditional seed production, selection and sharing mechanisms that shaped the crops that we currently know and that in several countries still supply the majority of seed for most crops. Formal seed systems are the regulated systems where specialized breeders, seed multipliers, and marketers operate in (quality) controlled chains, including variety trials, seed quality testing and certification.

However, many initiatives have been taken to integrate components of these two basic systems, such as quality-declared seed classes (FAO, 1993) and participatory plant breeding (Louwaars and de Boef, 2012). At the policy level such Integrated Seed Sector Development (ISSD) requires adaptation of various policies and regulations (Louwaars et al., 2013a).

1.2 Diversity

The genetic diversity of crops historically goes through a defined number of bottlenecks and subsequent increases due to different mechanisms after each major event (Louwaars, 2018). The wide natural diversity of plants is reduced during domestication, dispersal, and modernization bottlenecks (Tanksley and McCouch, 1997). Domestication is based on farmers' selection of plant species for their food and other needs. Within these crops, through many generations of selection, certain traits that have an evolutionary advantage in nature are changed in favour of traits that fit humankind. This is done through intentional selection and through natural selection in the agro-ecosystem. Typical examples are the increased seed size (intentional) and the reduced shattering behaviour (mainly natural selection) of cereal crops. This means that in the primary centres of diversity where such domestication has taken place new diversity has evolved in farmers' fields side by side to the wild populations in the natural environment.

The dispersal bottleneck is created by the introduction of crops outside their region (continent) of domestication. This is often based on a rather small genetic stock initially transported to the other continent (Pickersgill, 1998). After such an introduction, the widening of the genetic diversity starts through evolution, both intentional and natural selection in the new environments. This generally created a wide diversity of farmer varieties (landraces), especially in ecologically and culturally diverse regions. In mountainous regions, genetic differences within a crop between altitudes and valleys are likely higher than in uniform plains. Different cultures in the same country may have different food preferences, leading to crop diversity, for example, size and colour of dry beans (Kläsener et al., 2020).

The modernization bottleneck is based on the emergence of scientific plant breeding creating uniform varieties that replace the diverse farmers' varieties, and the introduction of single varieties in wide geographic areas. A typical example is that of rice variety IR8 developed at the International Rice Research Institute that was introduced throughout Southeast and South Asia, Latin America, and Africa. The international cooperation in plant breeding during the green revolution greatly increased food production but also vastly challenged crop genetic diversity (Evenson and Gollin, 2003), which resulted in the establishment of various types of ex situ conservation strategies.

The widest diversity on-farm can thus be expected in the centres of domestication, but significant diversity can also be found after introduction of the crops in other continents (secondary centres of diversity) and, as we will show, new diversity can also be found after the modernization bottleneck in areas with modern farming, which may be dubbed as tertiary centres of diversity.

2 Seed systems and diversity in farmers' fields

Seed systems can have a decisive influence on the genetic diversity in farmers' fields. Where crops are 'untouched' by modern breeding diversity is likely high, notably in the centres of domestication and next in ecologically or culturally diverse regions where the crop has been introduced a long time ago. In such areas, generations of farmers have shaped the genetic composition of their varieties, adding and loosing components of diversity over time as a result of exchanges, mutations and migration, and selection in changing (agro-) ecosystems. Natural and man-made disasters play a specific role in these dynamics (McGuire and Sperling, 2013). The varieties thus developed are called landraces or farmers' varieties that are commonly known and named by farmers based on a specific characteristic, or their (perceived) origin.

However, such 'untouched' conditions may be common only in underutilized crops that have undergone little formal research. Such situations are increasingly rare when we speak of major food crops. For such crops, next to the 'original' landraces, offspring of scientifically bred varieties may have reached farmers through variety demonstrations or farmer-to-farmer exchanges. Such materials are commonly either introduced into farmers' varieties, enriching their genetic diversity, or they have been kept as pure as possible, often for a specific purpose or market. Such may be the case with dry beans of a particular colour and shape. Also in this case, the seeds are multiplied, selected, and shared in the informal, farmers' seed systems. It is thus not true that farmers' seed systems only deal with genetically diverse landraces. Various types of contact of local seed systems with scientific breeders and their materials can enrich or reduce the crop genetic diversity in farmers' fields.

The promotion of modern varieties through the formal seed systems in farming systems using traditional farmers' varieties have indeed challenged local biodiversity leading to the modernization bottleneck. An effective formal seed system can however also increase diversity again, when a regular flow of new varieties from different sources reach farmers. Even though conventional plant breeding may operate within a fairly narrow genepool, various studies show that genetic diversity within crops has increased in European cereal crops over decades (Le Buanec, 2018; Malysheva-Otto et al., 2007; Law et al., 2001), where these fully depend on competitive breeding in formal seed systems. Van der Wouw et al. (2010, 2013) assign such increases of diversity to scientists increasingly making interspecific crosses and breeders using molecular biology tools that allow them from the 1990s onwards to use more distant genetic resources in their breeding programmes. Schouten et al. (2019) add market forces that stimulated breeders to broaden the diversity of vegetables. But breeding can also narrow diversity, for example, when a trait from a particular source is bred into all new varieties (Smith et al., 2010). However, given these results, one can safely say that more investments in plant breeding likely lead to more varieties and a wider diversity choice for farmers.

3 Supporting diversity in farmers' seed systems

On-farm management of plant genetic resources are a significant policy priority (Hammer et al., 2003). An important argument is that ex situ conservation 'freezes' the diversity, which can only go down during storage, whereas on-farm diversity continues to evolve (Kloppenburg, 1988; Fu, 2017). This means that we have to recognize that on-farm strategies by definition do not 'conserve' diversity in the literal sense, since during the evolutionary processes, and diversity may be gained but also lost (Seboka and Van Hintum, 2006). It is rather a strategy to manage diversity (FAO, 1997). This is even exacerbated in recent years due to quickly changing conditions in many regions (Meldrum et al., 2018). Climate change and intensified land use as a result of population pressure and urbanization challenge the traditional idea that farmers' varieties must be promoted because they are optimally adapted to the local conditions through many generations of selection. The conservation of genetic resources through the maintenance of traditional farming systems (Altieri and Merrick 1987) does not seem the right path when the slow genetic adaptation through natural and farmers' selection cannot keep up with the speed of ecological change.

A further basic challenge of such strategies is that it is not fair to make poor smallholder farmers responsible for the global challenge of plant genetic resource conservation. When introduced varieties appear to be better for their livelihoods, we should not stop such farmers from adopting such innovation.

Promoting diversity in farmers' fields only works when the diversity is functional in her conditions.

The management of genetic resources on-farm thus has to be dynamic, even though conservationists' basic attitude is likely more static (literally 'conservative'). Key in this respect is the variety of options that farmers can choose from: maintain their own landraces, to upgrade these, and to get access to landraces from other communities – but indeed only when these options are in their best personal interests.

3.1 Avoiding the loss of landraces: (community) seed banks

Landraces may disappear from farmers' fields when farmers lose interest, that is, when other materials are considered better, or through misfortune and disaster. Avoiding the first is, in radical terms, to make sure that farmers do not get access to other varieties. When indeed better varieties may be available for the farmer's particular conditions, then trying to do this is both morally wrong and quite impossible given the effectiveness of informal exchange systems (McGuire and Sperling 2016). Individual misfortune, such as expenses due to illness in the family, can lead to a farmer being unable to save seed for the next season. Commonly, the same landrace would then be available from other farmers in the community, but this may not always be the case. Community seed banks may guarantee that seeds of the right types will be available after such misfortune (Vernooy et al., 2015). When disaster affects the whole community though, neighbours may also have lost their seed and a seed bank may be equally affected. A severe drought or flood, a volcanic eruption, or conflict and war can lead to landraces getting lost on a larger scale. Community (or formal) genebanks may then harbour the landrace materials that may be reintroduced to the community, ideally backed up by a formal (inter)national genebank. In this way, the community genebank is both an ex situ conservation and local empowerment effort. Vernooy et al. (2015) combine these different functions under (the diversity of) the term community seed banks.

Reintroduction of local materials can also be done after extended periods of time. Varieties that have been grown in the past, which were replaced by modern varieties, and whose characteristics gain a renewed interest may be recovered from genebanks and reintroduced in the farming community (Melaku et al., 2000; Westengen et al., 2018).

3.2 Improving landraces: participatory plant breeding

Another aspect of dynamic adaptation is to support farmers in improving their own seed stock through participatory plant breeding (PPB) using farmers' own variety as a starting point (Almekinders and Louwaars, 2008). Such breeding would identify limitations of the local variety, for example, susceptibility to a disease, and crossing it with more resistant plants from a research programme. Such strategies of farmer-led PPB may enrich the diversity of the local stock and at the same time improve the variety for facing new challenges. They are, however, by definition local initiatives.

Various PPB strategies can be applied, including different distribution of tasks. The breeder may take the initiative, bringing in different materials based on farmers' priorities with regard to traits that they consider limiting in their farming practice. Alternatively, it is the farmer taking charge, requesting specific support. It is therefore not surprising that PPB has been initiated by either breeders, realizing that their on-station work had limitations (Witcombe et al., 1996; Ceccarelli and Grando, 1991), and by social scientists realizing that the locally available materials had limitations (Sperling et al. 1993). The outcome in terms of management of locally available diversity can differ among the variety of PPB strategies. These have been reviewed by Almekinders and Hardon (2006) and Ceccarelli et al. (2009). PPB does generally appear to increase diversity compared to on-station breeding, because different farmers may select different materials; it does however not necessarily include large parts of the original diversity in a given location.

PPB may be focusing on improving farmers' varieties, but also on farmers selecting from a wide range of introduced varieties (participatory variety selection). A hybrid between the two is that the breeder creates a wide diversity, including or not farmers' own landrace materials. The concept of 'evolutionary breeding' (Döring et al., 2012) is then based on allowing natural selection in the particular agro-ecological environment to optimally adapt the population to the specific local conditions. The authors report though that there is a risk that natural selection is likely to select for less preferred ('wild type') characteristics such as smaller seeds and shattering. Raggi et al. (2017) therefore put more emphasis in their work with barley on human selection in such populations alongside the selective effects by the environment. They stress that this strategy is particularly useful for breeding in low-productive environments.

3.3 Access to other landraces: seed fairs

A limitation of farmers seeds system is that even though exchange of seeds is generally open, there may be physical (different valleys having limited contact), and/or cultural barriers (Coomes et al., 2015). An important initiative to overcome such barriers is the seed fair where farmers are exposed to a diversity of options (Mazvimavi et al., 2012). Different communities may bring their varieties to the fair in normal times, but it can also be used to provide a choice for farmers after disaster (Sperling et al., 2008).

In all such options, farmers' seed system is the basis, but different components of the system are integrated with knowledge and materials from more formal origins. The concept of Farmer Field Schools is regularly applied, where farmers share knowledge about selection and storage of good seed, and about comparing different materials grown in their own farming conditions (Khisa, 2004). The diversity thus 'managed' is maintained by farmers, and flows within and among farming communities through traditional seed exchange mechanisms.

4 Supporting diversity in formal seed systems

The formal system, based on scientific breeding, official variety registration and quality-controlled seed production and marketing/distribution, does not necessarily favour diversity as it is an important cause of the modernization bottleneck. Eventually the diversity in farmers' fields depends on the number and genetic composition of varieties that farmers can choose from, also in mechanized agricultural systems.

4.1 Increasing the number of varieties

The choice for farmers depends primarily on the number of seed suppliers and the breeding programmes that they acquire their varieties from. Secondly, the investment that is made in breeding varieties relevant to the conditions of that farmer is a determining factor. The number of varieties that is actually available depends further on the variety registration system and the strategies of the seed producers; increasing the number of varieties increases costs in production planning and in promotion. The variety registration processes can prove to be a significant bottleneck, either because of excessive costs involved for each variety, or because registration decisions only favour varieties that present proven improvements over the existing varieties. When, furthermore, trial management is poor, statistical analysis of the trial results will not easily reveal significant improvements and thus reduce the number of varieties that get a positive release decision. Furthermore, when trial conditions are quite different from farmers' conditions, then the registration decision will have little relevance for the farmer (Tripp and Louwaars, 1997).

The more farmers' needs are taken into account, the better varieties can be developed that meet such needs. In commercial breeding this is done in the interplay between the seed marketers and the breeders within the company; in public breeding the extension service can be an important source of information. In the absence of an effective extension service – or inadequate extension-research linkages – breeding may not get the right focus. Involving farmers in the process directly can be valuable. Also in formal systems, PPB can contribute, as the potato-breeding experience in the Netherlands illustrates (Almekinders et al., 2014). The focus on the farmer-customer has a limitation though. Breeders have to take a large enough 'recommendation domain'

(Hildebrand et al., 1993) into account in order to warrant the investments. This is true for both private and public breeders. This by definition limits the number of varieties that a breeder can effectively release. One could furthermore say that, in general, when different breeding programmes operate in parallel such as in a competitive private breeding sector the farmer is likely to have a wider choice.

4.2 Increasing differences between varieties

An important positive aspect of breeding is that materials are introduced in any farming system from different regions. They thus solve one of the limitations of farmers' seed systems that these normally do not easily introduce fresh diversity from distant places. However, also the breeding process itself determines the options that are provided for farmers in terms of diversity. Breeders use diversity in order to breed different varieties. Two developments with opposing effects influence the diversity produced: technology and regulation. Technological developments have facilitated the use of more distant genetic resources in breeding. Marker-assisted breeding methods allow a breeder to a more focussed and quicker introgression of useful 'new' diversity into breeding programmes. Backcrossing using molecular data proves much more efficient than before, thus allowing breeders to use more 'distant' materials in their work.

Access to the genetic resources needed for crossing and selection has become, however, increasingly restricted during the past decades by global and national policies and agreements such as the Convention on Biological Diversity (CBD) and its Nagoya Protocol. Challenges in the implementation of these rules create uncertainty and administrative requirements introduce additional costs and liabilities which may make breeders often decide to stay within the genepool that they have known and used for a long time. Also the collection of genetic diversity by breeders and genebanks has greatly reduced since the CBD has come into force. The introduction of patents in the breeding realities during roughly the same period can also reduce the availability of genetic resources for breeders.

The International Treaty on Plant Genetic Resources for Food and Agriculture (ITPGRFA), which came into force in 2004, was an important attempt to facilitate both access and benefit sharing for major food crops. Recent attempts to improve this Treaty, both in terms of benefit sharing and widening its scope, have failed and new initiatives are needed to make sure that breeders continue to be able to create diversity for farmers.

4.3 Creating new diversity

Breeders not only recombine existing diversity from different origins, but they also have created new diversity, notably since the introduction of mutation breeding in the 1930s. Already in the 1950s important new traits were selected for crops like barley (Donini and Sonnino, 1998). Currently, mutation breeding using either physical (radiation) or chemical mutagenesis is commonly used in a wide variety of crops. Interspecific crosses also introduce new diversity into crops, which would not have happened through natural evolutionary forces (Van Tuyl and Lim, 2003). More recently, marker-assisted selection has made a much wider array of diversity of crops available to breeders. Backcrossing became much more efficient, which made the use of distant genebank material technically and economically feasible for breeders. Le Buanec (2018) presents an overview of studies indicating that plant breeding indeed has resulted in increased genetic diversity in farmers' field for a number of field crops.

Finally, important new opportunities are arising due to technological developments that will likely make gene editing applicable for a wide range of breeders. Due to the cost of research and deregulation, genetic transformation has not yielded a lot of new diversity (Potrykus, 2012), and early gene-editing techniques such as TALEN and ZincFinger have hardly led to impact outside the fundamental research laboratories. However, CRISPRcas9 and more recent uses of CRISPR technology have the potential to significantly change that situation (Voytas, 2013; Doudna and Charpentier, 2014), making targeted mutation breeding a reality for many breeders (Schaart et al., 2016). The extent to which that will come true will depend on regulatory decisions and consumer acceptance next to further technological developments.

4.4 Creating diversity within the variety

Some breeders see opportunities for heterogeneous varieties, notably with the aim to increase field resistance against diseases and pests by combining plants with different resistance genes in one population. Indeed, the development of an epidemic can be slowed down when not all plants are equally susceptible. This is achieved in many traditional agricultural systems through intercropping and also genetic diversity within landraces may contribute (Jarvis et al., 2010). It is a challenge to obtain similar results while responding to the modern efficient cultivation methods though. A key element is mechanized harvesting, which means that maturation of the crop has to be uniform. In several crops, also quality expectations of processors (wheat – baking quality) or consumers (vegetables) lead to more uniformity. In other crops, such as herbage and other fodder and catch crops (also sometimes called cover crops), such uniformity is less important, and varietal (and species) mixtures are common in the seed trade of such crops. However, for most row-crops and vegetables produced for the market, uniformity is more important.

An attempt to combine these diversity and uniformity demands was the creation of multiline varieties, consisting of a mixture of near-isogenic lines that are uniform except for their resistance genes (Groenewegen, 1977). The attempt in the 1970s to implement that in wheat against rust was technically successful, but because of the additional breeding time it took to create the multiline, variety 'Tumult' did not reach farmers' fields (Zeven and Waninge, 1985). With the development of 'Tumult' - also the regulatory system had challenges to respond to this new breeding strategy - it had to redefine its concept of uniformity in both variety registration and seed certification. It remains to be seen whether gene editing may create new opportunities to develop multiline varieties, that is, editing the target resistance genes after an otherwise good variety has been bred.

Recent calls, especially from the organic sector, to allow heterogeneous varieties, revive such discussions (Lammerts van Bueren amd Myers, 2012). Conventional variety registration policies are based on the idea that varieties must have an identity that farmers can rely on. This aspect of farmer protection should not be lost when heterogeneous varieties are to be marketed, as already now, with the current rules, fake seed is available in many markets. The introduction of heterogeneous varieties should not create additional opportunities for seedsmen to cheat farmers. Uniformity may not need to be the goal but maintaining a guaranteed identity should be a priority of the legislator. Creating opportunities for mixtures of (identifiable) varieties in the market may be a good opportunity to enhance biodiversity in the fields of productive agricultural systems next to other proposals such as strip-farming, biodiverse borders, and the promotion of catch crops in the rotation.

5 Policies and regulations

Supporting genetic diversity in agricultural systems as an important component of conservation policies for genetic resources faces a number of challenges from current policies and regulations.

5.1 Food security policies

First of all, securing the production of sufficient food creates pressures in making agriculture more productive, especially in areas where population growth and urbanization reduce the net available land area for agriculture. Adding climate change to the equation, this means that low-productive traditional farming systems that do not give a good livelihood for the farmers cannot be maintained. Creating 'museum-farmers' to maintain the old landraces is not an option as such landraces depend on the dynamic of farmer and natural selection.

5.2 Genetic resource policies

Reintroduction of genebank materials into farmers' fields may not be a fruitful strategy in regions where the agricultural production systems have significantly evolved due to population pressure and climate change. It may however provide interesting leads for PPB. Supporting farmers to further develop their varieties through the introduction of introduced diversity to enrich their varieties, with, for example, 'new' resistance is an important strategy to increase production and improve livelihoods while managing crop genetic diversity. Obviously, the introduction of new and well-adapted varieties is important in this respect as well. The wide use of introduced genetic resources in such breeding for informal seed systems is, however, limited by the Convention on Biological Diversity restricting the international movement of genetic resources, and its Nagoya Protocol which puts additional compliance obligations on breeders. Such administrative requirements are not compatible with introducing genetic diversity in breeding for farmers' seed systems. Biodiversity regulations can also limit the opportunity for breeders in the formal sector to support diversity. Where 'These regimes have (...) contributed to the restriction of access to genetic resources for food and agriculture' (Andersen, 2003), the diversity of the output of breeding is logically restricted. Such access restrictions have been felt by breeders over the last few decades.

5.3 Intellectual property policies

Access to genetic resources has also been an issue in intellectual property on plants since the early policy debates in the early twentieth century. Plants and plant varieties have been exempted from patent protection for the first century of harmonized patent systems since the Paris Convention of 1883. Only in very few countries, varieties can currently be patented. Maintaining free access to the genetic diversity has been a key policy priority when the concept of plant breeder's rights (PBR) was created during the 1930s. The 'breeder's exemption' is still a major component of the now widely introduced rights. However, since biotechnology has entered the sphere of plant breeding, plants (or plant traits) became patentable. This means that a plant that contains a patented trait cannot be used for further breeding without the consent of the patent holder (Louwaars et al., 2009). Breeders, notably in Europe increasingly became wary of these developments and specific policies have been approved there to avoid negative impact on access by breeders to the genetic resources embedded in varieties (Ghijsen, 2009).

First, the EU introduced a 'limited breeder's exemption' in its Agreement on the Unitary Patent Court (www.epo.org/law-practice/unitary/patent-court.html) allowing anybody to use for further breeding plants that fall under the scope of a patent. If the resulting variety does not contain the patented component, the breeder is free to register and market it without any need for a license from the patent holder. Secondly, the EU has adopted a policy not to allow patents on natural traits (products of essentially biological processes) any more (http://www.europarl.europa.eu/doceo/document/TA-9-2019-0020_EN.html). This is implemented by the European Patent Office. Other discussions have evolved in other jurisdictions.

Both PBR and patents do restrict the commercial multiplication and trade of protected plant varieties to the right holders and their licensees. This is necessary to maintain a productive plant breeding sector. This should however not affect the informal exchange of plant materials among farmers that mainly produce for family consumption (de Jonge et al., 2015; Louwaars et al., 2013b). Discussions are ongoing how to demarcate such non-commercial use of PBR-protected varieties and a concept has been developed a concept to implement that, which is widely accepted by both non-governmental organizations and the private seed sector, but which is awaiting formalization at the (inter-) governmental level (Oxfam et al., 2019).

5.4 Seed and genetic modification regulations

Further, seed laws may provide restrictions to supporting diversity in farmers' fields, notably PPB, or breeding for informal seed systems in general (Bocci, 2011). Results of PPB that enter the informal dissemination systems bypass the formal variety registration systems. Supporting informal farmers' seed systems is in some countries not very well received by Government as it bypasses official seed production rules and regulations. It requires policy makers to make a clear distinction between the formal seed systems where farmers should be able to rely on the label on the seed bag with regard to varietal identity and seed quality, and the informal systems where it is the farmer's own responsibility to trust the supplier - normally within the community. The biggest challenge is to avoid that unscrupulous businessmen misuse the grey area between those systems putting rubbish in serious-looking seed bags. Designing rules that properly regulate formal seed systems and that at the same time avoid unnecessarily regulating the farmers' seed systems is a major challenge when recognizing that different seed systems should be able to operate and thrive side by side (Louwaars et al., 2013a).

The most recent regulatory challenge for increasing diversity in breeding is the interpretation of genetic modification in relation to the new gene-editing technologies that contribute to genetic diversity in the formal seed systems. Regulations are an important reason why the application of transgenesis has been restricted to very few crops and traits. Gene editing, which is a less costly procedure with great promises towards breeding for improving agriculture and product qualities, may face the same consequence when the same rules apply as is currently the case in the EU following a Court of Justice case in 2018 declaring the products of targeted mutagenesis as regulated GMO (ECJ, 2018). In several other jurisdictions less restrictive procedures are being formulated.

6 Future trends in research

It would be important to quantify the impact of climate change on genetic diversity in farmers' varieties and the speed in which such varieties are able to adapt to climate change. Such data could guide on-farm management strategies for crop genetic resources.

Furthermore, further studies on trends in formal seed systems with respect to the number and diversity of crop varieties of different crops that are available to farmers would be useful in order to map diversity increases following the modernization bottleneck. Also, the impact on both technologies (e.g. marker-assisted selection and gene editing) and of policies (CBD - ITPGRFA and Nagoya Protocol) on the diversity of output from the scientific plant breeding sector would be important to see whether such technologies should be supported and whether such policies should be adapted.

7 Conclusions

It is the seed that 'carries' genetic diversity, and so it is the seed systems that have an important role in determining the diversity of crops sown. Crop diversity has gone through domestication, dispersal, and modernization bottlenecks. After each bottleneck diversity picks up due to different evolutionary and man-induced forces. It does not reach the original levels, but may well produce new diversity, beyond the natural crop genetic diversity.

Farmers' seed systems are known to harbour significant diversity. This can be supported by making sure that locally adapted, diverse seed is available, even when a farmer has lost her stock, for example, through community seed banks. However, due to climate change and population pressure the old 'landraces' may not be optimally adapted anymore. PPB may support diversity while at the same time supporting crop production to adapt to new realities.

Formal seed systems can much quicker produce positive results for farmers when these systems, including the marketing and distribution components have reached a certain level of maturity. Formal breeding can also increase genetic diversity in farmers' field among and within varieties. After the modernization bottleneck, scientific plant breeding creates new diversity in a crop through interspecific crosses and (targeted) mutagenesis, and makes diversity from distant places available to farmers.

Various policy areas can challenge the management of diversity in farmers' field and in breeding. Careful wording of seed regulations and IP-rights can provide for the necessary limits that operators in the formal systems have to obey, while avoiding unnecessary restrictions on the informal systems. Other policies, such as biodiversity and biotechnology policies are important in broadening the genetic base of breeding, producing optimal diversity levels for farmers to use. However, several of these policies prove to restrict rather than promote diversity.

8 Where to look for further information

Most of the literature about on-farm management of crop genetic diversity relates to biodiversity policy documents by international biodiversity policy bodies, such as the Convention on Biological Diversity (www.cbd.org) or the International Treaty on Plant Genetic Resources for Food and Agriculture (www/planttreaty.org). Basic documents are the reports on the state of the world's plant genetic resources for food and agriculture, available at:

- http://www.fao.org/agriculture/crops/thematic-sitemap/theme/seeds-pgr/sow/en/.
- Furthermore, the projects that the IT-PGRFA operates through its benefit sharing fund: http://www.fao.org/plant-treaty/areas-of-work/benefit-sharing-fund/overview/en/.
- The Treaty has also taken a seed systems perspective in its information document CGRFA-12/09/Inf.20 in 2009. It can be found at http://www.fao.org/tempref/docrep/fao/meeting/017/ak568e.pdf.
- In more formal settings, breeders are generally very concerned about broadening the genetic base of their programmes. A book, published back in 2001 gives the broadest overview, available at https://www.researchgate.net/publication/270581019_Broadening_the_Genetic_Base_of_Crop_Production.
- A more recent source is: https://bmcgenomics.biomedcentral.com/articles/10.1186/s12864-020-6756-0.
- And with a focus on modern technologies: https://www.researchgate.net/publication/284242519_MOLECULAR_MARKERS_AND_GENETIC_BASE_BROADENING_IN_LYCOPERSICON_SPP.

Literature on the impact of such strategies can be found in the references of this chapter.

9 References

Almekinders, C. and Hardon, J. (2006). *Bringing Farmers Back into Breeding: Experiences with Participatory Plant Breeding and Challenges for Institutionalisation, AgroSpecial* 5. The Netherlands, Agromisa Foundation.

Almekinders, C. J. M. and Louwaars, N. P. (2008). Supporting farmers in maintaining and selecting seeds of local varieties. In: Thijssen, M. H., Bishaw, Z., Beshir, A. and de Boef, W. S. (Eds) *Farmers, Seeds and Varieties: Supporting Informal Seed Supply in Ethiopia*. Wageningen, Wageningen International, pp 87–96.

Almekinders, C. J. M., Louwaars, N. P. and de Bruijn, G. H. (1994). Local seed systems and their importance for improved seed supply in developing countries. *Euphytica* 78(3): 207–216.

Almekinders, C. J. M., Mertens, L., van Loon, J. P. and Lammerts Van Bueren, E. T. (2014). Potato breeding in the Netherlands: a successful participatory model with collaboration between farmers and commercial breeders. *Food Security* 6(4): 515–524.

Altieri, M. A. and Merrick, L. C. (1987). In situ conservation of crop genetic resources through maintenance of traditional farming systems. *Economic Botany* 41(1): 86–96.

Andersen, R. (2003). FAO and the management of plant genetic resources. In: Schramm-Stokke, O. and Thommessen, Ø. (Eds) *Yearbook of International Co-Operation on Environment and Development 2003/2004*. London, Earthscan Publications, pp 43–53.

Bocci, R. (2011). Seed legislation and agrobiodiversity: conservation varieties. *Journal of Agriculture and Environment for International Development* 103(1/2): 31–49.

Ceccarelli, S. and Grando, S. (1991). Environment of selection and type of germplasm in barley breeding for low-yielding conditions. *Euphytica* 57(3): 207–219.

Ceccarelli, S., Guimaraes, E. P. and Weltzien, E. (Eds) (2009). *Plant Breeding and Farmer Participation*. Rome, Food and Agriculture Organization.

Coomes, O. T., McGuire, S. J., Garine, E., Caillon, S., McKey, D., Demeulenaere, E., Jarvis, D., Aistara, G., Barnaud, A., Clouvel, P., Emperaire, L., Louafi, S., Martin, P., Massol, F., Pautasso, M., Violon, C. and Wencélius, J. (2015). Farmer seed networks make a limited contribution to agriculture? four common misconceptions. *Food Policy* 56: 41–50.

Cromwell, E., Wiggins, S. and Wentzel, S. (1993). *Sowing beyond the State: NGOs and Seed Supply in Developing Countries*. London, Overseas Development Institute, 143 pp.

De Jonge, B., Louwaars, N. P. and Kinderlerer, J. (2015). A solution to the controversy on plant variety protection in Africa. *Nature Biotechnology* 33(5): 487–488.

Donini, P. and Sonnino, A. (1998). Induced mutation in plant breeding: current status and future outlook. In: Jain, S. M., Brar, D. S. and Ahloowalia, B. S. (Eds) *Somaclonal Variation and Induced Mutations in Crop Improvement. Current Plant Science and Biotechnology in Agriculture* (vol. 32). Dordrecht, Springer, pp 255–291.

Döring, T. F., Bocci, R., Hitchings, R., Howlett, S., Lammerts van Bueren, E. T., Pautasso, M., Raaijmakers, M., Rey, F., Stubsgaard, A., Weinhappel, M., Wilbois, K. P., Winkler, L. R. and Wolfe, M. S. (2012). The organic seed regulations framework in Europe–current status and recommendations for future development. *Org. Agr.* 2: 173–183. https://doi.org/10.1007/s13165-012-0034-7.

Doudna, J. A. and Charpentier, E. (2014). Genome editing. The new frontier of genome engineering with CRISPR-Cas9. *Science* 346(6213): 1258096.

ECJ–European Court of Justice. (2018). Website accessed 16.02.2020. Available at: http://curia.europa.eu/juris/document/document.jsf?docid=204387&doclang=EN.

Evenson, R. E. and Gollin, D. (2003). Assessing the impact of the green revolution, 1960 to 2000. *Science* 300(5620): 758–762.

FAO. (1993). Plant genetic resources service. Quality declared seed system. FAO Plant Production and Protection Paper 185. Rome, Food and Agriculture Organization of the United Nations.

FAO. (1997). The state of the world's plant genetic resources for food and agriculture. Rome, Food and Agriculture Organisation of the United Nations. 511 p. Available at: http://www.fao.org/3/a-w7324e.pdf.

Fu, Y. B. (2017). The vulnerability of plant genetic resources conserved ex situ. *Crop Science* 57(5): 2314–2328.

Ghijsen, H. (2009). Intellectual property rights and access rules for germplasm: benefit or straitjacket? *Euphytica* 170(1–2): 229–234.

Groenewegen, L. J. M. (1977). Multilines as a tool in breeding for reliable yields'. *Cereal Research Communications* 5: 12–132.

Hammer, K., Gladis, Th. and Diederichsen, A. (2003). In situ and on-farm management of plant genetic resources. *European Journal of Agronomy* 19(4): 509–517.

Hildebrand, P. E., Singh, B. K., Bellows, B. C., Campbell, E. P. and Jama, B. A. (1993). Farming systems research for agroforestry extension. *Agroforestry Systems* 23(2–3): 219–237.

Jarvis, D. I., Padoch, C. and Cooper, H. D. (Eds) (2010). *Managing Biodiversity in Agricultural Ecosystems*. New York, Columbia University Press, 512 pp.

Khisa, G. (2004). *Farmers Field School Methodology: Training of Trainers Manual* (1st edn.). Rome, Food and Agriculture Organization.

Kläsener, G. R., Dalfollo Ribeiro, N., Casagrande, C. R. and Daltrozo Arns, F. (2020). Consumer preference and the technological and nutritional quality of different bean colours. *Acta Scientiarum* 1807-8621. doi: 10.4025/actasciagron.v42i1.43689.

Kloppenburg, J. (Ed.) (1988). *Seeds and Sovereignty: Debate Over the Use and Control of Plant Genetic Resources*. Durham, Duke University Press, 369 p.

Lammerts van Bueren, E. T. and Myers, J. R. (2012). Organic crop breeding: Integrating agricultural approaches and traditional and modern plant breeding methods. In: Lammerts van Bueren, E. T. and Myers, J. R. (Eds) *Organic Crop Breeding*. Chichester, West Sussex, UK; Ames, IA, Wiley-Blackwell, pp 3–14.

Law, J. R., Donini, P., Koebner, R. M. D., Reeves, J. C., White, J. and, Cooke, R. J. (2001). Diversity in UK quality wheat. *Aspects of Applied Biology* 64: 143–144.

Le Buanec, B. (2018). La diversité génétique en agriculture. In: Reignault-Roger, C. (Dirs.): *Idées reçues en agriculture, parole à la science*. Académie d'Agriculture de France. Presses des Mines editor, Paris, pp 169–186.

Lipper, L., Leigh Anderson, C. and Dalton, T. J. (Eds) (2010). *Seed Trade in Rural Markets* (vol. 2010). Food and Agriculture Organization and Earthscan, Rome, Italy 223 pp.

Louwaars, N. P. (2018). Plant breeding and diversity a troubled relationship? *Euphytica: Netherlands Journal of Plant Breeding* 214(7): 114.

Louwaars, N. P. and de Boef, W. S. (2012). Integrated seed sector development in Africa: a conceptual framework for creating coherence between practices, programs and policies. *Journal for Crop Improvement* 26(1): 29–59.

Louwaars, N. P., de Boef, W. S. and Edeme, J. (2013a). Integrated seed sector development: a basis for seed policy and law. *Journal for Crop Improvement* 27: 186–214.

Louwaars, N. P., de Jonge, B. and Munyi, P. (2013b). Intellectual property rights in the plant sciences, and development goals in agriculture: an historical perspective.

In: Arapostathis, S. and Dutfield, G. (Eds) *Knowledge Management and Intellectual Property. Concepts, Actors and Practices from the Past to the Present*. Cheltenham, Edgar Elgar Publishing, pp 252–272.

Louwaars, N. P., Dons, H., van Overwalle, G., Raven, H., Arundel, A., Eaton, D. and Nelis, A. (2009). Breeding business. The future of plant breeding in the light of developments in patent rights and plant breeder's rights. CGN-Report 2009-14. Wageningen, Centre for genetic resources, The Netherlands. 65 p. Available at: http://documents.plant.wur.nl/cgn/literature/reports/BreedingBusiness.pdf.

Malysheva-Otto, L., Ganal, M. W., Law, J. R., Reeves, J. C. and Röder, M. S. (2007). Temporal trends in genetic diversity in European barley cultivars (Hordeum vulgare). *Molecular Breeding* 20(4): 309–322.

Mazvimavi, K., Pedzisa, T., Murendo, C., Minde, I. J. and Ndlovu, P. V. (2012). Cost effectiveness of seed fairs relative to direct relief distribution in Zimbabwe. *Development in Practice* 22(7): 978–990.

McGuire, S. and Sperling, L. (2013). Making seed systems more resilient to stress. *Global Environmental Change* 23(3): 644–653.

McGuire, S. and Sperling, L. (2016). Seed systems that smallholder farmers use. *Food Security* 8: 179–195.

Melaku, W., Awegechew, T. and Tesfaye, T. (2000). Participatory approaches linking farmer access to genebanks: Ethiopia. In: FriisHansen, E. and Sthapit, B. R. (Eds) *Participatory Approaches to the Conservation and Use of Plant Genetic Resources*. Rome, International Plant Genetic Resources Institute, pp 149–153.

Meldrum, G., Mijatović, D., Rojas, W., Flores, J., Pinto, M., Mamani, G., Condori, E., Hilaquita, D., Gruberg, H. and Padulosi, S. (2018). Climate change and crop diversity: farmers' perceptions and adaptation on the Bolivian Altiplano. *Environment, Development and Sustainability* 20(2): 703–730.

Oxfam, Plantum and Euroseeds. (2019). Can the exchange or sale of self-produced seed be allowed under UPOV 1991?. The Hague, OxfamNovib. Available at: https://oxfam.app.box.com/s/eo9d3cvaeffbtwdaqcvbvm1udyq4ynq5.

Pickersgill, B. (1998). Crop introductions and the development of secondary areas of diversity. In: Prendergast, H. D. V., Etkin, N. L., Harris, D. R. and Houghton, P. J. (Eds) *Plants for Food and Medicine*. Kew, UK, Royal Botanic Gardens, pp 93–105.

Potrykus, I. (2012). Unjustified regulation prevents use of GMO technology for public good.forum: science & society special issue: celebrating 30 years of biotechnology. *Trends in Biotechnology* 31(23): 131–133.

Raggi, L., Ciancaleoni, S., Torricelli, R., Terzi, V., Ceccarelli, S. and Negri, V. (2017). Evolutionary breeding for sustainable agriculture: selection and multi-environmental evaluation of barley populations and lines. *Field Crops Research* 204: 76–88.

Schaart, J. G., Van de Wiel, C. C. M., Lotz, L. A. P. and Smulders, M. J. M. (2016). Opportunities for products of new plant breeding techniques. *Trends in Plant Science* 21(5): 438–449. doi.org/10.1016/j.tplants.2015.11.006

Schouten, H. J., Tikunov, Y., Verkerke, W., Finkers, R., Bovy, A., Bai, Y. and Visser, R. G. F. (2019). Breeding has increased the diversity of cultivated tomato in the Netherlands. *Frontiers in Plant Science* 10: 1606. Doi: 10.3389.fpls.2019.01606.

Seboka B. and VanHintum, T. (2006). The dynamics of on-farm management of sorghum in ethiopia: implication for the conservation and improvement of plant genetic resources. *Genetic Resources and Crop Evolution* 53(7): 1385–1403.

Smith, S., Primomo, V., Monk, R., Nelson, B., Jones, E. and Porter, K. (2010). Genetic diversity of widely used U.S. Sorghum hybrids 1980-2008. *Crop Science* 50(5): 1664-1673.

Sperling, L., Cooper, H. D. and Remington, T. (2008). Moving towards more effective seed aid. *Journal of Development Studies* 44(4): 586-612.

Sperling, L., Loevinsohn, M. E. and Ntabomvura, B. (1993). Rethinking the farmer's role in plant breeding: local bean experts and on-station selection in Rwanda. *Experimental Agriculture* 29(4): 509-519.

Tanksley, S. D. and McCouch, S. R. (1997). Seed banks and molecular maps: unlocking genetic potential from the wild. *Science* 277(5329): 1063-1066.

Tripp, R. and Louwaars, N. P. (1997). The conduct and reform of crop variety regulation. In: Tripp, R. (Ed.) *New Seed and Old Laws; Regulatory Reform and the Diversification of National Seed Systems*. London, Intermediate Technology Publications, pp 88-120.

Van der Wouw, M. J., van Hintum, T., Kik, C., van Treuren, R. and Visser, B. (2010).Genetic diversity trends in twentieth century crop cultivars: a meta analysis. *TAG. Theoretical and Applied Genetics. Theoretische und Angewandte Genetik* 120(6): 1241-1252.

Van der Wouw, M. J., Van Treuren, R. and Van Hintum, T. J. L. (2013). A historical analysis of diversity trends in French and Dutch lettuce cultivars. *Euphytica* 190(2): 229-239.

Van Tuyl, J. M. and Lim, K.-B. (2003). Interspecific Hybridisation and polyploidisation as tools in ornamental plant breeding. *Acta Horticulturae* 612(612): 13-22. doi: 10.17660/ActaHortic.2003.612.1.

Vernooy, R., Shrestha, P. and Shtapit, B. (Eds) (2015). *Community Seed Banks, Evolution and Prospects*. Abingdon UK and New York, Routledge, 270 p.

Voytas, D. F. (2013). Plant genome engineering with sequence-specific nucleases. *Annual Review of Plant Biology* 64: 327-350.

Westengen, O. T., Skarbø, K., Berg, T. H. and Berg, T. (2018). Access to genes: linkages between genebanks and farmers' seed systems. *Food Security* 10(1): 9-25.

Witcombe, J. R., Joshi, A., Joshi, K. and Sthapit, B. (1996). Farmer participatory crop improvement. I. Varietal selection and breeding methods and their impact on biodiversity. *Experimental Agriculture* 32(04): 445-460.

Zeven, A. C. and Waninge, J. (1985). The yielding capacity of the winter wheat multiline Tumult, its components and recurrent parent Tadorna. *Cereal Research Communications* 13: 85-87.

Chapter 13

DNA-based screening of *Brassica* germplasm for sustainable and enhanced crop production

Yueqi Zhang, The University of Western Australia, Australia; Ting Xiang Neik, Sunway College Kuala Lumpur, Malaysia; and Junrey C. Amas, Aldrin Y. Cantila, Nur Shuhadah Mohd Saad, Tingting Wu and Jacqueline Batley, The University of Western Australia, Australia

1 Introduction

The Brassicaceae family is one of the largest eudicot families, and includes the model plant *Arabidopsis thaliana* (Al-Shehbaz, 2012). Members of the Brassicaceae form an important component of the human diet, as a major source of edible oil and vegetables. *Brassica* is a core genus of the Brassicaceae family and cultivated *Brassicas* consist of six main species: *B. napus* (oilseed rape/canola), *B. rapa* (Chinese cabbage), *B. oleracea* (cabbage, cauliflower, brussels sprouts, broccoli), *B. juncea* (Indian mustard), *B. nigra* (black mustard) and *B. carinata* (Ethiopian mustard) (U, 1935). Of these, *B. napus* is the most important source of edible oil and contributes approximately to 13% of the global edible oil supply (Raman et al., 2019b).

Plant genetic resources, or germplasm, supply essential raw materials to crop improvement programmes as they constitute reservoirs of genes which are useful for breeding, conservation of species and studying the origin and evolution of a crop species (Katche et al., 2019; Werner and Snowdon, 2018). In a broader sense, plant germplasm resources are the sum of genes in a crop

http://dx.doi.org/10.19103/AS.2020.0085.15

species. We can improve plant genetic resource management by ensuring proper documentation of germplasm for correct species identification, careful distribution of germplasm for research purposes and confirming accurate species annotation in research publications (Chen et al., 2017; El-Esawi, 2017; King and Baten, 2018). Germplasm should be evaluated for important morpho-agronomic traits under different agro-climatic conditions, which will result in the identification of donors for use in varietal improvement programmes. Worldwide, there are more than 40 256 accessions of *Brassica* conserved in over 450 germplasm banks around the world (CGIAR, 2020). The major world germplasm collections of *Brassica* include, but are not limited to, the Warwick Genetic Resource Unit (United Kingdom), Australia Grains Genebank (AGG, Australia), Northeast Regional Plant Introduction Station (United States) and Genebank, Leibniz Institute of Plant Genetics and Crop Plant Research (Germany) and the Nordic Genetic Resource Centre (NordGen) (CGIAR, 2020; Katche et al., 2019).

Compared with other marker systems, Single Nucleotide Polymorphism (SNP) markers are advantageous due to the bi-allelic nature of the marker and abundant distribution in plant and animal genomes, thus making them very effective markers in screening large populations for marker-trait association studies (Li et al., 2019a). SNP markers are robust and highly adapted to a variety of genotyping systems, and the straightforward allele calling coupled with the availability of large database resources from diverse crop species makes it convenient to conduct whole-genome analysis in many plant species (Thomson, 2014). SNP genotyping using the *Brassica* 60K and 90K Illumina Infinium™ SNP arrays (henceforth termed *Brassica* 60K/90K SNP array) provide rapid high-density genome scans and are cost-effective for large numbers of samples (Mason et al., 2017; Scheben et al., 2019). PCR-based fluorescently labelled SNP assays such as Kompetitive allele-specific PCR (KASP) markers screen one marker at a time (Thomson, 2014) and are ideal for marker-assisted selection studies. Sequencing-based SNP genotyping platforms including genotyping by sequencing (GBS) are an attractive option when no prior genomic information is available (Rasheed et al., 2017). Most GBS approaches involve reduced representation sequencing such as Restriction-site Associated DNA sequencing (RAD-seq) and Diversity Array Technology sequencing (DArTseq) (Rasheed et al., 2017). More recently, SNP genotyping based on bulked segregant analysis, such as BSR-seq using RNA-seq or QTL-seq using whole genome resequencing (WGR), allows efficient studies of marker-trait association through rapid identification of QTL and candidate genes, for example, in crop species including barley (Shi et al., 2019), peanut (Zhao et al., 2020) and *B. rapa* (Zhang et al., 2020). The approach of WGR at low coverage makes SNP genotyping of large populations viable (Scheben et al., 2017). Although the cost per data point is much lower than the reduced representation sequencing approaches, the cost per sample remains higher for WGR (Scheben et al., 2017). In *Brassicas*, SNP-based genotyping and screening is commonly used to assess the diversity of germplasm (Malmberg et al., 2018), as well as mapping for tolerance to biotic and abiotic stresses and enhancing yield and oilseed quality (Delourme et al., 2018).

2 Management and conservation of *Brassica* germplasm

Brassica species are well studied and extensively cultivated, with more than half a millennia of breeding history, which has resulted in narrower genetic diversity (Malmberg et al., 2018). These erosions in genetic diversity were mostly due to selection bias for breeding favourable traits such as disease resistance, climate adaptation and yield. This also includes intensive breeding selection for low erucic acid and glucosinolate content, resulting in inbreeding, most notably in Australian cultivars (Cowling, 2007). Intermixed gene pools also caused challenges for breeders to identify lines for successful hybrids (Luo et al., 2019). Past and current phenotype and genotype screening efforts have generated expansive information for elite and landrace *Brassicas*. A number of SNP-based screening techniques have been utilised to propose future management and conservation efforts for *Brassica* germplasm (Table 1). A well-informed management and conservation strategies could be applied to ensure diversity is sustainable within the global *Brassica* germplasm collections.

One of the important steps towards maintaining the *Brassica* germplasm is to ensure the validity of seed stocks species and absence of genotype contamination. Multiplex PCR approaches offer an easily accessible molecular option to distinguish between the A, B and C chromosomes that make up the six *Brassica* species in the U's triangle (Koh et al., 2017; U, 1935). While the method is optimised for less than 96 samples, it could still be utilised for preliminary analysis and to identify contamination in seed stocks, routinely and quickly (Koh et al., 2017). Alternatively, the *Brassica* 60K and 90K SNP arrays have been used extensively to screen more than 30 000 *Brassica* lines (Mason et al., 2017; Scheben et al., 2019). Complementarily, CropSNPdb (http://snpdb.appliedbioinformatics.com.au/) is an example of an accessible database and data management strategy, which hosts SNP array datasets from 533 *Brassica* lines, to date (Scheben et al., 2019). Sample misidentification from germplasm collections was also demonstrated by (Mason et al., 2015) using the *Brassica* 60K SNP array. Evidence of interspecific hybridisation events was detected, which likely resulted from cross-pollination during bulking processes. One such example includes presence of the C-genome in a *B. rapa* individual (Mason et al., 2015). It was further proposed that such tools are effective and could be routinely utilised by breeders and researchers to ensure seed socks were adequately and accurately catalogued after seed harvesting.

Table 1 Summary of recent findings in DNA-based screening of *Brassica* germplasm

Traits	DNA-based screening techniques	Germplasm characteristics	Key findings QTL and/or genes	References
Management and conservation of *Brassica* germplasm				
A and C genome composition	*Brassica* 60K SNP, principal component analysis (PCA), hierarchical clustering and chromosome count	180 Brassicaceae samples, including those suspected to be mislabelled	Species misclassification in 18% of the samples, which include 30% suspected misclassification samples and 9% randomly sourced samples. Robust amplification between samples for 11 983 A-genome, and 12,794 C-genome specific SNPs	Mason et al., 2015
Assess population history and diversity	*Brassica* 60K SNP, PCA, population structure and statistical analysis of molecular variance (AMOVA)	995 *B. napus* cultivars, predominantly from China	Identification of parental lines with ecological adaptability, diversity assessment and prediction of hybrid performance based on genetic distance	Luo et al., 2019
Assess population structure	SLAF-seq	300 *B. napus* accessions, predominantly from China	SNP mining and identification of 41 SNP rich regions. Population structure analysis identified richer variation in the A-genome over the C-genome	Zhou et al., 2017
Assess genetic diversity	KASP	Chinese cabbage including 231 *B. rapa* ssp. *pekinensis*, ssp. *rapifera*, ssp. *oleifera* (Su et al., 2018) and 70 *B. rapa* ssp. *chinensis* var. Makino and *communis* (Li et al., 2019a)	Identification of 1167 equally distributed SNPs generated by (Su et al., 2018), where 50 core SNPs were subsequently selected by Li et al., (2019a) accessible to breeders for rapid genetic diversity screening and genotyping purposes	Su et al., 2018; Li et al., 2019a

Assess population structure	GBS-transcriptomics (GBS-t) and AMOVA and WGR for a subset of 149 global diversity samples	633 accessions, predominantly Australian and European, with additional Asian *B. napus* varieties	Identification of 4 029 750 high confidence SNPs and structural variations in the genome. SNP list generated is more representative of the global germplasm, with suggestions to include more European and Canadian varieties in the future. Improved SNP list recommended for characterising SNP effects for canola genomics, proteomics and phenomics studies	Malmberg et al., 2018
Sustainable management of resistance to biotic stresses				
Blackleg resistance	GWAS and linkage mapping using *Brassica* 60K, 20K and 8K SNP arrays	166 winter *B. napus* varieties and mapping populations: DY (Darmor x Yudal), DS (Darmor x Samourai) and DB (Darmor x Bristol)	Identification of consensus QTL and significant SNP markers on chromosomes A08, A09/C09 and C04	Kumar et al., 2018
Blackleg resistance	GWAS using 37 346 SNPs GBS	139 global *B. napus* accessions	Identification of a major QTL on chromosome A01, and ten minor-effect QTL	Raman et al., 2016
Blackleg resistance	Linkage mapping using 18 851 DArTseq-SNPs and *in silico* DArTs	177 doubled haploid (DH) lines from an RP04/Ag-Outback cross	Identification of 19 QR QTL of which 2 were considered stable and candidate genes *BnaA03g26760D* (A03), *BnaA07g20240D* (A07) and *BnaC03g04240D* (C03)	Raman et al., 2020
Sclerotinia stem rot resistance	Linkage mapping using *Brassica* 60K SNP array	A natural population containing 521 *B. napus* inbred lines and a double haploid population containing 150 lines	Four QTL affecting both SSR resistance and flowering time on chromosomes A02, A06, C02 and C08	Zhang et al., 2019

(*Continued*)

Table 1 Summary of recent findings in DNA-based screening of *Brassica* germplasm (*Continued*)

Traits	DNA-based screening techniques	Germplasm characteristics	Key findings QTL and/or genes	References
Sclerotinia stem rot resistance	GBS and GWAS	A set of 96 *Brassica juncea–Erucastrum cardaminoides* introgression lines (ILs)	QTL for quantitative SSR resistance on chromosomes A03, A06 and B03	Rana et al., 2019
Clubroot resistance	*Brassica* 60K SNP array and GWAS analysis	A global population composed of 472 accessions	Nine QTL for resistance against the predominant pathotype 4 of *P. brassicae* in China	Li et al., 2016b
Clubroot resistance	ddRAD-seq and linkage mapping	An F2 mapping population of 348 lines derived from a Chinese cabbage line and a turnip line that are susceptible and resistant to a *P. brassicae* Korean Seosan-isolate, respectively	A novel qualitative resistance locus was identified on chromosome A08	Laila et al., 2019
Clubroot resistance	Using SNPs from BSR-Seq and linkage mapping	A cross between the resistant *B. nigra* line PI 219576 and susceptible line CR2748	The first qualitative clubroot resistance gene, *Rcr6*, was identified in the B-genome of *B. nigra*	Chang et al., 2019
Clubroot resistance	GBS and GWAS analysis	135 *B. oleracea* accessions	Ten QTL from six C-genome chromosomes for resistance to two recently evolved *P. brassicae* pathotypes, F3-14 (3A) and F-359-13 (5X L-G2)	Farid et al., 2020
Sustainable management of resistance to abiotic stresses				
Heat tolerance	GBS-SNP followed by GWAS	88 spring type *B. napus* from US Germplasm Resources Information Network (US GRIN)	Total 37 539 SNPs, 115 significant SNPs, 20 QTL associated with heat tolerance	Rahaman et al., 2018
Heat tolerance	GBS-SNP followed by QTL mapping	155 DH population of broccoli	Total 208 170 SNPs, five significant QTL, 82 candidate genes	Branham et al., 2017

Frost tolerance	GBS-SNP followed by GWAS	147 *B. napus* from US GRIN with representations from 15 countries worldwide and exhibiting three types of growth habits – spring, winter and semi-winter	Total 37 111 SNPs, 38 significant SNPs, 1 major QTL on chromosome C04	Wrucke et al., 2019
Drought tolerance	*Brassica* 60K SNP array followed by GWAS	140 *B. napus* accessions from 17 countries mostly from AGG, and the rest from Huazhong Agricultural University (HZAU) and Oil Crops Research Institute, Chinese Academy of Agriculture Science (OCRI-CAAS)	Total 25 495 SNPs, 16 significant SNPs, 79 candidate genes	Zhang et al., 2015
Drought tolerance	GWAS-SNP based on SLAF-seq, which is a modification of ddRADseq	228 *B. napus* accessions sourced mainly from China and the rest from other parts of the world	Total SNPs, 52 significant SNPs, 85 candidate genes	Khanzada et al., 2020
	SNP calling using WGR data of *B. napus* followed by GWAS	588 *B. napus* accessions originating in Asia, Europe, North America and Australia including old accessions from 1980 to 2000 and displaying all three growth habits – spring, winter and semi-winter	Total 670 028 SNPs, 60 significant loci associated with 10 important traits related to cultivar improvement, six candidate genes responsible for drought tolerance	Lu et al., 2019
Salinity tolerance	*Brassica* 60K SNP array followed by GWAS	368 diverse *B. napus* accessions from 10 countries across four continents mostly from China and included winter, semi-winter and spring varieties	Total 19 167 SNPs, 75 significant SNPs, 38 candidate genes	Wan et al., 2017

(*Continued*)

Table 1 Summary of recent findings in DNA-based screening of *Brassica* germplasm (*Continued*)

Traits	DNA-based screening techniques	Germplasm characteristics	Key findings QTL and/or genes	References
Enhance yield and nutritional related traits				
Seed yield	GWAS (3106 SNP bins and non-SNP markers and 11 phenotypic datasets)	348 *B. napus* DH lines from KenC-8 x N53-2	Of the 17 978 SNPs, 31 significant QTL	Chao et al., 2019
Seed yield	QTL alignment and Candidate gene identification	5,805 canola germplasm based on 22 different populations from America, China, and Germany	93 significant QTL, 16 candidate genes	Raboanatahiry et al., 2018
Silique number	GBS technology (1465 SNP bin markers) flowed by GWAS (two phenotypic datasets)	100 DH lines of P1 (01-20) 9 P2 (96-100) *B. oleracea*	One significant QTL, qSNS6.1 (C06)	Li et al., 2019b
Thousand seed weight	GWAS (1643 SNP markers and 11 phenotypic datasets), BLAST, fine-mapping	505 *B. napus* diverse GWAS panel	Six significant QTL in A03, A07, A08 and C02 which one QTL (fine-mapped) corresponded to *BnaA03G37960D*	Wang et al., 2020
Shattering resistance	Association mapping (SNP and DArT sequencing and two-year data for pod traits)	235 germplasm lines with 124 *B. juncea*, 90 *B. rapa* and 21 *B. nigra* from Australia, Canada, China, Central Asia, East Europe, Pakistan and India	In 23 associations, six shattering-related genes were significant	Kaur et al., 2020
Flowering time	ddRAD sequencing followed by association analysis	Three *B. juncea* Indian and three exotic germplasm	Of the 10 399 SNPs, two significant associations in A06, SNPs A06_6796237 and A06_23478761	Sudan et al., 2019
Flowering time	GWAS (Illumina customized Infinium platform with 5306 probes and 10 phenotypic datasets)	190 DH lines of *B. napus* from Tapidor×Ningyou7	Of 437 associated SNPs, 11 of which were linked to flowering time genes	Li et al., 2015

Flowering time	SNP/InDel analyses using WGRS data of *B. napus* followed by GWAS	991 *B. napus* germplasm accessions, including 658 winter, 145 semi-winter, and 188 spring types originating from 39 countries across the world	From more than 5 million SNPs and 1 million InDels, BnaA02g12130D was identified within the 6.37–6.38 Mb region on chromosome A02 and BnaA10g22080D within the region 14.99-15.01 Mb on chromosome A10 as significantly associated with the trait	Wu et al., 2019b
Flowering time	SNP/InDel/Copy number variation analyses after targeted deep sequencing	280 genetically diverse *B. napus* inbred lines (selfed ≥ 5 generations)	In 637 regions annotated as genes and 184 as target genes out of 1184 captured regions, 5216 SNPs and 569 INDELs were detected that revealed duplications and deletions in flowering time genes in different lines	Schiessl et al., 2017
Flowering time, plant height, branch initiation height, and stem diameter	QTL mapping (3073 SNP markers, six phenotypic datasets for plant height and five for branch initiation height, stem diameter and flowering time)	208 DH lines from Y689 (hybrid between *B. napus* cv. Zhongyou 821 and *Capsella bursa-pastoris*) and Westar	63, 36, 24 and 71 QTL for plant height, branch initiation height, stem diameter and flowering time, respectively	Shen et al., 2018
Plant height, branch initiation height and branch number	GWAS (*Brassica* 60K SNP array and four phenotypic datasets)	GWAS panel of 333 *B. napus* accessions	22 significant QTL for plant height, 12 for branch initiation height and 7 for branch number	Zheng et al., 2017
Dwarfism	WGR Library Construction and High-Throughput Sequencing	80 RILs (F2) from YA2016-12 x G184-189 population	In 12.4Mb genomic region (A03), nine candidate genes were found	Chai et al., 2019
Seed coat	GWAS (22 previously linked SNP marker for seed coat and two phenotypic datasets)	520 *B. napus* accessions GWAS panel	Of the 31 839 SNPs and one most significant candidate gene (*BnaC06g17050D*)	Wang et al., 2017

(*Continued*)

Table 1 Summary of recent findings in DNA-based screening of *Brassica* germplasm (*Continued*)

Traits	DNA-based screening techniques	Germplasm characteristics	Key findings QTL and/or genes	References
Oil qualities	Marker-trait association (1359 DArT and non-DArT markers, and two phenotypic datasets)	six populations of *B. napus* from 14 different researches	Flanking markers for the specific fatty acids in the oil, palmitic acid found on A01 (62.6 cM), oleic acid on A05 (94.9 cM), linoleic acid on A05 (94.2 cM) and linolenic acid on C04 (141.6 cM)	Raman et al., 2013b
Low erucic and glucosinate contents	GWAS (*Brassica* 60K SNP array and two phenotypic datasets)	472 accessions association panel	Of the total 26 841 SNPs, two SNPs were linked to two key genes for low erucic content, and five SNPs linked to genes governing glucosinolate biosynthesis	Li et al., 2014
Low glucosinate contents	High-throughput sequencing and GWAS	238 *B. napus* accessions	Of the 4 542 011 SNPs and 628 666 INDELs, 27 candidate genes were identified and involved in glucosinolate biosynthesis	Wang et al., 2018

Analysing genotype information from the germplasm could help inform our understanding on the domestication history and assess diversity between closely related gene pools in canola and other modern *Brassicas*. This will also help to distinguish new sources to integrate into highly intermixed gene pools. For example, Zhou et al. (2017) used Specific-Locus Amplified Fragment Sequencing (SLAF-seq) to analyse ecotype diversity for the validation of breeding history between spring and early winter canola. *B. rapa* is a source of rich diversity, as a progenitor of modern *B. napus*-derived canola, as well as being an agronomically significant vegetable crops commercially valued worldwide. KASP is another rapid and valuable tool for SNP validation (Li et al., 2019a; Su et al., 2018). This is especially crucial for germplasm characterisation as a consequence of genetic diversity losses following decades of artificial selection. A total of 568 SNPs generated from Su et al. (2018) were proposed for germplasm characterisation and were densely informative for future genetic mapping, QTL identification and association studies. These markers are useful for tracing domestication events including differentiating modern varieties.

The events leading to *B. oleracea* domestication were previously inferred with minimal molecular information like restriction fragment length polymorphisms (RFLPs) (Song et al., 1988) and morphological observations (Stansell et al., 2018). Using the *Brassica* 60K SNP array Luo et al. (2019) identified broccoli (*B. oleracea* var. *italica*) as the most genetically diverse *B. oleracea* morphotype, over cauliflower (var. *botrytis*), and Chinese kale (var. *alboglabra*). Although the study included a categorically disproportionate sample size, they managed to identify more genetic diversity in landrace broccoli and cauliflower, over their modern and well-cultivated counterparts. Additionally, Malmberg et al. (2018) utilised GBS-transcriptomics (GBS-t) and a WGR approach to assess SNP diversity in a global canola germplasm dataset. Both studies found higher intra-specific genetic variability than between populations, thus highlighting a favourable avenue to be explored in future breeding programs.

3 Sustainable management of resistance to biotic stress

Blackleg, *Sclerotinia* stem rot (SSR) and clubroot are the three major diseases affecting *Brassicas* (Neik et al., 2017). The most effective measure to control the diseases is to deploy host resistance. In *Brassicas* both qualitative resistance, controlled by a single gene, and quantitative resistance, controlled by multiple genes, have been described (Neik et al., 2017). To date, at least 18 qualitative blackleg *R* genes have been identified (Van de Wouw and Howlett, 2019) while several QTL for quantitative resistance have been mapped in various germplasm (Kumar et al., 2018; Raman et al., 2016; Raman et al., 2020). Almost all mapping of SSR resistance focuses on *B. napus* and only QR has been characterized in the A and C genomes (Lv et al., 2020). Abundant qualitative resistance genes have been identified in *B. rapa* for clubroot resistance. Array-based and sequencing-based SNP discovery and association studies have allowed identification of abundant QTL and candidate genes for disease resistance in *Brassica* germplasm (Table 1).

3.1 Blackleg resistance

Most of the major *Brassica R* genes against blackleg have been identified from the A genome (*Rlm1, Rlm2, Rlm3, Rlm4, Rlm7, Rlm9, Rlm12, LepR1, LepR2, LepR3, LepR4, BLMR1, BLMR2, LmFr1* and *LEM1*), while a few have been sourced from the B genome (*Rlm6, Rlm10, LMJR1, LMJR2*) (Raman et al., 2016; Raman et al., 2013b). These genes were mapped using early generation molecular markers generated through different marker systems using diverse genetic populations. These differences make these markers less useful for comparative mapping and precise location of causative genes (Raman et al., 2013a). The establishment of genome references permitted a comprehensive study of *R* genes in *Brassica* and facilitated the identification and cloning of specific genes controlling three (*Rlm2, LepR3* and *Rlm9*) of the 18 known blackleg *R* genes (Larkan et al., 2013, 2019). Sequence-specific primers are currently being routinely used for marker-assisted breeding and confirmation of gene content in released varieties.

A number of germplasm have been evaluated using genome-wide SNPs to validate blackleg *R* gene content and/or to identify candidate genes and closely linked markers. GWAS and linkage mapping using *Brassica* 60K, 20K and 8K SNP arrays for genotyping 166 winter *B. napus* varieties and genetic mapping populations allowed the identification of consensus QTL for blackleg resistance on chromosomes A08, A09/C09 and C04 (Kumar et al., 2018). Raman et al. (2016) screened 139 global *B. napus* accessions obtained from Germplasm Resources Information Network (GRIN) to map for blackleg QTL, using GWAS with 37,346 SNPs generated using GBS. The association analysis revealed one major QTL on chromosome A01, and ten minor-effect QTL in all the chromosomes in the A genome, except for A05 and on chromosomes C01, C03, C06, and C08. Four SNP markers were found to be in perfect LD with the major QTL, and this haplotype information was used to predict the blackleg resistance of 22 *B. napus* cultivars. Phenotypic analysis showed that these markers could accurately predict the resistance response to the blackleg pathogen. Further alignment of significant markers to the reference genome revealed colocalization of these genomic regions with orthologs of *Arabidopsis thaliana* hydroxyproline-rich glycoprotein family protein, acyl-CoA oxidase 1 and cysteine-/histidine-rich C1 domain family protein – all of which have been previously linked to plant defence mechanism.

In a recent paper from Raman et al. (2020), 177 doubled haploid (DH) lines derived from an RP04/Ag-Outback cross were screened for blackleg resistance to discover new alleles for QR. The genome-wide scan of a set of 18,851 DArTseq-SNPs and *in silico* DArTs yielded 19 significant QTL associated with QR, of which two were consistently detected across environments. Further alignment of significant markers to the reference genome enabled the identification of three candidate genes: *BnaA03g26760D* (A03), *BnaA07g20240D* (A07) and *BnaC03g04240D* (C03), the sequences of which were highly identical to disease response genes in *A. thaliana*. All of these genes have been previously described to have direct or indirect roles in plant disease immunity.

3.2 SSR resistance

Screening with the *Brassica* 60K SNP array of a natural population containing 521 *B. napus* inbred lines and a DH population containing 150 lines, all of which displayed great variation in flowering time, identified four QTL affecting both SSR resistance and flowering time on chromosomes A02, A06, C02 and C08 (Zhang et al., 2019). Furthermore, co-localisation of the four QTL and those from previous studies for these two traits, based on the physical location of markers on *B. napus* reference genomes, permitted the identification of QTL hot spots on chromosomes A02, A03, C02 and C06 (Zhang et al., 2019). This study provided a better understanding of the genetic linkage between SSR resistance and flowering time and will help in breeding varieties with both early maturation and SSR resistance (Zhang et al., 2019).

A set of 96 *Brassica juncea-Erucastrum cardaminoides* introgression lines (ILs) were developed from introgression of a genomic region associated with SSR resistance from the wild relative *E. cardaminoides* (Rana et al., 2019). Genotyping with transferable genetic markers and association mapping allowed the identification of six resistance loci in both A and B genomes and subsequent GBS and GWAS analysis enabled the localisation of the QTL on chromosomes A03, A06 and B03 (Rana et al., 2019). Candidate genes underlying the QTL were involved in signal transduction pathways, hypersensitive responses, and production of anti-fungal proteins and metabolites, as well as LRR-RLK and TIR-NBS-LRR type resistance genes (Rana et al., 2019). SNP markers in close linkage with the candidate genes will facilitate introgression of resistance sources into *B. juncea* and other crop *Brassica* species (Rana et al., 2019).

3.3 Clubroot resistance

A global population composed of 472 accessions was screened with the *Brassica* 60K SNP array for resistance against the predominant pathotype 4 of *Plasmodiophora brassicae* in China. GWAS analysis led to the discovery of nine QTL, including seven novel ones which have additive effects on the genetic control of clubroot resistance, and candidate gene analysis suggested that the TIR-NBS gene family might play an important role (Li et al., 2016b). An F_2 mapping population of 348 lines derived from a Chinese cabbage line and a turnip line that were susceptible and resistant to a *P. brassicae* Korean Seosan-isolate, respectively, was genotyped using double digest restriction site-associated DNA sequencing (ddRAD-seq). Through this a novel qualitative resistance locus was identified on chromosome A08 (Laila et al., 2019). Four out of seventeen validated SNP markers cover a 0.8-Mb region flanking three novel NBS-LRR candidate genes (Laila et al., 2019). The closely linked markers could be used to genotype Seosan-resistant lines and assist in the development of resistant Chinese cabbage cultivars (Laila et al., 2019).

The first qualitative clubroot resistance gene in the B-genome of *B. nigra*, *Rcr6*, was identified through mapping with SNPs from BSR-Seq, in a cross between the resistant *B. nigra* line PI 219576 and susceptible line CR2748. It was further fine mapped using KASP assays onto *B. rapa* A08 and *B. nigra* B07. Only the *Rcr6* candidate gene, *BniB015819* encoding a TIR-NBS-LRR type resistance protein, had a homolog Bra010663 in the target region (Chang et al., 2019). GBS and GWAS analysis on 135 *B. oleracea* accessions allowed the identification of 10 QTL from six C-genome chromosomes for resistance to two recently evolved *P. brassicae* pathotypes, F3-14 (3A) and F-359-13 (5X L-G2) (Farid et al., 2020). Each of the QTL accounted for 8–18% of the phenotypic variance and thus could be deployed for enhancing clubroot resistance in *Brassica* crops (Farid et al., 2020). Three QTL carried more than 3 SNP makers which will facilitate marker-assisted selection on the QTL (Farid et al., 2020).

3.4 Insect resistance

Insect pests such as cabbage seedpod weevils (*Ceutorhynchus obstrictus*), crucifer flea beetles (*Phyllotreta cruciferae*), or pollen beetles (*Brassicogethes aeneus*) cause damage to *Brassica* flower buds (Friedt et al., 2018). Insecticides cause environmental pollution and damage ecosystems (Mpumi et al., 2020). Restrictions on insecticide usage make genetic approaches to insect resistance increasingly important (Friedt et al., 2018). Although substantial progress has been achieved on improving disease resistance in *Brassicas*, less progress has been made regarding resistance to insect pests (Delourme et al., 2018). Five QTL were associated with resistance to diamondback moth (*Plutella xylostella*) using a mapping population derived from crossing between cv. 'SLMO46' and cv. 'Quantum' (Asghari et al., 2012). The QTL mapping was based on 70 microsatellite and randomly amplified polymorphic DNA (RAPD) markers. A further QTL was identified on chromosome A07, associated with kaempferol 3-O-sinapoyl-sophoroside 7-O-glucoside (KSSG) metabolite accumulation, using microsatellite markers, which was correlated with reduction of cabbage seedpod weevil (*Ceutorhynchus obstrictus*) larval infestation in a *Sinapis alba* and *B. napus* mapping population (Lee et al., 2014). In another study, SNP genotyping of the cabbage whitefly (*Aleyrodes proletella*) resistant line Rivera and susceptible line Christmas Drumhead identified two QTL on chromosomes C02 and C09 in the modern white cabbage (*Brassica oleracea* var. *capitata*), of which several genes were involved in abscisic acid (ABA) signalling (Broekgaarden et al., 2018).

4 Sustainable management of resistance to abiotic stresses

The genetic basis of tolerance towards abiotic factors in plants is often polygenic or genetically controlled by more than one gene (Delourme et al., 2018). Diverse genetic resources can be tapped from various *Brassica* germplasm collections that include wild relatives, landrace, breeding lines and cultivars (Delourme et al., 2018). SNP-based screening on *Brassica* germplasm has greatly contributed to the understanding of the genetic basis of tolerance to heat, frost, drought and salt stresses and has facilitated breeding of resistant/tolerant crops (Table 1).

4.1 Heat tolerance

Heat stress in *Brassica* has a significant impact on the reproduction of the plant. A study conducted on heat tolerance in *B. rapa* showed a reduction in the development of mature seeds in pods during heat stress, and that *B. rapa* is a good source of heat tolerance for breeding of *B. napus* (Chen et al., 2013). Another study in *B. napus* showed similar findings in that heat stress severely impacts *B. napus* during the early flowering stage (Rahaman et al., 2018), where GWAS analysis on 88 spring type *B. napus* revealed 20 QTL regions associated with flowering traits such as pollen sterility, sterile/aborted pods and number of pods on main raceme with candidate genes identified, such as a pentatricopeptide repeat (PRR) protein involved in RNA processing and F-box family protein involved in protein degradation. The SNP markers were mostly concentrated on the A genome (Rahaman et al., 2018), suggesting that the A genome harbours greater genetic diversity for heat tolerance and serves as a good source of germplasm for breeding heat-tolerant *B. napus* varieties. *B. oleracea* var. italica (broccoli) also provides a good source of heat tolerance, as shown from a study applying GBS-SNPs to 155 DH lines of *B. oleracea* resulting in identification of five significant QTL associated with heat tolerance and 82 candidate genes encoding for floral development proteins, heat-shock proteins, antioxidant enzymes and transcription factors (Branham et al., 2017).

4.2 Frost tolerance

To investigate the frost tolerant trait in *B. napus*, a diverse set of 147 *B. napus* germplasm accessions obtained from GRIN, with global origins, was subjected to GBS. This leads to the identification of one significant major QTL, containing five candidate genes located on chromosome C04, including a protein phosphatase 2C family protein and ubiquitin-conjugating enzyme, which are involved in abscisic acid (ABA) signalling and abiotic stress response pathway (Wrucke et al., 2019). Unlike the heat tolerance candidate genes, the frost tolerance candidate genes were located mostly in the C genome. Further frost-tolerant screening experiments in *B. napus* incorporating larger sample size, higher population genetic diversity and higher quality *B. napus* reference genome assembly are needed to confirm the accuracy of the major QTL on chromosome C04.

4.3 Drought tolerance

Drought stress has been shown to affect *B. napus* plant at all stages including germination, flowering or seed setting, as well as vegetative growth stages (Sabagh et al., 2019; Zhu et al., 2016), seriously impacting the productivity of the crop. A GWAS study conducted by Zhang et al. (2015) using the *Brassica* 60K SNP array to screen for drought stress response in *B. napus* lines identified 16 significant loci associated with drought tolerance, mostly within the A sub-genome. One of the candidate genes was located on chromosome A03, encoding for a late embryogenesis abundant (LEA) protein, and another candidate gene on chromosome A05 encoded a carboxylate synthase enzyme involved in proline biosynthesis, both of which were previously known to play a role in drought tolerance (Zhang et al., 2015). Khanzada et al. (2020) reported 85 candidate genes associated with drought tolerance trait in *B. napus,* mainly being transcription factors such as the homeodomain-leucine zipper subfamily, distributed across both A and C genome chromosomes, with more candidate genes found on the A genome.

In a recent large-scale study using 588 worldwide *B. napus* accessions to investigate the origin of *B. napus* and to identify the candidate genes associated with 11 important traits that are favourable for cultivar improvement, six candidate genes responsible for drought tolerance genes were identified in the A sub-genome of *B. napus*, and these genes were important in ABA biosynthesis during water deficit conditions, highlighting that the A sub-genome is strongly selected for adaptation to environmental stress (Lu et al., 2019). The implication of this finding is such that A genome of the *Brassica* species harbours more genes responsible for abiotic stress compared to the C sub-genome; hence, focusing on DNA screening in *B. rapa* for novel alleles is a good strategy for *Brassica* crop improvement in abiotic stress.

4.4 Salinity tolerance

Research on tolerance towards salinity in *B. napus* is relatively extensive, covering proteomics, transcriptomics, gene expression and molecular breeding, including developing transgenic *B. napus* with salt tolerance (Shokri-Gharelo and Noparvar, 2018). Screening methods for salinity tolerance in *B. napus* have also been developed using mathematical models which can reliably predict the physiological response of *B. napus* towards different levels of salinity during germination (Wu et al., 2019a). However, this method of physiological screening using mathematical models requires intensive labour and time. Screening *B. napus* germplasm for salinity tolerance using marker-assisted methods provides a more accurate and cost-effective method for breeding. In the GWAS-SNP study screening for salt tolerance trait across 368 *B. napus* lines conducted by Wan et al. (2017), 75 significant SNP markers, distributed across 14 chromosomes, were associated with salt tolerance. There were 38 candidate genes underlying the salt tolerance QTL region with a biased distribution within the A genome of *B. napus* and the gene families include transcription factor, transporter proteins for water and solutes, and enzymes involve in salt and osmotic stress biochemical pathway (Wan et al., 2017).

Besides *B. napus*, *B. oleracea* germplasm potentially offers a good genetic source for the development of salt tolerance in *Brassica* crops. Morphological and physiological studies under controlled environments of six accessions of *B. oleracea* var. *capitata* (cabbage) and *B. oleracea* var. *acephala* (kale) involving wild type, landraces and cultivars showed that the wild *B. oleracea* adapted better to high salinity conditions compared to the other *B. oleracea* varieties (Lema et al., 2019). In another biochemical study of *B. oleracea*, which did not include wild germplasm but different morphotypes of *B. oleracea*, comparing the root response towards salt stress in *Brassica rapa* ssp. *pekinensis* (Chinese cabbage), *B. oleracea* var. *capitata* (white cabbage) and *B. oleracea* var. *acephala* (kale), kale was identified as the most salt tolerant *Brassica* species amongst the other *Brassica* groups (Pavlović et al., 2019). These findings suggest that *B. oleracea* is an important genetic source of salt-tolerant genes and the application of DNA-based screening across all the *B. oleracea* germplasm including the wild, landrace and cultivar should be the focus in future *Brassica* salinity tolerance studies.

5 Enhancing yield and nutritional related traits

5.1 Seed yield

Seed yield and oil quality are affected by numerous traits including silique number, seed number per silique, thousand seed weight, flowering time, plant height, branch initiation height, branch number, shattering resistance, seed coat, fatty acids (palmitic, oleic, linoleic and linolenic acids), glucosinolate and erucic acid content (Table 1). A high density SNP-based map using 3106 SNP-bins (17 978 SNPs) and 101 non-SNP markers in 348 *B. napus* DH lines (KenC-8 x N53-2) linked to seed yield in 11 environments of diverse area conditions (semi-winter, spring, and winter-types) resolved into 31 QTL (Chao et al., 2019). A consensus physical map of *B. napus* Darmor-*bzh* based on 22 different populations with diverse germplasm from America, China and Germany was used to identify 93 QTL, including the identification of candidate genes in seed yield (Raboanatahiry et al., 2018). Interestingly, there was a common candidate gene: *BnaA02g02560D* (*Bra023513*, gene in *B. rapa*) (Chao et al., 2019; Raboanatahiry et al., 2018; Zhao et al., 2016), which was also identified using GWAS and RNA-seq tools in 520 *B. napus* accessions (Lu et al., 2017). The marker derived from this candidate gene could be a helpful tool in screening for high-yielding *Brassicas*.

Silique number, seed number per silique and thousand seed weight have direct effects on seed yield (Yang et al., 2012) while flowering time, plant height, branch initiation height, branch number and shattering resistance have indirect effects on seed yield (Hu et al., 2012; Liu et al., 2016; Raman et al., 2019a; Shi et al., 2009). In *B. oleracea*, GBS technology was applied to construct a high-density SNP-based map with 1465 bin markers resolved into one significant QTL, qSNS6.1 (C06), which was consistent for two years based on 100 DH lines (Li et al., 2019b). Using 1643 SNP markers in a 505 *B. napus* diverse GWAS panel, the basic local alignment search tool (BLAST) identified significant hits/six QTL in thousand seed weight on four chromosomes, in which one QTL (fine-mapped) corresponding to *BnaA03G37960D* on chromosome A03 was consistent across six environments (Wang et al., 2020). In shattering resistance, large sets of germplasm lines of *B. juncea* (124), *B. rapa* (90) and *B. nigra* (21) (from Australia, Canada, China, Central Asia, East Europe, Pakistan and India) were phenotypically evaluated with pod traits for two years and genotyped with SNP and DArT sequencing, which led to the identification of six shattering-related genes (*SHP1* and *SHP2* in *B. rapa*, *B. nigra* and *B. juncea*, *RPL* in *B. rapa*, and *NAC* in *B. juncea*) through the association mapping (Kaur et al., 2020).

5.2 Flowering time

DdRAD sequencing led to the identification of 10 399 SNPs with two highly significant associations, on chromosome A06 at 6.8 MB (A06_6796237) and 23.4 MB (A06_23478761), with flowering time using six diverse *B. juncea* germplasm (Sudan et al., 2019). GWAS was also done in 190 DH lines of *B. napus* and identified 437 SNPs associated with flowering time, which includes *UQnapus5172* on chromosome A06, homologous to *AT2G01820* (a flowering time gene) and a receptor-like kinase *TMK3* gene (primarily for cell expansion or growth and proliferation) in *A. thaliana* (Dai et al., 2013; Li et al., 2015). Using higher coverage with 991 *B. napus* accessions from 39 countries identified more than 6 million SNPs/INDELs, which when used for GWAS identified *BnaA02g12130D* on chromosome A02 and *BnaA10g22080D* on chromosome A10 as significantly associated with the flowering time (Wu et al., 2019b). In addition, targeted deep sequencing of flowering time pathways, including genes in circadian clock regulators, the input pathways for vernalisation, photoperiod sensitivity and downstream signal transducers, revealed extensive nucleotide and copy number variation using 280 diverse *B. napus* accessions (Schiessl et al., 2017). Of the genes showing copy number variation, a total of 259 genes on the A-subgenome and 189 genes on the C-subgenome were found to be the result of duplication and 139 on the A-subgenome and 351 on the C-subgenome were the result of deletion, where the highest rate of variation was found between A02/C02 and A03/C03 (Schiessl et al., 2017). Most of the variable genes on chromosomes A02/C02 were involved in the vernalisation pathway, while a single variable gene was identified as a downstream signal transducer. On chromosomes A03/C03, all of these variable genes are involved in pathways for vernalisation.

Flowering time, plant height, branch initiation height and stem diameter were significantly and positively correlated to each other, which had been co-localised in chromosomes A02 and A07, and highlight *BnaA02g02560* as a candidate gene for plant height, stem diameter, and flowering time, and *BnaA02g08490* was associated with plant height, branch initiation height and flowering time (Shen et al., 2018). A diverse panel of 333 *B. napus* accessions were genotyped using the *Brassica* 60K SNP array and phenotypically evaluated over four years, which identified 22 significant QTL for plant height, 12 for branch initiation height and 7 for branch number (Zheng et al., 2017), of which a few of the QTL were also found in previous studies (Chen et al., 2007; Ding et al., 2012; Li et al., 2013, 2016a; Mei et al., 2009; Shi et al., 2011). Dwarfism makes *Brassica* show lodging resistance over tall plants. The region conferring dwarfism in *B. napus* was also identified with nine candidate genes on chromosome A03, involved in gibberellin or cytokinin signalling, using whole genome re-sequencing and association analysis (Chai et al., 2019).

5.3 Seed coat

Seed coat is a significant breeding trait for *Brassica* crops, especially for seed quality (Wang et al., 2017), and yellow-seeded *Brassica* oilseeds had desirable agronomic properties (Mahmood et al., 2006). A diverse set of 520 *B. napus* accessions showing different seed coat colour intensity based on near-infrared reflectance spectroscopy was assessed using 22 previously known SNP markers for the trait, along with a further 31 839 SNPs. Through this, one significant candidate gene (*BnaC06g17050D*), 43.42 kb from the *TRANSPARENT TESTA 12 (TT12)* gene, was identified (Wang et al., 2017). SRAP markers developed from chromosome walking, including SA12BG18388, SA7BG29245 and DC1GA27197, could distinguish seed coat colour, especially the latter that can determine dark to light yellow seed coat in *B. napus* (Rahman et al., 2010). In *B. rapa*, intron polymorphism markers (IP) 1, IP2 and IP3 linked to *Bra036827*, *Bra036828* and *Bra036829* on chromosome A09, respectively, co-segregated with seed coat colour and could accurately distinguish three genotypes at the yellow seed locus in separate populations (Huang et al., 2016).

5.4 Seed quality

To determine oil qualities in *Brassicas*, a consensus map in *B. napus* (in six populations) consisting of 1359 DArT and non-DArT markers where its result (sequences) was aligned with genomic scaffolds from *B. rapa* and *B. oleracea* (Raman et al., 2013a). The study derived flanking markers for the specific fatty acids; palmitic acid on chromosome A01, oleic acid and linoleic acid on chromosome A05, and linolenic acid on chromosome C04 (Raman et al., 2013a). The information in these DArT markers were utilised in several oil quality research studies not only in *Brassica* but also in other species with oil potentials (Gacek et al., 2016; Jiang et al., 2014; Raboanatahiry et al., 2017; Shasidhar et al., 2017; Wang et al., 2013). The low erucic acid content in *B. napus* was identified through mutation at two additive loci: the genes *BnFAE1.1* (chromosome A08) and *BnFAE1.2* (chromosome C03) (Basunanda et al., 2007). Using the *Brassica* 60K SNP array in 472 accessions, Bn-A08-p12599446 (on chromosome A08) and Bn-scaff_15794_3-p29807 (on chromosome C03) SNPs were found to be located 233 and 128 kb away from the key genes *BnFAE1.1* and *BnFAE1.2*, respectively, (Li et al., 2014). Using *B. juncea* (Pusa Mustard 30, a low erucic acid genotype) sequence paralogs, a functional CAPS markers for the *fatty acid elongase (FAE)* genes was used to identify two SNPs that can distinguish low and high erucic acid genotypes (Saini et al., 2016). These markers had also been utilized by the *Brassica* team of Punjab Agricultural University (PAU), Ludhiana, India, to expedite the *B. napus* breeding programme (Sandhu and Singh, 2018).

Reducing the glucosinolate content is an important goal in oilseed *Brassicas* (Walker and Booth, 2001). Using 24 256 SNPs in a GWAS panel, different copies of MYB28, a transcription factor regulating biosynthesis of glucosinolate in *A. thaliana* (Hirai et al., 2007), have been found on chromosomes A09, C02, C07 and C09 responsible for glucosinolate content (Li et al., 2014). A similar result was also found for glucosinate content based on 4 542 011 SNPs and 628 666 INDELs, 27 candidate genes were identified and involved in glucosinolate biosynthesis on chromosomes A09, C02 and

C09 (Wang et al., 2018). Associative transcriptomics, which uses transciptome sequencing to identify markers representing variation in gene expression and gene sequences, has been used to identify deletions underlying QTL for glucosinolate content (Harper et al., 2012). In this study the deleted regions contained orthologs of an Arabidopsis transcription factor (*HAG1*) which controls aliphatic glucosinolate biosynthesis.

6 Conclusion and future trends

SNP marker-based genotyping technologies, aimed at rapid and high-throughput detection of SNP markers along with discovery of large numbers and genome-wide availability of SNPs without compromising on data accuracy and cost effectiveness, have revolutionised DNA-based screening methods in *Brassica* species. Improvements on the current SNP genotyping methods in *Brassica* research to achieve greater accuracy in the detection of significant marker-trait association are ongoing, for example incorporating long-range sequencing approach or continuous refinement of bioinformatics scripts for SNP data analysis. With these improvements in place, we will expect increasing amounts of high-quality marker-trait association studies to be reported for *Brassica* species. In the future, in addition to the well-studied and managed *B. rapa*, *B. oleracea* and *B. napus* germplasm, agronomically desirable traits could be sourced from under-represented *Brassica* germplasms such as *B. juncea* (Nanjundan et al., 2020), *B. carinata* (Ethiopian mustard) (Thakur et al., 2019), and *Brassica* wild relatives (Panjabi et al., 2019) for breeding of improved varieties of *Brassica* crop.

7 Where to look for further information

The following articles provide a good overview of the subject:

- Lv, H., Fang, Z., Yang, L., Zhang, Y. and Wang, Y. (2020). 'An update on the arsenal: mining resistance genes for disease management of *Brassica* crops in the genomic era'. *Horticulture Research*, 7(1), 1–18.
- Delourme, R., Laperche, A., Bouchet, A.-S., Jubault, M., Paillard, S., et al. (2018). 'Genes and quantitative trait loci mapping for major agronomic traits in *Brassica* napus L'. In Liu, S., Snowdon, R. and Chalhoub, B. (Eds), *The Brassica Napus Genome*, Springer, Switzerland, pp. 41–85.8

8 References

Al-Shehbaz, I. A. (2012). A generic and tribal synopsis of the Brassicaceae (Cruciferae). *Taxon* 61(5), 931–954.

Asghari, A., Asghar Fathi, A., Mohammadi, S. and Mohammaddust, H. (2012). QTL analysis for diamondback moth resistance in canola (*Brassica napus* L.). *International Journal of Plant Production* 3(3), 29–34.

Basunanda, P., Spiller, T. H., Hasan, M., Gehringer, A., Schondelmaier, J., Lühs, W., Friedt, W. and Snowdon, R. J. (2007). Marker-assisted increase of genetic diversity in a double-low seed quality winter oilseed rape genetic background. *Plant Breeding* 126(6), 581–587.

Branham, S. E., Stansell, Z. J., Couillard, D. M. and Farnham, M. W. (2017). Quantitative trait loci mapping of heat tolerance in broccoli (*Brassica oleracea* var. italica) using genotyping-by-sequencing. *TAG – Theoretical and Applied Genetics (Theoretische und Angewandte Genetik)* 130(3), 529–538.

Broekgaarden, C., Pelgrom, K. T. B., Bucher, J., van Dam, N. M., Grosser, K., Pieterse, C. M. J., van Kaauwen, M., Steenhuis, G., Voorrips, R. E., de Vos, M., Vosman, B., Worrich, A. and van Wees, S. C. M. (2018). Combining QTL mapping with transcriptome and metabolome profiling reveals a possible role for ABA signaling in resistance against the cabbage whitefly in cabbage. *PLoS ONE* 13(11), e0206103.

CGIAR (2020), CGIAR genebank platform. Available at: https://www.cgiar.org/research/program-platform/genebank-platform/.

Chai, L., Li, H., Zhang, J., Wu, L., Zheng, B., Cui, C., Jiang, J., Zuo, S. and Jiang, L. (2019). Rapid identification of a genomic region conferring dwarfism in rapeseed (*Brassica napus* L.) YA2016-12. *Agronomy* 9(3), 129.

Chang, A., Lamara, M., Wei, Y., Hu, H., Parkin, I. A. P., Gossen, B. D., Peng, G. and Yu, F. (2019). Clubroot resistance gene *Rcr6* in *Brassica nigra* resides in a genomic region homologous to chromosome A08 in *B. rapa*. *BMC Plant Biology* 19(1), 224.

Chao, H., Raboanatahiry, N., Wang, X., Zhao, W., Chen, L., Guo, L., Li, B., Hou, D., Pu, S., Zhang, L., Wang, H., Wang, B. and Li, M. (2019). Genetic dissection of harvest index and related traits through genome-wide quantitative trait locus mapping in *Brassica napus* L. *Breeding Science* 69(1), 104–116.

Chen, R., Hara, T., Ohsawa, R. and Yoshioka, Y. (2017). Analysis of genetic diversity of rapeseed genetic resources in Japan and core collection construction. *Breeding Science* 67(3), 239–247.

Chen, S., Turner, N. C., and Cowling, W. A. (2013). Genetic variation for heat tolerance during the reproductive phase in *Brassica rapa*. *Journal of Agronomy and Crop Science* 199(6), 424–435.

Chen, W., Zhang, Y., Liu, X., Chen, B., Tu, J. and Tingdong, F. (2007). Detection of QTL for six yield-related traits in oilseed rape (*Brassica napus*) using DH and immortalized F(2) populations. *TAG – Theoretical and Applied Genetics (Theoretische und Angewandte Genetik)* 115(6), 849–858.

Cowling, W. A. (2007). Genetic diversity in Australian canola and implications for crop breeding for changing future environments. *Field Crops Research* 104(1–3), 103–111.

Dai, N., Wang, W., Patterson, S. E. and Bleecker, A. B. (2013). The TMK subfamily of receptor-like kinases in *Arabidopsis* display an essential role in growth and a reduced sensitivity to auxin. *PLoS ONE* 8(4), e60990.

Delourme, R., Laperche, A., Bouchet, A.-S., Jubault, M., Paillard, S., Manzanares-Dauleux, M. J., and Nesi, N. (2018). Genes and quantitative trait loci mapping for major agronomic traits in *Brassica napus* L. In: Liu, S., Snowdon, R. and Chalhoub, B. (Eds), *The Brassica Napus Genome*, Springer: Switzerland, pp. 41–85.

Ding, G., Zhao, Z., Liao, Y., Hu, Y., Shi, L., Long, Y. and Xu, F. (2012). Quantitative trait loci for seed yield and yield-related traits, and their responses to reduced phosphorus supply in *Brassica napus*. *Annals of Botany* 109(4), 747–759.

El-Esawi, M. A. (2017). Genetic diversity and evolution of *Brassica* genetic resources: from morphology to novel genomic technologies–a review. *Plant Genetic Resources* 15(5), 388–399.

Farid, M., Yang, R. C., Kebede, B. and Rahman, H. (2020). Evaluation of *Brassica oleracea* accessions for resistance to *Plasmodiophora brassicae* and identification of genomic regions associated with resistance. *Genome* 63(2), 91–101.

Friedt, W., Tu, J. and Fu, T. (2018). Academic and economic importance of *Brassica napus* rapeseed. In: Liu, S., Snowdon, R. and Chalhoub, B. (Eds), *The Brassica napus Genome*, Springer: Switzerland, pp. 1–20.

Gacek, K., Bayer, P. E., Bartkowiak-Broda, I., Szala, L., Bocianowski, J., Edwards, D. and Batley, J. (2016). Genome-wide association study of genetic control of seed fatty acid biosynthesis in *Brassica napus*. *Frontiers in Plant Science* 7, 2062.

Harper, A. L., Trick, M., Higgins, J., Fraser, F., Clissold, L., Wells, R., Hattori, C., Werner, P. and Bancroft, I. (2012). Associative transcriptomics of traits in the polyploid crop species Brassica napus. *Nature Biotechnology* 30, 798–802.

Hirai, M. Y., Sugiyama, K., Sawada, Y., Tohge, T., Obayashi, T., Suzuki, A., Araki, R., Sakurai, N., Suzuki, H., Aoki, K., Goda, H., Nishizawa, O. I., Shibata, D. and Saito, K. (2007). Omics-based identification of *Arabidopsis* Myb transcription factors regulating aliphatic glucosinolate biosynthesis. *Proceedings of the National Academy of Sciences of the United States of America* 104(15), 6478–6483.

Hu, Z., Hua, W., Huang, S., Yang, H., Zhan, G., Wang, X., Liu, G. and Wang, H. (2012). Discovery of pod shatter-resistant associated SNPs by deep sequencing of a representative library followed by bulk segregant analysis in rapeseed. *PLoS ONE* 7(4), e34253.

Huang, Z., Liu, L., Lu, H., Lang, L., Zhao, N., Ding, J. and Xu, A. (2016). Development of IP and SCAR markers linked to the yellow seed color gene in *Brassica juncea* L. *Breeding Science* 66(2), 175–180.

Jiang, C., Shi, J., Li, R., Long, Y., Wang, H., Li, D., Zhao, J. and Meng, J. (2014). Quantitative trait loci that control the oil content variation of rapeseed (*Brassica napus* L.). *TAG - Theoretical and Applied Genetics (Theoretische und Angewandte Genetik)* 127(4), 957–968.

Katche, E., Quezada-Martinez, D., Katche, E. I., Vasquez-Teuber, P. and Mason, A. S. (2019). Interspecific hybridization for *Brassica* crop improvement. *Crop Breeding, Genetics and Genomics* 1(1), e190007.

Kaur, J., Akhatar, J., Goyal, A., Kaur, N., Kaur, S., Mittal, M., Kumar, N., Sharma, H., Banga, S. and Banga, S. S. (2020). Genome wide association mapping and candidate gene analysis for pod shatter resistance in *Brassica juncea* and its progenitor species. *Molecular Biology Reports* 47(4), 2963–2974.

Khanzada, H., Wassan, G. M., He, H., Mason, A. S., Keerio, A. A., Khanzada, S., Faheem, M., Solangi, A. M., Zhou, Q., Fu, D., Huang, Y. and Rasheed, A. (2020). Differentially evolved drought stress indices determine the genetic variation of *Brassica napus* at seedling traits by genome-wide association mapping. *Journal of Advanced Research* 24, 447–461.

King, G. J. and Baten, A. (2018). *Brassica napus* genomic resources. In: Liu, S., Snowdon, R. and Chalhoub, B. (Eds), *The Brassica Napus Genome*, Springer: Switzerland, pp. 233–244.

Koh, J. C. O., Barbulescu, D. M., Norton, S., Redden, B., Salisbury, P. A., Kaur, S., Cogan, N. and Slater, A. T. (2017). A multiplex PCR for rapid identification of *Brassica* species in the triangle of U. *Plant Methods* 13(1), 49.

Kumar, V., Paillard, S., Fopa-Fomeju, B., Falentin, C., Deniot, G., Baron, C., Vallée, P., Manzanares-Dauleux, M. J. and Delourme, R. (2018). Multi-year linkage and association mapping confirm the high number of genomic regions involved in oilseed rape quantitative resistance to blackleg. *TAG - Theoretical and Applied Genetics (Theoretische und Angewandte Genetik)* 131(8), 1627–1643.

Laila, R., Park, J. I., Robin, A. H. K., Natarajan, S., Vijayakumar, H., Shirasawa, K., Isobe, S., Kim, H. T. and Nou, I. S. (2019). Mapping of a novel clubroot resistance QTL using ddRAD-seq in Chinese cabbage (*Brassica rapa* L.). *BMC Plant Biology* 19(1), 13.

Larkan, N. J., Lydiate, D. J., Parkin, I. A., Nelson, M. N., Epp, D. J., Cowling, W. A., Rimmer, S. R. and Borhan, M. H. (2013). The *Brassica napus* blackleg resistance gene *LepR3* encodes a receptor-like protein triggered by the *Leptosphaeria maculans* effector *AvrLm1*. *New Phytologist* 197(2), 595–605.

Larkan, N. J., Ma, L., Haddadi, P., Buchwaldt, M., Parkin, I. A. P., Djavaheri, M. and Borhan, M. H. (2019). The *Brassica napus* Wall-Associated Kinase-Like (WAKL) gene *Rlm9* provides race-specific blackleg resistance. *The Plant Journal* 104(4), 892–900.

Lee, R. W. H., Malchev, I. T., Rajcan, I. and Kott, L. S. (2014). Identification of putative quantitative trait loci associated with a flavonoid related to resistance to cabbage seedpod weevil (*Ceutorhynchus obstrictus*) in canola derived from an intergeneric cross, *Sinapis alba* × *Brassica napus*. *Theoretical and Applied Genetics* 127(2), 419–428.

Lema, M., Ali, M. Y. and Retuerto, R. (2019). Domestication influences morphological and physiological responses to salinity in *Brassica oleracea* seedlings. *AoB Plants* 11(5), plz046.

Li, F., Chen, B., Xu, K., Gao, G., Yan, G., Qiao, J., Li, J., Li, H., Li, L., Xiao, X., Zhang, T., Nishio, T. and Wu, X. (2016a). A genome-wide association study of plant height and primary branch number in rapeseed (*Brassica napus*). *Plant Science: An International Journal of Experimental Plant Biology* 242, 169–177.

Li, F., Chen, B., Xu, K., Wu, J., Song, W., Bancroft, I., Harper, A. L., Trick, M., Liu, S., Gao, G., Wang, N., Yan, G., Qiao, J., Li, J., Li, H., Xiao, X., Zhang, T. and Wu, X. (2014). Genome-wide association study dissects the genetic architecture of seed weight and seed quality in rapeseed (*Brassica napus* L.). *DNA Research* 21(4), 355–367.

Li, L., Long, Y., Zhang, L., Dalton-Morgan, J., Batley, J., Yu, L., Meng, J. and Li, M. (2015). Genome wide analysis of flowering time trait in multiple environments via high-throughput genotyping technique in *Brassica napus* L. *PLoS ONE* 10(3), e0119425.

Li, L., Luo, Y., Chen, B., Xu, K., Zhang, F., Li, H., Huang, Q., Xiao, X., Zhang, T., Hu, J., Li, F. and Wu, X. (2016b). A genome-wide association study reveals new loci for resistance to clubroot disease in *Brassica napus*. *Frontiers in Plant Science* 7(1483), 1483.

Li, P., Su, T., Wang, H., Zhao, X., Wang, W., Yu, Y., Zhang, D., Wen, C., Yu, S. and Zhang, F. (2019a). Development of a core set of KASP markers for assaying genetic diversity in *Brassica rapa* subsp. chinensis Makino. *Plant Breeding* 138(3), 309–324.

Li, X., Kong, C., Yu, H., Liu, X., Fang, Z., Liu, Y., Yang, L., Zhuang, M., Wang, Y., Lv, H. and Zhang, Y. (2019b). Identification of a major QTL for seed number per silique in cabbage (*Brassica oleracea* L. var. capitata) using genotyping by sequencing. *Euphytica* 215(7), 1–13.

Li, X., Ramchiary, N., Dhandapani, V., Choi, S. R., Hur, Y., Nou, I. S., Yoon, M. K. and Lim, Y. P. (2013). Quantitative trait loci mapping in *Brassica rapa* revealed the structural and functional conservation of genetic loci governing morphological and yield component traits in the A, B, and C subgenomes of *Brassica* species. *DNA Research: An International Journal for Rapid Publication of Reports on Genes and Genomes* 20(1), 1–16.

Liu, J., Wang, J., Wang, H., Wang, W., Zhou, R., Mei, D., Cheng, H., Yang, J., Raman, H. and Hu, Q. (2016). Multigenic control of pod shattering resistance in Chinese rapeseed germplasm revealed by genome-wide association and linkage analyses. *Frontiers in Plant Science* 7(1058), 1058.

Lu, K., Peng, L., Zhang, C., Lu, J., Yang, B., Xiao, Z., Liang, Y., Xu, X., Qu, C., Zhang, K., Liu, L., Zhu, Q., Fu, M., Yuan, X. and Li, J. (2017). Genome-wide association and transcriptome analyses reveal candidate genes underlying yield-determining traits in *Brassica napus*. *Frontiers in Plant Science* 8(206), 206.

Lu, K., Wei, L., Li, X., Wang, Y., Wu, J., Liu, M., Zhang, C., Chen, Z., Xiao, Z., Jian, H., Cheng, F., Zhang, K., Du, H., Cheng, X., Qu, C., Qian, W., Liu, L., Wang, R., Zou, Q., Ying, J., Xu, X., Mei, J., Liang, Y., Chai, Y. R., Tang, Z., Wan, H., Ni, Y., He, Y., Lin, N., Fan, Y., Sun, W., Li, N. N., Zhou, G., Zheng, H., Wang, X., Paterson, A. H. and Li, J. (2019). Whole-genome resequencing reveals *Brassica napus* origin and genetic loci involved in its improvement. *Nature Communications* 10(1), 1154.

Luo, X., Tan, Y., Ma, C., Tu, J., Shen, J., Yi, B. and Fu, T. (2019). High-throughput identification of SNPs reveals extensive heterosis with intra-group hybridization and genetic characteristics in a large rapeseed (*Brassica napus* L.) panel. *Euphytica* 215(10), 157.

Lv, H., Fang, Z., Yang, L., Zhang, Y. and Wang, Y. (2020). An update on the arsenal: mining resistance genes for disease management of *Brassica* crops in the genomic era. *Horticulture Research* 7(1), 34.

Mahmood, T., Rahman, M. H., Stringam, G. R., Yeh, F. and Good, A. G. (2006). Identification of quantitative trait loci (QTL) for oil and protein contents and their relationships with other seed quality traits in *Brassica juncea*. *TAG – Theoretical and Applied Genetics (Theoretische und Angewandte Genetik)* 113(7), 1211–1220.

Malmberg, M. M., Shi, F., Spangenberg, G. C., Daetwyler, H. D. and Cogan, N. O. I. (2018). Diversity and genome analysis of Australian and global oilseed *Brassica napus* L. germplasm using transcriptomics and whole genome re-sequencing. *Frontiers in Plant Science* 9, 508.

Mason, A. S., Higgins, E. E., Snowdon, R. J., Batley, J., Stein, A., Werner, C. and Parkin, I. A. (2017). A user guide to the *Brassica* 60 K Illumina Infinium™ SNP genotyping array. *TAG – Theoretical and Applied Genetics (Theoretische und Angewandte Genetik)* 130(4), 621–633.

Mason, A. S., Zhang, J., Tollenaere, R., Vasquez Teuber, P., Dalton-Morgan, J., Hu, L., Yan, G., Edwards, D., Redden, R. and Batley, J. (2015). High-throughput genotyping for species identification and diversity assessment in germplasm collections. *Molecular Ecology Resources* 15(5), 1091–1101.

Mei, D. S., Wang, H. Z., Hu, Q., Li, Y. D., Xu, Y. S. and Li, Y. C. (2009). QTL analysis on plant height and flowering time in *Brassica napus*. *Plant Breeding* 128(5), 458–465.

Mpumi, N., Machunda, R. S., Mtei, K. M. and Ndakidemi, P. A. (2020). Selected insect pests of economic importance to *Brassica oleracea*, their control strategies and the potential threat to environmental pollution in Africa. *Sustainability* 12(9), 3824.

Nanjundan, J., Radhamani, J., Thakur, A. K., Berliner, J., Manjunatha, C., Sindhu, A., Aravind, J. and Singh, K. (2020). Utilization of rapeseed-mustard genetic resources for *Brassica* improvement: a retrospective approach. In: Wani, S. H., Thakur, A. K. and Khan, Y. J. (Eds), *Brassica Improvement*, Springer: Switzerland, pp. 1–30.

Neik, T. X., Barbetti, M. J. and Batley, J. (2017). Current status and challenges in identifying disease resistance genes in *Brassica napus*. *Frontiers in Plant Science* 8, 1788.

Panjabi, P., Yadava, S. K., Kumar, N., Bangkim, R. and Ramchiary, N. (2019). Breeding *Brassica juncea* and *B. rapa* for sustainable oilseed production in the changing climate: progress and prospects. In: *Genomic Designing of Climate-Smart Oilseed Crops*, Springer, pp. 275–369.

Pavlović, I., Mlinarić, S., Tarkowská, D., Oklestkova, J., Novák, O., Lepeduš, H., Bok, V. V., Brkanac, S. R., Strnad, M. and Salopek-Sondi, B. (2019). Early *Brassica* crops responses to salinity stress: a comparative analysis between Chinese cabbage, white cabbage, and kale. *Frontiers in Plant Science* 10, 450.

Raboanatahiry, N., Chao, H., Dalin, H., Pu, S., Yan, W., Yu, L., Wang, B. and Li, M. (2018). QTL alignment for seed yield and yield related traits in *Brassica napus*. *Frontiers in Plant Science* 9(1127), 1127.

Raboanatahiry, N., Chao, H., Guo, L., Gan, J., Xiang, J., Yan, M., Zhang, L., Yu, L. and Li, M. (2017). Synteny analysis of genes and distribution of loci controlling oil content and fatty acid profile based on QTL alignment map in *Brassica napus*. *BMC Genomics* 18(1), 776.

Rahaman, M., Mamidi, S. and Rahman, M. (2018). Genome-wide association study of heat stress-tolerance traits in spring-type *Brassica napus* L. under controlled conditions. *The Crop Journal* 6(2), 115–125.

Rahman, M., Li, G., Schroeder, D. and McVetty, P. B. E. (2010). Inheritance of seed coat color genes in *Brassica napus* (L.) and tagging the genes using SRAP, SCAR and SNP molecular markers. *Molecular Breeding* 26(3), 439–453.

Raman, H., Raman, R., Coombes, N., Song, J., Diffey, S., Kilian, A., Lindbeck, K., Barbulescu, D. M., Batley, J., Edwards, D., Salisbury, P. A. and Marcroft, S. (2016). Genome-wide association study identifies new loci for resistance to *Leptosphaeria maculans* in canola. *Frontiers in Plant Science* 7, 1513.

Raman, H., Raman, R., Kilian, A., Detering, F., Long, Y., Edwards, D., Parkin, I. A., Sharpe, A. G., Nelson, M. N., Larkan, N., Zou, J., Meng, J., Aslam, M. N., Batley, J., Cowling, W. A. and Lydiate, D. (2013a). A consensus map of rapeseed (*Brassica napus* L.) based on diversity array technology markers: applications in genetic dissection of qualitative and quantitative traits. *BMC Genomics* 14, 277.

Raman, H., Raman, R. and Larken, N. (2013b). Genetic dissection of blackleg resistance loci in rapeseed (*Brassica napus* L.). In: Andersen, S. B. (Ed), *Plant Breeding from Laboratories to Fields*, InTech: Croatia, pp. 86–120.

Raman, H., Raman, R., Qiu, M., Yadav, A. S., Sureshkumar, S., Borg, L., Rohan, M., Wheeler, D., Owen, O., Menz, I. and Balasubramanian, S. (2019a). GWAS hints at pleiotropic roles for FLOWERING LOCUS T in flowering time and yield-related traits in canola. *BMC Genomics* 20(1), 636.

Raman, H., Uppal, R. K. and Raman, R. (2019b). Genetic solutions to improve resilience of canola to climate change. In: Kole, C. (Ed), *Genomic Designing of Climate-Smart Oilseed Crops*, Springer: Switzerland, pp. 75–131.

Raman, R., Diffey, S., Barbulescu, D. M., Coombes, N., Luckett, D., Salisbury, P., Cowley, R., Marcroft, S. and Raman, H. (2020). Genetic and physical mapping of loci for resistance to blackleg disease in canola (*Brassica napus* L.). *Scientific Reports* 10(1), 4416.

Rana, K., Atri, C., Akhatar, J., Kaur, R., Goyal, A., Singh, M. P., Kumar, N., Sharma, A., Sandhu, P. S., Kaur, G., Barbetti, M. J. and Banga, S. S. (2019). Detection of first marker trait associations for resistance against *Sclerotinia sclerotiorum* in *Brassica juncea-Erucastrum cardaminoides* introgression lines. *Frontiers in Plant Science* 10, 1015.

Rasheed, A., Hao, Y., Xia, X., Khan, A., Xu, Y., Varshney, R. K. and He, Z. (2017). Crop breeding chips and genotyping platforms: progress, challenges, and perspectives. *Molecular Plant* 10(8), 1047–1064.

Sabagh, A. E. L., Hossain, A., Barutçular, C., Islam, M. S., Ratnasekera, D., Kumar, N., Meena, R. S., Gharib, H. S., Saneoka, H. S. and Teixeira da Silva, J. A. (2019). Drought and salinity stress management for higher and sustainable canola (*Brassica napus* L.) production: a critical review. *Australian Journal of Crop Science* 13, 88–97.

Saini, N., Singh, N., Kumar, A., Vihan, N., Yadav, S., Vasudev, S. and Yadava, D. K. (2016). Development and validation of functional CAPS markers for the *FAE* genes in *Brassica juncea* and their use in marker-assisted selection. *Breeding Science* 66(5), 831–837.

Sandhu, S. K. and Singh, G. (2018). Genomic-assisted breeding in oilseed *Brassicas*. In: Gosal, S. S. and Wani, S. H. (Eds), *Biotechnologies of Crop Improvement (Vol. 3): Genomic Approaches*, Springer: Switzerland, pp. 317–348.

Scheben, A., Batley, J. and Edwards, D. (2017). Genotyping-by-sequencing approaches to characterize crop genomes: choosing the right tool for the right application. *Plant Biotechnology Journal* 15(2), 149–161.

Scheben, A., Verpaalen, B., Lawley, C. T., Chan, C. K. K., Bayer, P. E., Batley, J. and Edwards, D. (2019). CropSNPdb: a database of SNP array data for *Brassica* crops and hexaploid bread wheat. *The Plant Journal: For Cell and Molecular Biology* 98(1), 142–152.

Schiessl, S., Huettel, B., Kuehn, D., Reinhardt, R. and Snowdon, R. J. (2017). Targeted deep sequencing of flowering regulators in *Brassica napus* reveals extensive copy number variation. *Scientific Data* 4, 170013.

Shasidhar, Y., Vishwakarma, M. K., Pandey, M. K., Janila, P., Variath, M. T., Manohar, S. S., Nigam, S. N., Guo, B. and Varshney, R. K. (2017). Molecular mapping of oil content and fatty acids using dense genetic maps in groundnut (*Arachis hypogaea* L.). *Frontiers in Plant Science* 8, 794.

Shen, Y., Xiang, Y., Xu, E., Ge, X. and Li, Z. (2018). Major co-localized QTL for plant height, branch initiation height, stem diameter, and flowering time in an alien introgression derived *Brassica napus* DH population. *Frontiers in Plant Science* 9(390), 390.

Shi, J., Li, R., Qiu, D., Jiang, C., Long, Y., Morgan, C., Bancroft, I., Zhao, J. and Meng, J. (2009). Unraveling the complex trait of crop yield with quantitative trait loci mapping in *Brassica napus*. *Genetics* 182(3), 851–861.

Shi, J., Li, R., Zou, J., Long, Y. and Meng, J. (2011). A dynamic and complex network regulates the heterosis of yield-correlated traits in rapeseed (*Brassica napus* L.). *PLoS ONE* 6(7), e21645.

Shi, L., Jiang, C., He, Q., Habekuß, A., Ordon, F., Luan, H., Shen, H., Liu, J., Feng, Z., Zhang, J. and Yang, P. (2019). Bulked segregant RNA-sequencing (BSR-seq) identified a novel rare allele of *eIF4E* effective against multiple isolates of BaYMV/BaMMV. *TAG - Theoretical and Applied Genetics (Theoretische und Angewandte Genetik)* 132(6), 1777–1788.

Shokri-Gharelo, R. and Noparvar, P. M. (2018). Molecular response of canola to salt stress: insights on tolerance mechanisms. *PeerJ* 6, e4822–e4822.

Song, K. M., Osborn, T. C. and Williams, P. H. (1988). *Brassica* taxonomy based on nuclear restriction fragment length polymorphisms (RFLPs). *Theoretical and Applied Genetics* 75(5), 784–794.

Stansell, Z., Hyma, K., Fresnedo-Ramírez, J., Sun, Q., Mitchell, S., Björkman, T. and Hua, J. (2018). Genotyping-by-sequencing of *Brassica oleracea* vegetables reveals unique phylogenetic patterns, population structure and domestication footprints. *Horticulture Research* 5(1), 38.

Su, T., Li, P., Yang, J., Sui, G., Yu, Y., Zhang, D., Zhao, X., Wang, W., Wen, C., Yu, S. and Zhang, F. (2018). Development of cost-effective single nucleotide polymorphism marker assays for genetic diversity analysis in *Brassica rapa*. *Molecular Breeding* 38(4), 42.

Sudan, J., Singh, R., Sharma, S., Salgotra, R. K., Sharma, V., Singh, G., Sharma, I., Sharma, S., Gupta, S. K. and Zargar, S. M. (2019). ddRAD sequencing-based identification of inter-genepool SNPs and association analysis in *Brassica juncea*. *BMC Plant Biology* 19(1), 594.

Thakur, A. K., Singh, K. H., Sharma, D., Parmar, N. and Nanjundan, J. (2019). Breeding and genomics interventions in Ethiopian mustard (*Brassica carinata* A. Braun) improvement – A mini review. *South African Journal of Botany* 125, 457–465.

Thomson, M. J. (2014). High-throughput SNP genotyping to accelerate crop improvement. *Plant Breeding and Biotechnology* 2(3), 195–212.

U, N. (1935). Genome analysis in Brassica with special reference to the experimental formation of B. napus and peculiar mode of fertilization. *Japanese Journal of Botany* 7, 389–452.

Van de Wouw, A. P. and Howlett, B. J. (2019). Advances in understanding the *Leptosphaeria maculans-Brassica* pathosystem and their impact on disease management. *Canadian Journal of Plant Pathology* 42(2), 1–15.

Walker, K. C. and Booth, E. J. (2001). Agricultural aspects of rape and other *Brassica* products. *European Journal of Lipid Science and Technology* 103(7), 441–446.

Wan, H., Chen, L., Guo, J., Li, Q., Wen, J., Yi, B., Ma, C., Tu, J., Fu, T. and Shen, J. (2017). Genome-wide association study reveals the genetic architecture underlying salt tolerance-related traits in rapeseed (*Brassica napus* L.). *Frontiers in Plant Science* 8, 593.

Wang, B., Wu, Z., Li, Z., Zhang, Q., Hu, J., Xiao, Y., Cai, D., Wu, J., King, G. J., Li, H. and Liu, K. (2018). Dissection of the genetic architecture of three seed-quality traits and consequences for breeding in *Brassica napus*. *Plant Biotechnology Journal* 16(7), 1336–1348.

Wang, H., Yan, M., Xiong, M., Wang, P., Liu, Y., Xin, Q., Wan, L., Yang, G. and Hong, D. (2020). Genetic dissection of thousand-seed weight and fine mapping of cqSW.A03-2 via linkage and association analysis in rapeseed (*Brassica napus* L.). *TAG – Theoretical and Applied Genetics (Theoretische und Angewandte Genetik)* 133(4), 1321–1335.

Wang, J., Xian, X., Xu, X., Qu, C., Lu, K., Li, J. and Liu, L. (2017). Genome-wide association mapping of seed coat color in *Brassica napus*. *Journal of Agricultural and Food Chemistry* 65(26), 5229–5237.

Wang, X., Wang, H., Long, Y., Li, D., Yin, Y., Tian, J., Chen, L., Liu, L., Zhao, W., Zhao, Y., Yu, L. and Li, M. (2013). Identification of QTLs associated with oil content in a high-oil *Brassica napus* cultivar and construction of a high-density consensus map for QTLs comparison in *B. napus*. *PLoS ONE* 8(12), e80569.

Werner, C. and Snowdon, R. (2018). Genome-facilitated breeding of oilseed rape. In: Liu, S., Snowdon, R. and Chalhoub, B. (Eds), *The Brassica Napus Genome*, Springer: Switzerland, pp. 245–269.

Wrucke, D. F., Mamidi, S. and Rahman, M. (2019). Genome-wide association study for frost tolerance in canola (*Brassica napus* L.) under field conditions. *Journal of Plant Biochemistry and Biotechnology* 28(2), 211–222.

Wu, H., Guo, J., Wang, C., Li, K., Zhang, X., Yang, Z., Li, M. and Wang, B. (2019a). An effective screening method and a reliable screening trait for salt tolerance of *Brassica napus* at the germination stage. *Frontiers in Plant Science* 10, 530.

Wu, D., Liang, Z., Yan, T., Xu, Y., Xuan, L., Tang, J., Zhou, G., Lohwasser, U., Hua, S., Wang, H., Chen, X., Wang, Q., Zhu, L., Maodzeka, A., Hussain, N., Li, Z., Li, X., Shamsi, I. H., Jilani, G., Wu, L., Zheng, H., Zhang, G., Chalhoub, B., Shen, L., Yu, H. and Jiang, L. (2019b). Whole-Genome Resequencing of a Worldwide Collection of Rapeseed Accessions Reveals the Genetic Basis of Ecotype Divergence. *Molecular Plant* 12(1), 30–43.

Yang, P., Shu, C., Chen, L., Xu, J., Wu, J. and Liu, K. (2012). Identification of a major QTL for silique length and seed weight in oilseed rape (*Brassica napus* L.). *TAG - Theoretical and Applied Genetics (Theoretische und Angewandte Genetik)* 125(2), 285–296.

Zhang, F., Huang, J., Tang, M., Cheng, X., Liu, Y., Tong, C., Yu, J., Sadia, T., Dong, C., Liu, L., Tang, B., Chen, J. and Liu, S. (2019). Syntenic quantitative trait loci and genomic divergence for *Sclerotinia* resistance and flowering time in *Brassica napus*. *Journal of Integrative Plant Biology* 61(1), 75–88.

Zhang, J., Mason, A. S., Wu, J., Liu, S., Zhang, X., Luo, T., Redden, R., Batley, J., Hu, L. and Yan, G. (2015). Identification of putative candidate genes for water stress tolerance in canola (*Brassica napus*). *Frontiers in Plant Science* 6, 1058.

Zhang, X., Zhang, K., Wu, J., Guo, N., Liang, J., Wang, X. and Cheng, F. (2020). QTL-seq and sequence assembly rapidly mapped the gene *BrMYBL2.1* for the purple trait in *Brassica rapa*. *Scientific Reports* 10(1), 2328.

Zhao, W., Wang, X., Wang, H., Tian, J., Li, B., Chen, L., Chao, H., Long, Y., Xiang, J., Gan, J., Liang, W. and Li, M. (2016). Genome-wide identification of QTL for seed yield and yield-related traits and construction of a high-density consensus map for QTL comparison in *Brassica napus*. *Frontiers in Plant Science* 7(17), 17.

Zhao, Y., Ma, J., Li, M., Deng, L., Li, G., Xia, H., Zhao, S., Hou, L., Li, P., Ma, C., Yuan, M., Ren, L., Gu, J., Guo, B., Zhao, C. and Wang, X. (2020). Whole-genome resequencing-based QTL-seq identified *AhTc1* gene encoding a R2R3-MYB transcription factor controlling peanut purple testa colour. *Plant Biotechnology Journal* 18(1), 96–105.

Zheng, M., Peng, C., Liu, H., Tang, M., Yang, H., Li, X., Liu, J., Sun, X., Wang, X., Xu, J., Hua, W. and Wang, H. (2017). Genome-wide association study reveals candidate genes for control of plant height, branch initiation height and branch number in rapeseed (*Brassica napus* L.). *Frontiers in Plant Science* 8, 1246.

Zhou, Q., Zhou, C., Zheng, W., Mason, A. S., Fan, S., Wu, C., Fu, D. and Huang, Y. (2017). Genome-wide SNP markers based on SLAF-seq uncover breeding traces in rapeseed (*Brassica napus* L.). *Frontiers in Plant Science* 8, 648.

Zhu, M., Monroe, J. G., Suhail, Y., Villiers, F., Mullen, J., Pater, D., Hauser, F., Jeon, B. W., Bader, J. S., Kwak, J. M., Schroeder, J. I., McKay, J. K. and Assmann, S. M. (2016). Molecular and systems approaches towards drought-tolerant canola crops. *New Phytologist* 210(4), 1169–1189.

Index